Springer Proceedings in Physics

Volume 287

Indexed by Scopus

The series Springer Proceedings in Physics, founded in 1984, is devoted to timely reports of state-of-the-art developments in physics and related sciences. Typically based on material presented at conferences, workshops and similar scientific meetings, volumes published in this series will constitute a comprehensive up to date source of reference on a field or subfield of relevance in contemporary physics. Proposals must include the following:

– Name, place and date of the scientific meeting
– A link to the committees (local organization, international advisors etc.)
– Scientific description of the meeting
– List of invited/plenary speakers
– An estimate of the planned proceedings book parameters (number of pages/articles, requested number of bulk copies, submission deadline).

Please contact:

For Americas and Europe: Dr. Zachary Evenson; zachary.evenson@springer.com
For Asia, Australia and New Zealand: Dr. Loyola DSilva; loyola.dsilva@springer.com

Luisa Bonolis · Luciano Maiani · Giulia Pancheri
Editors

Bruno Touschek 100 Years

Memorial Symposium 2021

 Springer

Editors
Luisa Bonolis
Max Planck Institute for the History
of Science
Berlin, Germany

Luciano Maiani
Department of Physics
Sapienza University of Rome
Rome, Italy

Giulia Pancheri
INFN Frascati National Laboratories
Frascati, Italy

ISSN 0930-8989   ISSN 1867-4941 (electronic)
Springer Proceedings in Physics
ISBN 978-3-031-23041-7   ISBN 978-3-031-23042-4 (eBook)
https://doi.org/10.1007/978-3-031-23042-4

This Springer imprint is published by the registered company Springer Nature Switzerland AG
The registered company address is: Gewerbestrasse 11, 6330 Cham, Switzerland

Preface

On March 7, 1960, in a seminar at Frascati National Laboratories, Bruno Touschek illustrated the importance of high-energy electron-positron collisions and proposed to create a *Storage Ring*, a new type of machine in which it would be possible to produce this type of reactions, never studied before in a systematic way.

The idea of circulating beams of electrons and positrons in a single ring had its physical basis in the fundamental symmetry between matter and antimatter discovered by Dirac in the 1920s, but clashed with technological difficulties unknown until then (high vacuum and beam electronics among many others). The challenge was welcomed with enthusiasm by the young and very young researchers of the Laboratory, recruited by Giorgio Salvini in the years of construction of the Electron Synchrotron, which had just entered into operation. In two years, under the guidance of Bruno Touschek, the construction of the first electron-positron storage ring (AdA = *Anello di Accumulazione* in Italian) was completed in Frascati. Subsequent developments were carried out by a Franco-Italian collaboration at the Linear Accelerator Laboratory of Orsay, near Paris.

AdA was the beginning of a story that had a deep influence on the course of modern particle physics, with the construction of collision rings of larger dimensions and higher energies, in Italy and in the world. The discovery of new phenomena at colliders has contributed in a fundamental way to the modern theory of elementary particles.

To remember Bruno Touschek and his scientific heritage in theoretical and experimental physics, in Italy and elsewhere, we organized an International Conference on the 100th anniversary of his birth, which was held on December 2–4, 2021 in the three places where Touschek mainly carried out his Roman activity: the Physics Departments of the University of Rome La Sapienza, the National Laboratories of INFN in Frascati, and the National Academy of Lincei.

The present Volume collects contributions presented in the three days. In addition to historical speeches on the life of Bruno Touschek and on the creation of AdA and ADONE, the first major storage ring for physics experiments carried out in Frascati in the 1970s, the volume presents a panorama of present Theoretical and Experimental Physics and prospects for future developments. Among these, the projects of very high-energy colliders and the prospects for the new astronomy opened by the observation of gravitational waves coming from the Cosmos.

The Conference would not have taken place without the strong support of the institutions involved. For this, we thank Rahatlu Shahram, Director of the Physics Department; Aleandro Nisati, Director of the Roma1 Section of INFN; Fabio Bossi, Director of the Frascati Laboratories; Antonio Zoccoli, President of INFN; and Roberto Antonelli, President of the Academy of Lincei. We thank the Austrian Embassy in Rome for sponsorship and participation in the Symposium.

Finally, we thank Giovanni Jona-Lasinio, Giovanni Gallavotti, Daniele del Re, Luca Silvestrini, Stefano Giagu, Pierluigi Campana, and Paola Gianotti, for their participation in the Scientific Committee of the Conference. Ilaria Bonincontro and Maria Cristina D'Amato provided precious technical support and Sonia Mozzillo took care of the correspondence with the Authors.

Berlin, Germany   Luisa Bonolis
Rome, Italy   Luciano Maiani
Frascati, Italy   Giulia Pancheri

Contents

Contributors

Ugo Amaldi TERA Foundation, Novara, Italy

Gianni Battimelli INFN Sezione di Roma, c/o Dipartimento di Fisica, Università di Roma "La Sapienza", Rome, Italy

Luisa Bonolis Max Planck Institute for the History of Science, Berlin, Germany

Marica Branchesi Gran Sasso Science Institute, INFN, L'Aquila, Italy; Laboratori Nazionali del Gran Sasso, Assergi, Italy

Franco Buccella INFN, Sezione di Napoli, Napoli, Italy

Luisa Cifarelli University of Bologna, Bologna, Italy

Carlo di Castro Dipartimento di Fisica, Università di Roma "La Sapienza", Roma, Italy

Paolo Di Vecchia Nordita, KTH Royal Institute of Technology, University of Stockholm, Stockholm, Sweden

Giovanni Gallavotti Dipartimento di, Fisica and INFN, Università di Roma "La Sapienza", Roma, Italia

Andrea Ghigo National Laboratories of INFN, Frascati, Italy

Gian Francesco Giudice Theoretical Physics Department, CERN, Geneva, Switzerland

Mario Greco Department of Mathematics and Physics and INFN, University of Roma Tre, Rome, Italy

Jacques Haïssinski Laboratoire IJCLab Orsay, Paris, France

Giovanni Jona-Lasinio Dipartimento di Fisica and INFN, Università di Roma "La Sapienza", Roma, Italy

Luciano Maiani Dipartimento di Fisica, Università di Roma La Sapienza and INFN, Sezione di Roma1, Roma, Italy; TH Division, CERN, Geneve 23, Switzerland

Guido Martinelli Dipartimento di Fisica, Università di Roma La Sapienza, Roma, Italy; INFN, Roma, Italy

Giulia Pancheri Laboratori Nazionali di Frascati dell'INFN, Frascati, Italy

Giorgio Parisi Department of Physics, Rome University La Sapienza, Roma, Italy

Claudio Pellegrini SLAC National Accelerator Laboratory, Menlo Park, CA, USA

Luciano Pietronero Enrico Fermi Research Center, Roma, Italy

Giancarlo Rossi Dipartimento di Fisica, Università di Roma "Tor Vergata", Roma, Italy; Sezione di Roma "Tor Vergata", INFN, Roma, Italy; Museo Storico della Fisica e Centro Studi e Ricerche Enrico Fermi, Roma, Italy

Lucio Rossi Dipartimento Di Fisica, LASA Laboratory, Università degli Studi di Milano, INFN-Sezione di Milano, Milan, Italy

Carlo Rubbia CERN, Geneva, Switzerland

Steinar Stapnes ATS-DO, CERN, Geneva, Switzerland

Achille Stocchi Université Paris-Saclay, CNRS/IN2P3, IJCLab, Orsay, France

Francis Touschek Rome, Italy

Gabriele Veneziano Theory Department, CERN, Geneva 23, Switzerland; Collège de France, Paris, France

Yifang Wang Institute of High Energy Physics, Beijing, P.R. China

Part I
Talks Given at the Physics Department, Università di Roma La Sapienza

Chapter 1
Bruno "Burl" Touschek (1921–1978)

Francis Touschek

I was kindly invited to give a small contribution to this symposium which I gladly accepted, fully aware I was surrounded by the gotha of the physics world. So, I spent a long time brooding over what might be appropriate, my first instinct was to denounce the ugliness and brutality my father had to endure throughout his whole life. I finally decided on a softer approach, impromptu and recount some of the more amusing moments I was lucky enough to share and enjoy with him. He was an extraordinary man, with a vicious sense of humour. I have never met anyone quite like him.

In the inner circle of his family and friends, my father was known as **Burl**. He was born an only child, into a modest family, father Franz (after whom I was named) a retired officer of the Austrian army and mother, Camilla (née Weltmann) a talented artist who was struck by the Spanish flu and eventually died of it in 1931 when **Burl** was only 10. This had a huge impact on his future life and drew him closer and closer to his father.

The short, brilliant and tragic life of **Burl** can be broken down simply, into five stages. His childhood in Vienna, the war years, his time in Glasgow, Italy and finally his short spell in Geneva.

He was born in Vienna in 1921, in a city which was a cradle of culture (despite the ravages and consequences of the first world war). Camilla, his mother came from a very interesting Jewish family, with very close connections to the art and cultural world of the time. Klimt, Schiele, Kokoschka, Kandinski, Gropius, Kraus and Mahler (and his very interesting wife, Alma) to mention a few, were indeed known acquaintances to the Weltmann family. It is in this extraordinary environment that my father grew up. He must have gone through, what I refer to as the cultural osmosis of his time. He not only soaked the knowledge but most of all the spirit of

F. Touschek (✉)
Rome, Italy
e-mail: frtousch@msn.com

© The Author(s) 2023
L. Bonolis et al. (eds.), *Bruno Touschek 100 Years*,
Springer Proceedings in Physics 287,
https://doi.org/10.1007/978-3-031-23042-4_1

his surroundings. He rapidly learned there were no borders to knowledge and thrived in this awareness.

I am not going to dwell into his biography, Dr. Pancheri and Dr. Bonolis will do this in much greater detail and with greater ability than I but, there are a few aspects that I did share with my father and which are beyond the boundaries of biographies or biographers. I am obviously referring to the period he spent in Italy. My brother (Steven) and I, were born in Rome, possibly during the happiest moments of his life. He came to Rome in the early '50s, met my mother (Elspeth, a brilliant artist) in Naples. They got married in Glasgow right next to the well known Ballantine whisky distillery (there is a sense of foreboding). They then settled in Rome, I was born in 1958 and my brother Steven in 1961. The first 15 years in Italy must have been sheer bliss for both of them. A beautiful couple with a brilliant future.

Sadly, this moment came to a sudden end towards the end of the 60s. Two episodes were to change his life dramatically. The first event was the "caso Ippolito", the second the global student revolution which, in Italy turned into a nasty political affair. One day he came home, with a photograph of a "graffiti" on the walls of the faculty of physics, portraying the phrase 'Touschek=nazista". After what he had endured during the war, this was the final straw. He started drinking heavily, he detached himself from the projects he was involved in. He sought employment at the University of Vienna. I went with him for his interview. We were met by Walter Thirring just outside the faculty of physics. I was left outside the hall where the interview was to take place, my father entered the room and came out a few minutes later, ashen. One of the interviewers was his torturer in the concentration camp where he was interned during the war.

Burl was never the same again. We returned to Italy with a sense of gloom, my father took to drinking even more heavily and after a series of hepatic comas died in 1978 (same year curiously, a few months apart, as Aldo Moro).

This brief synopsis is a gentle introduction to the life of Bruno Touschek and during this Symposium the details of his scientific life and his contribution to the world of physics will be better described by members of a much higher standard than myself.

There are though, quite a few amusing moments I recall, all of which took place during the "happy" years. Our flat in Rome was a Mecca for many friends and colleagues, I was privileged to meet some extraordinary people. I recall one instance, we had just moved into a new flat, Jerry O'Neill came to visit us, escorted by CIA agents together with Gersh Budker escorted by KGB agents. It was a surreal evening. After a few drinks the atmosphere became quite merry. My father noticed a hole in the ceiling (about the size of a 100 lira coin) and the discussion among the three rapidly became quite idiotic (or at least in my mind). They spent a long time discussing the possible solution to close the offending hole. Then one of them, I can't remember who, came up with the idea of putting a post stamp adequately moistened on a coin and then proceeded to throw in the general direction of the hole in the hope that the stamp would remain attached to the ceiling sealing the hole. This exercise went on for hours, becoming rowdier and rowdier. My mother ushered my brother and myself rapidly to our room before things got out of hand. Being a pragmatic, rational and logical person I was extremely perplexed and curious to see the result of all the combined efforts. So, early next morning I went into the sitting room and the hole was still there. For years I was haunted by the thought, "why did they not use a ladder"? I was not a physicist!

Professor Palma, of the University of Catania, invited my father to give a talk on AdA. We stayed at the hotel Jolly in Catania. One evening, in the hotel foyer, a "bunch" of professors, including my father, started assembling mini rockets made out of the local Italian matches, the "cerini". The exercise involved rapping the tip of the matches with silver foil (in this case using the rapping of a packet of cigarettes). You then splayed the matches in a vague semblance of a tripod and then lit the heads of the matches and these lethal objects would then start flying across the hotel foyer to the utter dismay of staff and guests. Inevitably, and I am convinced it was my fathers rocket that hit a curtain and set fire to it. As you can imagine, total chaos ensued. By some quirk of fate, 30 or so years later I was to stay in the very same hotel. When I checked in, I prayed none of the staff would recognize me as the son of the man who set the hotel ablaze!

The Touschek family always loved the sea, whenever we could we enjoyed the beautiful coastlines of Italy and not only. Finally, the day came that my father decided to acquire a boat. Heaven forbid. He bought a 15 foot "pilotina" and at this stage Prof. Touschek rose to the self appointed rank of admiral. Our life became hell overnight. As the boat was being assembled, Bruno decided he wanted a toilet on board, not a marine toilet but a proper ceramic one with all the amenities including a hand pump needed to clear the offending remains. Now you have to realize the very limited space that was available in the cabin, roughly two and a half square meters and the toilet stood proudly as a throne taking up much of the space. The builder suggested sheepishly, that it should be covered so as not to be seen during normal hours. My father rapidly agreed and the builder concocted something that looked eerily like a guillotine and turned out to be just that. In order to use the shameful object, you had to lift the cover, make sure you hooked it carefully or else, as you were busy pursuing your ablution the lid could swiftly decapitate you. Well, having achieved his lifelong dream of having a toilet on board a tiny boat, we set sail followed by the raucous laughter of the builder and the casual passerby. It was humiliating. The boat turned out to be bow heavy due to the ceramic wc. So, when we turned on the outboard engine, the propeller was halfway out of the water barely pushing the boat forward. Now any sensible person would at this stage have eliminated the offending object. Not my father, he persevered in his criminal intent and decided against all logic to extend the drive shaft of the engine by about ten inches, this way the propeller was fully submerged. The maiden voyage was something out of a play of Max Frisch and the theatre of the absurd. Admiral Touschek and his crew sets sail for the Argentario from the harbour of Civitavecchia a distance of roughly 30 nautical miles. Fully equipped making the boat even more bow heavy. We leave the harbour in a cloud of shame whilst my father stood proudly at the helm of what he felt was the sister ship of the Queen Elizabeth cruise ship. And off we go with a boat that defeated all standards of nautical engineering. Ten hours later we reach Porto Ercole at sunset, after a gruelling 10 h of navigation which had taken its toll on the Admiral, use of the toilet on board proved to be impossible during navigation and at this stage my mother was definitely feeling uncomfortable and commenting on the mental instability of her husband, the Admiral. At this stage, the Touschek crew was faced with another gruelling prospect. Berthing the boat in the harbour of Porto Ercole. You have to understand my father was still convinced he had the sister ship of the Queen Elizabeth and would not take berth next to the other boats, he opted for a space in between two oil tankers. After having successfully moored the boat, my father realised there was a vertical wall nearly twenty feet high, how on earth would we get off the boat. Worry not, my father had a solution, by means of a series of pulleys...... Well, I leave you to draw the inevitable conclusion. After this experience, my brother and I were completely put off by the world of physics. Logic and good common sense eluded the mind of the Admiral.

My memories of Frascati were also happy ones. My brother and I enjoyed the hustle and bustle of the building of AdA and Adone. We frequently played in the building that was to house Adone. There was an atmosphere of excitement that permeated the whole establishment. **Burl** had a great gift, he was a catalyst, people were drawn to him and as children we enjoyed the limelight reflected on us. During lunch we would take the short drive to lake Albano and enjoyed swimming and snorkelling for a few hours. It was a blissful childhood until it lasted.

In his final years, I enjoyed a brief spell with him, when he spent more and more of his time at the Accademia dei Lincei. He came up with the brilliant idea of setting up "lectio magistralis" recordings. In the late 1970s video recordings were rare and expensive. The Accademia provided us with sufficient funds to buy video recording equipment. We set up the camera in one of the lecture halls of the Accademia and I recorded my father giving a brilliant lecture followed a few days later by Paul Dirac and many others thereafter. There are over a hundred recordings available to the general public.

I then returned to the UK to pursue my own studies in the meantime he moved to Geneva. He died a few months later. Aged 57.

My mother and I went to collect his remains in Innsbruck. They could not find his body. Together with an orderly we found his remains on the floor of the cellar of the hospital. And then began the ordeal of trying to obtain a pension from the Italian government. It took eleven years to obtain to 262 euro a month for my mother.

This short introduction of mine is dedicated to the extraordinary figure of my mother Elspeth Jennifer Yonge in Touschek who stood by his side in good and bad health and my dear brother who died prematurely, aged 48 of the same ailment that killed my father.

A very special thank you goes to Prof. Luciano Maiani who made all this possible and to two special ladies for whom I have a great deal of respect, Dr. Lia Pancheri (a former student of my fathers) and Dr. Luisa Bonolis whose kindness has become a rare commodity.

To conclude, my gratitude to all the organizing committee who have helped me metabolise the premature loss of my father.

Chapter 2
Bruno Touschek (1921–1978). The Path to Electron-Positron Collisions

Luisa Bonolis

Abstract The 100th anniversary of Bruno Touschek's birth also marks 60 years since the first beams of electrons and positrons circulated in AdA, the first ever matter-antimatter collider, built in Frascati National Laboratories following Touschek's visionary proposal of February 1960. Touschek's path to such idea is briefly outlined, beginning with his early years as a student—first in Vienna and later in Germany—through his first experiences with Rolf Widerøe's betatron and the electron synchrotron accelerator in Glasgow after the war, along with his relationships with the fathers of modern physics in Europe, to his arrival in Italy, where crucial reflections during the 1950s led him to the profound belief that matter-antimatter annihilations should become a primary goal for future physics. Based on these premises, Touschek and his collaborators dared to take on the challenge of realizing what at the time seemed an "unthinkable idea": keep beams of electrons and positron circulating for hours in the vacuum chamber of a storage ring and making them collide.

2.1 A European Scientist

The centennial of Bruno Touschek's birth inaugurates a new phase in the historical studies on one of the most original figures of 20th century physics. Touschek's scientific path is closely intertwined with his personality and of course his life—a life in many respects quite out of the ordinary. Like many of his generation, Touschek went through a dramatic period of the last century, but he also experienced the enthusiasm and excitement of struggling for the reconstruction and revival of European physics after the tragedy of World War II. And indeed, Touschek's life both as a scientist and an intellectual, unfolded across Europe in space and time in different phases during which he had the chance to come in contact with some of the most influential European physicists and more in general with different scientific communities that greatly enriched his cultural background and scientific thought. His mentors were

L. Bonolis (✉)
Max Planck Institute for the History of Science, Boltzmannstrasse 22, Berlin, Germany
e-mail: lbonolis@mpiwg-berlin.mpg.de

© The Author(s) 2023
L. Bonolis et al. (eds.), *Bruno Touschek 100 Years*,
Springer Proceedings in Physics 287,
https://doi.org/10.1007/978-3-031-23042-4_2

the main protagonists of 20th century physics, among the founding fathers of quantum theory and of the new quantum mechanics. Touschek was essentially self-taught until graduation, and yet had the opportunity to interact with many eminent German scientists and to gain practical experience as a theoretical physicist working on the project for the construction of a betatron during the war. Precisely because of his peculiar training as a physicist, who did not follow a standard path, he was able to conquer his own very special scientific style, resulting from the interweaving of his human and scientific vicissitudes against the backdrop of the annexation of Austria to Hitler's Reich, the war years and the early post-war period. He eventually landed in Italy, where his creative potential was able to flourish in contact with a dynamic scientific reality at the reconquest of excellence after the dramatic consequences of Mussolini's racist laws and the war. In Italy, working at the Physics Institute of Sapienza University of Rome and the INFN Frascati National Laboratories, Touschek conceived and built AdA, the first matter-antimatter collider, and in France, at the Laboratoire de l'Accélérateur Linéaire, he finally proved with his Franco-Italian team that the collisions had taken place.

2.2 Intellectual and Family Roots. Childhood and Early Youth in Vienna

Bruno Touschek was born to Camilla Weltmann and Franz Xaver Touschek on February 3rd, 1921, in Vienna, where he spent his childhood and early youth. Between the end of the 19th century and the beginning of the First World War, the Austrian capital was a highly cosmopolitan city and one of the most important centers of scientific advancement. But it was equally a center for the creation of modernity, the cradle for a number of ideas that shaped the whole 20th century and flourished in art, architecture, design, literature, science, philosophy and music. Among other political and artistic movements, the Austrian capital was a home to psychoanalysis, but also to Nazi ideology [1]. At the epicenter of this multifaceted world was the writer and essayist Karl Kraus. His caustic satirical spirit and his cultural engagement in the fundamental ideological issues of his time, had a profound influence on Touschek's intellectual formation.

Touschek's own family was actively involved in this scenario. His mother Camilla Weltmann and his aunt Ella—and Ella's own husband, the architect Josef Margold—were active in the circle of the Wiener Werkstätte, the association evolving from the Vienna Secession movement, founded by Josef Hoffmann and Koloman Moser as an alliance of artists, architects, designers and artisans that pioneered modern design and eventually influenced the Bauhaus movement as well as the Art Deco style. His maternal uncle Oskar was also an important reference figure for Bruno in his early years. He committed suicide in 1933, while Bruno's mother, Camilla, had already died, when he was only 9 years old (Fig. 2.1).

Fig. 2.1 Left: Bruno as a child with his mother Camilla Weltmann. Right: "Russia supreme war council", drawing made by Touschek when he had just turned 6 years old (title and date on the back: "Obersterkriegsrat Russland", March 31, 1927). *Credit* Francis Touschek

Touschek grew up in such amazing cauldron of great avant-garde cultural movements. His precocious talent for drawing, was influenced by the innovative expressionism of Egon Schiele and the famous psychological portraits of Oskar Kokoschka, artists who challenged established ideals of 'beauty' and shaped new ways to look at art seeking novel subjects for representation in their work. The style of Touschek's own drawings—well-known among friends and colleagues—testifies the persistence during his whole life of such strong and lively bonds with the rich cultural and intellectual world of his home town, that subtly blended in his personal and very original style.

2.3   University Studies in Vienna After the *Anschluss*

He had just turned seventeen years old on March 15, 1938, when Hitler announced the *Anschluss*, in Heiden Square, in front of an oceanic crowd. Due to his Jewish origin on the maternal side, the annexation of Austria into Nazi Germany completely turned his life upside down and dramatically affected his future forever. At that age, he was deeply aware of and intensely suffered in experiencing the dramatic events that were happening around him.

He was no longer able to attend classes as a regular student at school, but in 1939, after taking his Matura, his final examination, he enrolled in physics at the University of Vienna whose tradition included scientists such as Ludwig Boltzmann and Ernst Mach, who had influenced the turn-of-the century new generation of physicists, such as Lise Meitner and Paul Ehrenfest. Pauli himself, born in 1900, was raised among the intellectual elite of Vienna and Ernst Mach had been his godfather and first mentor. However, in June 1941, Touschek was expelled from the University for racial and political reasons and could only continue his studies privately, helped by

Fig. 2.2 Bruno Touschek's passport photo at 18 years old. *Credit* Francis Touschek

Paul Urban, who had received his Ph.D. in theoretical physics under the supervision of Hans Thirring, and had been his assistant at the Institute for Theoretical Physics at the University of Vienna (Fig. 2.2).[1]

2.4 Moving to Germany Protected by Arnold Sommerfeld

Paul Urban put Touschek in contact with Arnold Sommerfeld, who had educated and mentored a whole generation of young physicists and students (notably Werner Heisenberg and Wolfgang Pauli) who had a key role in the new era of theoretical physics. Sommerfeld had also openly supported Einstein and his work when the latter had been attacked by Philipp Lenard and Johannes Stark, who labeled relativity and quantum mechanics as *Jewish Physics*. For defending his Jewish colleagues Sommerfeld was forced into retirement when the Nazis came to power in 1933. Touschek had carefully studied Sommerfeld's classical treatise *Atombau und Spektrallinien*, finding some small errors. The beginning of their correspondence dates back to that time. In the years following the advent of the Nazi regime, Sommerfeld became increasingly concerned about the fate of physics in Germany—in particular theoretical physics. With his well known ability in the discovery of talents, he saw in such a gifted student a promise for the future. By the early 1940s, German physics had already lost so many brilliant scientists and there were no longer any Jewish physics professors left after the Nuremberg Laws. Somerfeld helped Touschek to find a work in Hamburg in an electronic firm led by Günther Jobst, a former student of his. In early 1942 Touschek abandoned Vienna and with great courage continued to pursue his passion for physics during the war years. Despite still being unable to

---

[1] For further details on Touschek's life and science see Giulia Pancheri's contribution in these proceedings and her volume *Bruno Touschek's Extraordinary Journey—from death-rays to antimatter* (Springer 2022) [3].

Fig. 2.3 Bruno Touschek's drawing in a letter written to his father on September 11, 1944. The drawing depicts one of his journeys from Vienna to Berlin and the makeshift seat he arranged by means of his own and another suitcase and his faithful typewriter that he always carried with him. *Credit* Francis Touschek

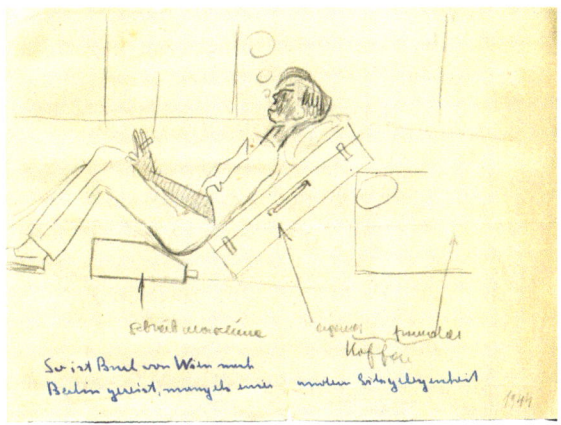

attend classes as a regular student, he continued his studies in Germany protected by Sommerfeld's colleagues and friends in Hamburg and later in Berlin, and at the same time worked to support himself. Hamburg University had risen to international fame during the era of the Weimar Republic thanks to its outstanding scholars, such as Ernst Cassirer, Erwin Panofsky, Otto Stern. As a Jew, after Nazi's seizure of power in 1933, Stern had been forced to resign from his post, similarly to many other colleagues, notably Panofsky, who was Pauli's and Einstein's friend. Both Panofsky and Stern found refuge in the United States, and the latter resided there when he was awarded the 1943 Nobel Prize in Physics. Panofsky's younger son, Wolfgang, became a renowned physicist and later played a relevant role in the art of particle accelerators. In the 1950s, his path eventually crossed Touschek's own way to new physics with colliding beams.

In Hamburg, Touschek established a strong relationship with Wilhelm Lenz (who had been Sommerfeld's student and his assistant in Munich), and was Director of the Institute of Theoretical Physics. Lenz had trained Ernst Ising and his assistants had included Pascual Jordan and Wolfgang Pauli. In Hamburg Touschek also became friends with Hans Jensen and Paul Harteck (of Austrian origin), both members of the *Uranium Club*, the German nuclear project led by Werner Heisenberg since 1942.

In late 1942, Touschek moved to Berlin and began to work at the Löwe Opta Radio company. In Berlin and Hamburg, Touschek experienced the heavy bombing raids and in particular the horror of firebombing of which he gave a chilling but very lucid and detailed description in his letters to his father and stepmother, often enriched with vivid drawings to complement the rich account of his daily life. This large group of letters constitutes a precious and irreplaceable documentation, a direct record of Touschek's life starting from 1939, when he was forced to abandon his native city (Fig. 2.3).

In Berlin he met Werner Heisenberg, Karl von Weizsäcker and Max von Laue, who showed great interest in his studies. Planck himself was still alive. Touschek followed their lessons and participated in seminars. But in 1944, when bombing began to menace the Max Planck Institute for Physics in Dahlem, Heisenberg and the others moved to Hechingen, in southern Germany, where they continued to work on the German nuclear project.

Those were still the months of Hitler's conquest of Europe; German troops had even occupied Paris in 1940. But since 1942, as the war situation worsened for Germany, Hermann Göring, high commander of the Air Force, fostered great hopes in the saving power of wonder weapons, the so called *Wunderwaffen*, which actually were also part of the propaganda disseminated by the government to keep up the morale of the population and instill confidence in the war resources of the Reich [4].

2.5   Building a 15-MeV Betatron with Rolf Widerøe

In early 1943, Touschek read an article submitted to the journal *Archiv für Elektrotechnik* by the Norwegian electric engineer Rolf Widerøe and describing a project for a betatron, a new kind of accelerator built by Donald Kerst in US, inspired by Widerøe's own ideas of several years before [5].

Widerøe's article entitled *Der Strahlentransformator*, was actually prepared for publication, but it was never included in the volume 37, issue 8, of the *Archiv für Elektrotechnik*. A copy is preserved in the Max Planck Archives in Berlin, together with a second typewritten manuscript with plans for a more powerful betatron. Following Touschek's indication of the interest of the proposal, the betatron immediately became a secret project of the Luftwaffe, as the military expected it would be able to produce powerful death rays to be used against enemy aircraft during air battles.[2] The possibility of producing high-energy X-rays for such aims was of course immediately set aside on scientific grounds, but nevertheless resulted in funding the construction of a 15-MeV betatron by the Reich Aviation Ministry, in view of a 100- and even 200-MeV machine.

Touschek was involved in this project as a theoretician. For the first time he found himself facing as a physicist both the challenge and the responsibility of the theoretical part of a relevant project, in addition to having to deal with a completely new field. This implied understanding what happens to charged particles circulating in an accelerator such as the betatron, subjected to electric and magnetic fields, as well as studying a series of phenomena that only in those years began to be tackled. This involvement temporarily saved him from being deported to forced labor by the Todt Organization, the OT, which he often mentioned in his letters home. This would have meant to be treated as a slave, doing extremely hard work in terrible conditions, which were often impossible to survive.

However, in the meantime, the Gestapo continued to keep an eye on him. In mid-March 1945, after having helped to move the completed betatron away from Hamburg to escape the advance of Russian troops, Touschek travelled back to Hamburg, where

---

[2] See B.I.O.S. Final Report No. 201, Visit to C.H.F. Müller A.G., Hamburg, p. 3. https://www.cdvandt.org/bios-201.htm.

something terrible happened, as he wrote in the first letter to his father after the war: "*I went back to sleep to be waken up at 7.30 in the morning by two gentlemen. I was so sleepy that when they said: 'Secret state police!' I answered: 'Yes, but at midnight?' [...]*. Despite the dramatic vicissitudes he had experienced, which only after many months he was able to tell his loved ones, he managed to minimize them in their eyes with the power of his extraordinary sense of humor.

2.6   Surviving Gestapo Captivity and a Death March to Kiel

During the weeks following his arrest, Touschek went through a horrible experience, languishing in a Gestapo prison and being shot twice during a march to the concentration camp of Kiel since he had fallen to the ground having a very high fever and under the weight of a large quantity of books that the prisoners had been forced to carry. Around 200.000 people died during such so-called death marches. Orders were to shoot the weakest prisoners and all those lagging behind or attempting to flee [6]. By that time, his maternal grandmother had already died at the concentration camp of Theresienstadt.

After being shot, Touschek was left there on the side of the street, but he was still alive and was imprisoned again and bounced around between different prisons until he was finally released a few days before the end of the war on the initiative of the

Fig. 2.4 "Round stamp of liberated Austria", drawing in a letter written to his father in March 1950. The drawing is created as a graphic extension of the family nickname "Burl", his usual signature in letters to the family. *Credit* Francis Touschek

betatron team. Otherwise, he would have certainly been shot by the German troops retreating during the last hours of the war (Fig. 2.4).

2.7 In Göttingen with Heisenberg and Other Members of the *Uranium Club*

He miraculously survived such dreadful events and after the war, he finally obtained his degree in Physics at the University of Göttingen with a dissertation on the theory of the betatron based on his reports to British Intelligence,[3] which testify how crucial were his participation and his role as a theoretical physicist in Widerøe's betatron project.[4] In Göttingen, Touschek came in contact with many members of the *Uranium Club* who had been released from their internment in UK at the beginning of 1946. The integrity of scientists such as Max Planck, Otto Hahn, and Max von Laue had never been in doubt. With their help and the support of the British authorities it became possible to initiate the reconstruction of fundamental scientific activity in Germany from the ashes and destruction of Nazi's regime. In the meantime, as Touschek wrote to Sommerfeld, he made "a bit of neutrino theory", "a bit of radiation damping," as well as "betatron calculations".[5] For some time, he was Werner Heisenberg's assistant at the Max Planck Institute for Physics in Göttingen, continuing his formation under the influence of the great German theoretical school.[6]

2.8 In Glasgow (1947–1952): The Making of a Theoretical Physicist

On the whole, such a difficult period in his life turned out to be a major step along the way to the first matter-antimatter collider. Despite having lost several years of his

---

[3] Copies of such reports on the theory of the betatron written by Touschek (Zur Theorie der Strahlentransformators; On the Starting of Electrons in the Betatron; Die magnetische Linsenstrasse und ihre Anwendung auf den Strahlen-Transformator, 1945; Zur Frage der Strahlungsdämpfung im Betatron, 1945) can be found in Rolf Widerøe's papers at the Eidgenossischen Technischen Hochschule (ETH) in Zurich.

[4] Touschek's contribution is clearly acknowledged also in a detailed British Intelligence Objective Sub-Committee (B.I.O.S.) report on *European Electron Induction Accelerators* prepared in October 1945 by the U.S. Naval Technical Mission in Europe: "In collaboration with the design work of Wideroe, a considerable amount of theoretical work was carried out by Touschek which was known to have been of invaluable aid in the development of the 15-MV accelerator. Further theoretical work has been done by Touschek on the starting of electrons in the accelerator" (Miscellaneous Report No. 77, Technical Report No. 331–345 prepared by H. F. Kaiser. https://www.cdvandt.org/bios-miscell-77.htm, p. 6).

[5] Arnold Sommerfeld's papers, Archives Deutsches Museum, Munich.

[6] On this period of Touschek's life see [7–9].

Fig. 2.5 Bruno Touschek with the physicist Samuel C. Curran, his colleague at Glasgow University. *Credit* Francis Touschek

youth, he re-emerged from the war and early-post war years as one of the first physicists in Europe endowed with a unique expertise about the theory and functioning of accelerators. And so, in early 1947, the British Intelligence arranged for his transfer to Glasgow University, in the Physics Department directed by Philip Dee, where they had built a 30-MeV electron synchrotron as a testing ground for the planned 300-MeV machine. While being involved in theoretical studies and in the building of the 300-MeV electron synchrotron (Fig. 2.5), Touschek was also consulted as a betatron expert by other research centers in UK. In 1949 Touschek was awarded a Ph.D. in Physics with a dissertation entitled *Collisions between electrons and nuclei*. His external supervisor was Rudolf Peierls, who had studied under Sommerfeld and Heisenberg, and had been Pauli's assistant in Zurich. While in UK with a post-doc grant, Peierls had decided not to return home after Hitler's rise to power in 1933 because of his Jewish background.

In UK Touschek came also in contact with Max Born, one of the fathers of the new quantum mechanics, and collaborated with him writing an appendix for a new edition of Born's book *Atomic Physics*. Born himself, being Jewish, had been suspended from his professorship at the University of Göttingen in 1933 and had moved to UK. Physicists such as Heisenberg, Pascual Jordan, Pauli, Edward Teller, Eugene Wigner, Viktor Weisskopf, had been his Ph.D. students or assistants. Born received the Nobel Prize in 1954.

In Glasgow, Touschek became a full-fledged theoretical physicist. He did theoretical work on the phenomenology of meson physics at accelerators, and began to work in quantum field theory. He thus experienced the initial phase of particle physics with accelerators, the transition from cosmic rays to more systematic studies with artificial beams of particles as one can see from Touschek's published works between 1948 and 1951.[7]

At that time he collaborated with Walter Thirring on the article *A covariant formulation of the Block–Nordsieck method* [12]. They both shared a growing interest in quantum electrodynamics which was becoming a hot topic and led Touschek to

---

[7] See [10] for a complete list of Touschek's works.

a friendship with the well-known Italian physicist Bruno Ferretti, who was in UK around 1948, and who eventually became the intermediary for Touschek's transfer to Italy.

As we know from letters to his father and especially to Arnold Sommerfeld, Touschek soon began to feel that in Glasgow he was rather far from the mainstream of theoretical physics. He was unhappy, and soon after being awarded the Ph.D., he felt that it was time to consider other possible positions in Europe. He missed Germany, and was particularly looking forward to working with Heisenberg again, as he wrote to Sommerfeld in October 1950: "*I have found that I am more comfortable in Germany and Austria than in Scotland, and that I have learned more in Göttingen in a month than in Glasgow in a year [...] If Professor Heisenberg comes to Munich and wants me, then I want him too, if it can be done somehow financially.*"[8]

In the very early 1950s, he shared lively discussions on quantum field theory with Bruno Ferretti, from whom he also heard about the great plans and expectations related to the rebuilding and relaunching of Italian physics, after the disaster of World War II. Both agreed that Touschek would obtain a leave of absence to be spent in Italy, a country he knew since his childhood, when he visited his second maternal aunt Ada, married with an Italian. In fall of the following year he wrote to his father: "*I have applied for a job in Rome [...] Only masochists can live in England in the long run [...]*".[9]

2.9   Participating in the Reconstruction and Revival of Italian Physics

Like Heisenberg in Germany, Patrick Blackett in UK, Pierre Auger in France—and similarly to other physicists who had experienced the flowering of European physics before the war—Edoardo Amaldi in Italy was deeply frustrated by the passage of leadership from Europe to the United States. Physics in Rome had a long-standing tradition that began immediately after the unification of Italy. The new Institute of Physics at Sapienza University, inaugurated in Via Panisperna in 1881 with first director Pietro Blaserna, who had studied physics in Vienna with Andreas von Ettingshausen, was the most modern in Italy for its research laboratories and for the quality of teaching [13]. Thanks to its second director, Orso Mario Corbino, the first chair of Theoretical Physics was established and Enrico Fermi, with his friend and collaborator Franco Rasetti, an excellent experimentalist, since 1927 opened the way to modern physics in Italy. While Fermi's group in Rome was making its well-known major contributions to nuclear physics, in Florence Bruno Rossi and his colleagues, which included Gilberto Bernardini, Giulio Racah, and Daria Bocciarelli were building a further novel research tradition which for several years would be a well established expertise of Italian physics: the study of cosmic rays.

---

[8] Arnold Sommerfeld's papers, Archives Deutsches Museum, Munich.

[9] On Touschek's stay in Glasgow in the years 1947–1952 see [11].

In 1937 the Physics Institute moved from via Panisperna to the new campus of Sapienza University, in the exemplary rationalist building designed by architect Giuseppe Pagano and named after Guglielmo Marconi. Soon after, in 1938, the Roman group definitely disbanded when Fermi decided to move to US and did not come back after being awarded the Nobel Prize. Bruno Rossi, being of Jewish family, was expelled from the Institute in Padua, which he directed—and whose design and construction he had personally supervised—and eventually emigrated to US.

The Italian physics community—and not only—was decimated by the racist policies implemented by Mussolini's fascist government, and deprived of some of its most influential and prestigious members. With them, many others left Italy because of political or racial reasons. After the isolation of wartime, Italian physicists were left with the difficult task of relaunching the field on a new basis, in line with the latest developments but at the same time taking into account the difficult economic situation of the country, half-destroyed by the consequences of the conflict. Edoardo Amaldi and Gilberto Bernardini, heirs of the fathers of modern physics in Italy and bridging that glorious tradition over the war years, initiated an intensive program for the revival of physics in the country and in parallel began promoting an international strategy to relaunch physical sciences in Europe. Amaldi became a key figure in the birth of CERN and the European Space Agency, and Bernardini was CERN'S first Director of Research and first president of the European Physical Society, of which he had strongly advocated the foundation.

Deeply aware of Touschek's potential, which was particularly suited to such ambitious plans, Edoardo Amaldi invited him officially in Rome with an INFN contract.[10]

At the end of 1952, when Touschek arrived in Italy, the National Institute for Nuclear Physics had just been founded with first president Bernardini and Italian physicists were deciding to establish a national laboratory for high energy physics in order to host an electron synchrotron, a new-generation machine they had planned to build as a powerful tool for elementary particle physics. The direction of the project was given to the 33-year-old Giorgio Salvini. After having been for many years at the frontier of nuclear and cosmic-ray research, Italy would have been able to regain a prominent international position, in particular in the sub-nuclear realm. And indeed, Touschek's unique expertise found a very fertile ground and turned out to be destined to have a profound influence on the future of this field, both theoretically and experimentally (Fig. 2.6).

---

[10] As emphasized by his son Francis, Touschek had always a special relationship with Amaldi, who showed his affection and esteem by writing with great commitment an accurate biography after Touschek's death [10].

Fig. 2.6 Left: Bruno Touschek with Edoardo Amaldi in the 1950s. Right: Touschek's portrait of Giorgio Salvini, director of Frascati National Laboratories. *Credit* E. Amaldi Archives, Physics Department, Sapienza University, Rome; Francis Touschek

2.10   Towards Matter-Antimatter Physics

During the 1950s, Touschek further evolved as a theoretician, giving relevant contributions to the study of discrete symmetries in particle physics, time reversal and neutrino physics also collaborating with several Italian physicists. Let me note in passing that, during the war, we know from the correspondence that he studied Eugene Wigner's book *Group Theory and its Application to the Quantum Mechanics of Atomic Spectra* (published for the first time in 1931) at a time when there wasn't great interest in the subject. But he was aware of its importance, and commented in a letter that it was a kind of "treasure" that he would put aside for later times...

In the 1950s, Touschek had a scientific correspondence with Pauli, whose interests were since some time centered on quantum field theory—and had already resulted in two fundamental pillars of the theory: the spin-statistics theorem and the CPT theorem. The exchange of letter intensified between 1957 and 1958. At that time, the shocking discovery of Parity violation in weak interactions was increasing interest in the discrete symmetry operations, a topic which both actually discussed since 1954. Pauli and Touschek also wrote a joint paper, published in 1959 as a contribution to the *International School of Physics "Enrico Fermi"* (8th Course: "Mathematical problems of the quantum theory of particles and fields") and which appeared only when Pauli had already passed away, in December 1958 [14]. After the funeral, Touschek wrote in a letter to his father: "*Without him, for me physics is only half interesting [...]*".

Nevertheless, all this was instrumental in preparing his mind for further crucial reflections (Fig. 2.7).

Fig. 2.7 Left: Bruno Touschek with Tsung Dao Lee and Wolfgang Pauli in Venice at the International Conference on *Mesons and recently discovered particles* in September 1957. Right: Touschek's joking portrait of T. D. Lee alluding to the recent discovery of parity violation, widely discussed during the conference. *Credit* Francis Touschek

2.11   Colliding Beams in the 1950s: e^-e^- versus e^+e^-

During the 1950s, while being actively involved in the life of the Italian scientific community, brilliantly integrated into such lively academic and scientific environment, Touschek closely experienced the birth and development of Frascati National Laboratories that had been established to host the brand new accelerator. Towards the end of 1959, Touschek was around in the Labs directed by Giorgio Salvini, where the 1100 MeV electron synchrotron had just gone on line. Touschek had frequent conversations with Carlo Bernardini on the research perspectives: "*Bruno was continuously exploding in his picturesque Austro-Italian, being rather unsatisfied with the experimental opportunities. He insisted that experiments had to be clean in channels with very definite quantum numbers, thus excluding the overcrowded proton reactions. Electrons appeared to him as 'gentle probes'*" [15, p. 8]. In Touschek's own words, "*On hitting their target [...] a beam of protons loses its identity [...] Protons are a rich source of events, which are difficult to interpret because the witnesses are too much involved. Electrons peering gently at their targets rarely produce spectacular events, but what they produce can be more easily interpreted*".[11]

Since the early 1950s, an extraordinary progress in high-energy physics was mainly due to the availability of large proton accelerators such as the Cosmotron (3 GeV, 1952, Brookhaven), the Bevatron (6.2 GeV, 1954, Berkeley), and the CERN Proton-Synchrotron (28 GeV, 1959, Geneva).

On the other hand, when it started to work the Frascati machine was one of the three biggest of its kind in the world, the other two were in the USA, at Cornell and at

---

[11] B. Touschek, *Ada and Adone are storage rings* (incomplete manuscript, Bruno Touschek Archive, Physics Department, Sapienza University of Rome, from now on B. T. A., Folder 11, 3.92.4, p. 5).

Caltech. *"That the machine which would bring Italy to a level with international and in particular U.S. high energy physics should be an electron accelerator*—recalled Bruno Touschek later—*was a courageous choice if confronted with a general tendency of physicists who at the time were bent on producing proton accelerators."*[12] *"At the same time, however*—Touschek further remarked—*new preoccupations arose. All over the world newer and bigger machines were being built and planned and it was felt that if Frascati wanted to keep abreast something big and new had to be planned."*[13]

Since the second half of the 1950s, US physicists had been proposing to exploit the colliding beam technique to obtain a larger center-of-mass energy and to carry out high-precision experiments to test the predictions of QED, in particular to investigate "the breakdown of electrodynamics" [15, p. 4]. In parallel, also Gersh Budker with his team in USSR was planning an e^-e^- experiment, Vep-1, apparently also discussing the interest of e^+e^- [16, 17]. Thus, at the end of the 1950s, the storage-ring and the colliding beam ideas were not new.[14]

As recalled by Touschek himself, *"The interest in storage rings at Frascati was started by a visit of Dr. Panofsky in the autumn of 1959"* [19]. In fall 1959, Pief Panofsky, who at the time, was already planning the future 2-Mile accelerator at Stanford, gave a seminar in Rome presenting the US Princeton-Stanford e^-e^- project. Raul Gatto and Nicola Cabibbo, who were both present, well remembered how Touschek immediately reacted proposing a different idea, based on quite different scientific aims, involving the physics of particles and antiparticles: *"It was after the seminar that Bruno Touschek came up with the remark that an e^+e^- machine could be realized in a single ring, 'because of the CTP theorem'* [...]" [20, p. 219]. To support his view on scientific grounds, *"Bruno kept insisting on CPT invariance, which would grant the same orbit for electrons and positrons inside the ring!"*[15]

We also have Touschek's own recollections about the late 1950s, the years ushering the transition to a new phase of his scientific adventure: *"At the time I felt rather exhausted from an overdose of work which I had been trying to perform in the most abstract field of theoretical research: the discussion of symmetries which had been opened up by the discovery of the breakdown of one of them, parity, by Lee and Yang. I therefore wanted to get my feet out of the clouds and onto the ground again, touch things (provided there was no high tension on them) and take them apart and get back to what I thought I really understood: elementary physics [...]"*.[16]

---

[12] B. Touschek, *AdA and Adone are storage rings* (manuscript, B. T. A, Folder 11, 3.92.4, p. 4).

[13] B. Touschek, *A Brief Outline of the Story of AdA* (manuscript, B. T. A., Folder 11, 3.92.5, p. 3).

[14] See [18] for a discussion on related US projects during the 1950s.

[15] R. Gatto, personal communication, January 15, 2004.

[16] B. Touschek, *AdA and Adone are storage rings*, manuscript, B.T.A., Box 11, Folder 3.92.4, p. 7.

2.12   From CPT to AdA, the First Particle-Antiparticle Collider

Then, in February 1960, during a meeting in Frascati Touschek surprised everybody proposing to go far beyond experiments with gamma beams obtained by hitting electrons against a fixed target inside the synchrotron, or even experiments such as those US physicists were scheduling at Stanford with two colliding beams of electrons stored in two tangent rings.[17]

According to Touschek, what would really be worth exploring, instead, was the physics of electron-positron annihilations, which would allow *to open a channel into the hadronic world through the quantum numbers of* e^+e^-. He tried to convince Giorgio Salvini, director of the National Laboratories, to re-convert the electron-synchrotron—that had just gone into operation—into an electron-positron collider. During the discussion that followed, Giorgio Ghigo suggested to build a small dedicated machine in order to perform the experiment proposed by Touschek.[18]

The following day, Touschek started to make calculations on a new notebook, which he named "SR", for Storage Ring, since it was clear that the beam-storage problem would be the most serious one, as he himself recalled some time later: *"The challenge of course consists in having the first machine in which particles which do not naturally live in the world that surrounds us can be kept and conserved."*[19]

During a seminar held the following March 7, 1960, Touschek emphasized the creative character of e^+e^- collisions, i.e. the possibility of a complete transformation of the collision energy in the creation of new particles, and this through a channel with well-defined quantum numbers—those of a photon—and proposed as an example muon and pion pairs.

The colliding beam technique, which other physicists were planning to exploit both in USA and USSR—basically to obtain a larger center-of-mass energy or to perform high-precision experiments to test the predictions of QED—was definitely moving towards a conceptually novel stage. Because of the CPT symmetry, insisted Touschek, electrons and positron could circulate in the same orbit, in opposite directions, and eventually collide: *"One of the leading motivations for planning e^+e^- colliding beam experiments (rather than e^-e^- or $p - p$) was that in such an experiment one could 'observe' the virtual time-like photon [...]"*[20]

As recalled by Nicola Cabibbo, *"F. Calogero, R. Gatto. C. Zemachs, L. Brown (the two were spending their sabbatical in Rome) and myself rushed to compute the relevant cross-sections"* [20, p. 2].

---

[17] For a list of Kerst's, O'Neill's et al. works on the e^-e^- storage ring idea and related colliding beams project see [21, footnote 1].

[18] See preliminary draft of his proposal: B. Touschek, "Proposta d'esperienza", two manuscript pages, B.T.A., Box 11, Folder 3.87.

[19] B. Touschek, *A brief outline of the story of AdA*, excerpts from a talk delivered by Touschek at the Accademia dei Lincei on May 24, 1974 (B. T. A., Box 11, Folder 3.92.5, p. 8).

[20] B. Touschek, *The time-like photon* (manuscript, B. T. A., Box 11, Folder 3.92.9, p. 1).

Fig. 2.8 Left: Bruno Touschek on his boat on the Lake of Albano. Right: Touschek's drawing, part of the famous series of his surreal bicycles. *Credit* Francis Touschek

Carlo Rubbia has well expressed such conceptual leap: *"[…] in his mind electron-positron collisions were nothing else than the way of realizing in practice the idea of symmetry between matter and antimatter, in the deep sense of the Dirac equation […] His boundless enthusiasm for particle-antiparticle collisions was dominated by a sense of perfect and intellectual esthetics…"* [22, p. 59].

The response of the Laboratory was enthusiastic. The project was fully approved and immediately funded (Fig. 2.8).

2.13 The Garden of Eden of Electron-Positron Annihilations

Following his challenging ideas—based on his firm belief in CPT and QED—the first matter antimatter collider AdA (for 'Anello di Accumulazione', Storage Ring), a 4-m perimeter ring suited for electrons and positrons of up to 250 MeV, was built. It inaugurated a brand-new research line at Frascati National Laboratories, heralding a new era in high-energy physics. The team led by Touschek, including Carlo Bernardini, Giorgio Ghigo, Gianfranco Corazza, Ruggero Querzoli and Giuseppe Di Giugno, was able already in February 1961 to observe the light signal from a single circulating electron after its capture in the ring as a pulse in the phototube output, or even as a white-bluish spot that could be seen with the naked eye through a small porthole. Electrons lived for 30 or 40 hours thanks to an unprecedented vacuum of 10^{10} torr, obtained by Corazza, a leader in the field of vacuum technology. In parallel, together with Raul Gatto, Nicola Cabibbo—who had graduated in 1958 with Touschek as supervisor—fully explored the annihilation processes of interest that were presented in their a seminal paper *Electron-Positron Colliding Beam Experiments*, universally known as the 'Bible' in Frascati circles [23]. As recalled by Cabibbo, *"The result of this explorative work confirmed beyond the wildest dreams the intuitions of Bruno Touschek".* And emphasized how, *"While doing this work we had the exhilarating experience of expanding into a vacuum: for a few years the only theoretical papers on the physics of e^+e^- were those issuing out of Rome or in Frascati"* [20, p. 221].

2.14 AdA in Orsay, at LAL

In Orsay, at the Laboratoire de l'Accélérateur Linéaire, where AdA was moved in 1962 to benefit from the Linac high particle injection rates, Touschek and his team, now including Pierre Marin and Jacques Haïssinski, were able to demonstrate that the two beams had actually collided, through the observation of the single bremsstrahlung as a monitoring reaction. At the time, the overlapping of the two beams was often put in doubt, as Carlo Bernardini repeatedly recalled: *"How can you be sure that electrons and positrons will meet?"* To such a question Touschek invariably replied: *"Obviously, TCP theorem! Actually, CP is enough!"* [24, p. 170]. This epochal achievement of the Franco-Italian collaboration definitely proved the feasibility of this type of machine [25]. They also discovered an unexpected effect, a loss of particles from the stored beams reducing their lifetime, whose origin was immediately explained by Touschek as an intra-bunch scattering. But luckily, it turned out that the so-called Touschek effect scales sharply with energy and does not seriously affect more powerful colliders [26].[21]

In the meantime, the French started their own e^+e^- project, the collider ACO [29].

2.15 1960s–1970s: ADONE and the Formation of a Theoretical School in Rome and Frascati

In November 1960, even before AdA had showed the feasibility of electron-positron collisions, opening the way to higher energy and luminosity, Touschek prepared a draft plan for a bigger and more powerful storage ring for electrons and positrons, ADONE.[22] The official proposal was presented in January 1961 [30].

ADONE (1.5 GeV per beam, 105 m in circumference), which became operational in 1969, eventually discovered the multi hadron production making electron-positron physics a field of major interest, further encouraging the construction of a large family of high-energy colliders all over the world where new types of elementary constituents of matter were detected. In 1974, ADONE confirmed the existence of the J/ψ, a bound state of a charm quark and a charm anti-quark, discovered at

---

[21] For a reconstruction of the birth and development of the Franco–Italian collaboration see [27, 28].

[22] *ADONE—a draft proposal for a colliding beam experiment*, typescript, B. T. A., Box 12, Folder 3.89.

Fig. 2.9 Bruno Touschek
with his dog Lola. *Credit*
Francis Touschek

the Brookhaven National Laboratory and at the Stanford Linear Accelerator Center with the e^+e^- collider SPEAR. It was the first breakthrough discovery of the colliding beam technique and the first firm experimental evidence for the charm quark—establishing the quark model as a credible description of nature—for which Samuel Ting and Burton Richter were awarded the 1976 Nobel Prize in Physics.[23] As Touschek had predicted, e^+e^- physics was showing its intrinsic simplicity and power.

With AdA and ADONE Touschek created a brand new, major research line at Frascati Laboratories and made a further fundamental contribution with the formation of a theoretical school in Rome and Frascati. Touschek, who was particularly valued also as an extremely brilliant, fascinating and inspiring teacher,[24] was invited by Luigi Radicati to give lessons at the Specialization School of the Scuola Normale Superiore in Pisa.[25] In 1972 he was elected as Foreign Member of the Accademia Nazionale dei Lincei, while in 1975 he was awarded the prestigious Matteucci Medal by the Italian National Academy of Sciences. A glance at the list of distinguished physicists to whom the Prize has been awarded since 1870 places Touschek at the highest level of world physics (Fig. 2.9).

---

[23] See contributions by Pancheri, Pellegrini and Greco in these Proceedings.

[24] See contribution by G. Rossi in these Proceedings.

[25] In an interview Luigi Radicati recalled his friendship with Touschek: "*He was my closest friend, the person I was closest to among the physicists who were in Italy, without any doubt [...] A bizarre character, perhaps a little crazy, but very cultured and extremely intelligent*" (L. Radicati, interview by L. Bonolis, Rome, June 6, 1997).

Fig. 2.10 *Physics with Intersecting Storage Rings*, 'Enrico Fermi Summer School' directed by Bruno Touschek, held in Varenna in 1969. *Credit* Italian Physical Society

2.16   1977–1978: At CERN in Geneva

Between 1977 and 1978, Touschek spent the last months of his life as visiting scientist at CERN, at a time when early plans for a giant electron-positron collider were being discussed. However, when LEP eventually came into operation in 1989, Touschek was no more there. He had prematurely passed away, on the 25th of May 1978, while he was participating in the planning of the proton-antiproton collider ($Sp\bar{p}S$) proposed by Carlo Rubbia, with whom in the late 1960s he had long discussions about the possibility of transforming a conventional accelerator into a proton-antiproton collider [22, pp. 59–60]: "*Clearly in his and in our mind at the time the proton-antiproton option was the logical continuation of the AdA–Adone line.*"

According to Salvini, during a conference in Saclay in September 1966, a session was dedicated to Novosibirsk and the method of cooling antiprotons, as suggested by Budker [2, p. 62]: "*[...] Budker was only at the beginning of his report, and Bruno Touschek had understood everything; he was getting excited, could not keep himself [...] Bruno told us that morning: 'We cannot get highest energies with electrons, but we'll get them by proton-antiproton collisions. It is a most important development, and probably this is not the only way to tame antiproton beams.*"

About ten years later, further recalled Rubbia [22, pp. 59–60],"*The fire of the proton-antiproton collision was still burning in the back of my mind, and I must say that so it was in the mind of Bruno [...] As soon as he knew that the proton-antiproton collision adventure at last was actually going to start—although already terribly affected by his illness—Bruno decided to move immediately to CERN. I remember having long discussions with him first at CERN and then, toward the end, at the nearby Hospital de La Tour [...]*".

At that time also Giorgio Salvini, who was taking part in the preparation of Rubbia's UA1 experiment, visited Touschek quite often to keep him informed [2, p. 65]: "*[...] almost every day [we] discussed the developments of UA1 in detail.*"

Fig. 2.11 Bruno Touschek
during his last days, when he
was visiting scientist at
CERN. *Credit* Francis
Touschek

Rubbia concluded his remembrance of Touschek by saying [22, p. 60]: "*I have learned from Bruno how to love matter-antimatter reactions. Without this fact, my own scientific career would certainly have been very different. So I believe it is the case for many of us.*"

Unfortunately Touschek did not live long enough to see his ideas triumph and to witness the discovery of the $W^{\pm}$ and Z^0 vector bosons and the related 1984 Nobel Prize in Physics to Rubbia and Simon van der Meer.

2.17 Conclusion

Bruno Touschek's small AdA (about 1.3 m in diameter, storing beams of 250 MeV) has opened the way to new bigger matter-antimatter colliders and precision measurements which have been instrumental in confirming our understanding of the basic building blocks of matter in the Universe and the fundamental forces that operate between them. The detection of the long-sought Higgs boson at the Large Hadron Collider at CERN in 2012, has eventually completed the Standard Theory of particle physics. The fundamental contribution of AdA as a progenitor of entire generations of colliders was recognized on 5 December 2013, when the world's first particle-antiparticle accelerator—still visible on the grounds of INFN Frascati National Laboratories—was declared an Historic Site by the European Physical Society. This important recognition has definitely marked AdA's role as a milestone in the Italian and European scientific heritage.

As historians, we are still left with the task of in-depth investigations related to the evolution of Touschek's theoretical thought during the twenty years or so between the war years—and his work on the betatron theory—to the late 1950s, when such a long process finally materialized into his daring and drastic proposal, that he considered "*the future goal*" of Frascati Laboratories: transform the electron synchrotron, that had just begun to function, into an electron-positron collider and explore the physics of matter-antimatter annihilations. His bold idea was wisely and enthusiastically converted into the decision to build a dedicated small prototype, AdA, the first ever matter-antimatter machine, which in the early 1960s set the stage for a new era in particle physics.

The period from Touschek's arrival in Italy at the end of 1952, to the end of the 1950s, during which he fully developed into a mature theoretical physicist dialoguing with prominent theoreticians of his time, has not yet been thoroughly studied. In particular his scientific production as well as his scientific correspondence with Heisenberg, which dates back to the early post-war period, and continued during the 1950s, has yet to be analyzed, as well as his letters with Wolfgang Pauli, himself born in Vienna from a prominent Jewish family, of whose work Touschek had always been an attentive follower since his early youth. They had an intense exchange of ideas during 1957–1958, at a time when much of Pauli's work was still centered on quantum field theory.[26] Such a dialogue with Pauli and with other theorists (notably Charles Enz, Gerhart Lüders, Markus Fierz, Kurt Symanzik, Luigi Radicati, Giacomo Morpurgo, Marcello Cini) was instrumental in the development of his ideas on QED and discrete symmetries, as well as in stimulating his own reflections on the CPT theorem, the solid conceptual base for AdA, as can be derived from correspondence of the period preserved within his papers at 'Edoardo Amaldi Archives' at the Physics Department of Sapienza University in Rome, and in published articles.

The analysis of Touschek's scientific life in the 1950s is thus to be pursued as one of the main keys to a deeper understanding of all the implications of his unique path towards what became a standard practice: using matter-antimatter annihilations to probe the ultimate nature of the basic building blocks of the Universe and their interactions.

References

1. G. Steiner, Hitler's Vienna. Salmagundi **139/140**, 63–71 (2003). http://www.jstor.org/stable/40549618
2. G. Pancheri, *Bruno Touschek's Extraordinary Journey. From Death Rays to Antimatter* (Springer, Cham, 2022) https://doi.org/10.1007/978-3-031-03826-6
3. P. Waloschek, *Death-Rays as Life-Savers in the Third Reich* (DESY, 2012). http://www-library.desy.de/preparch/books/death-rays.pdf

---

[26] See Pauli's archive at CERN and Touschek's papers in Rome.

4. R. Widerøe, *The Infancy of particle accelerators. Life and work of Rolf Widerøe*, ed. by P. Waloschek (Vieweg+Teubner Verlag, Braunschweig, Germany, 1994). https://doi.org/10.1007/978-3-663-05244-9

5. U. Fentsham, Der "Evakuierungsmarsch" von Hamburg–Fuhlsbüttel nach Kiel–Hassee (12.–15. April 1945). Informationen zur Schleswig-Holsteinischen Zeitgeschichte **44**, 66–105 (2004)

6. L. Bonolis, G. Pancheri, Bruno Touschek: particle physicist and father of the e^+e^- collider. Eur. Phys. J. H **36**(1), 1–61 (2011). https://doi.org/10.1140/epjh/e2011-10044-1

7. G. Pancheri, L. Bonolis, The path to high-energy electron-positron colliders: from Widerøe's betatron to Touschek's AdA and to LEP (2018). https://doi.org/10.48550/arXiv.1710.09003. arXiv:1710.09003 [physics.hist-ph]

8. L. Bonolis, G. Pancheri, Bruno Touschek in Germany after the War: 1945-46 (2019). https://doi.org/10.48550/arXiv.1910.09075. arXiv:1910.09075 [physics.hist-ph]

9. E. Amaldi, The Bruno Touschek Legacy (Vienna 1921 – Innsbruck 1978). CERN Yellow Reports, No. 81-19 (1981). http://cdsweb.cern.ch/record/135949/files/CERN-81-19.pdf

10. G. Pancheri, L. Bonolis, Bruno Touschek in Glasgow. The making of a theoretical physicist (2020). https://doi.org/10.48550/arXiv.2005.04942. arXiv:2005.04942 [physics.hist-ph]

11. B. Touschek, W. Thirring, A covariant formulation of the Bloch-Nordsieck method. Lond. Edinb. Dublin Philos. Mag. **42**(326), 244–249 (1951). https://doi.org/10.1080/14786445108561260

12. M. Focaccia, *Pietro Blaserna and the Birth of the Institute of Physics in Rome. A Gentleman Scientist at Via Panisperna* (Springer, Cham, 2019). https://doi.org/10.1007/978-3-030-10825-0

13. W. Pauli, B. Touschek, Report and comment on F. Gürsey's. Group structure of elementary particles. Il Nuovo Cimento **14**(1), 205–211 (1959). https://doi.org/10.1007/BF02724849

14. C. Bernardini, From the frascati electron synchrotron to ADONE, in *Present and Future of Collider Physics, Conference in honour of Giorgio Salvini's 70th birthday, Italian Physical Society*, ed. by Bacci et al. (Editrice Compositori, Bologna, 1990), pp. 3–15

15. V.N. Baier, Forty years of acting electron-positron colliders (2006). https://doi.org/10.48550/arXiv.hep-ph/0611201. arXiv:hep-ph/0611201

16. A. Skrinsky, Accelerator field development at Novosibirsk (history, status, prospects), in *Proceedings of the 16th Particle Accelerator Conference and International Conference on High-Energy Accelerators, HEACC 1995, Dallas, USA*, May 1–5 (IEEE, 1996), pp. 14–26. https://accelconf.web.cern.ch/p95/ARTICLES/MAD/MAD04.pdf

17. L. Bonolis, Bruno Touschek vs. machine builders: AdA, the first matter-antimatter collider. La Rivista del Nuovo Cimento **28**(11), 1–60 (2005). https://doi.org/10.1393/ncr/i2005-10006-x

18. B. Touschek, The Italian storage rings, in *Proceedings of the 1963 Summer Study on Storage Rings, Accelerators and Experimentation at Super-High Energies*, June 10 to July 19, 1963, vol. C630610, ed. by J.W. Bittner (Brookhaven National Lab, Upton, NY, 1963), pp. 171–208 https://inspirehep.net/files/847d16bc41da3d4d8e099f8ed89f4a3e

19. N. Cabibbo, e^+e^- Physics - a View from Frascati in 1960's, in *Adone a Milestone on the Particle Way*, ed. by V. Valente (INFN Frascati National Laboratories, Frascati Physics Series, Frascati, 1997), pp. 219–225

20. L. Bonolis, Bruno Touschek Remembered. 1921–2021. Bibliography and Sources (2021). https://doi.org/10.48550/arXiv.2111.00625. arXiv:2111.00625 [physics.hist-ph]

21. C. Rubbia, The role of Bruno Touschek in the realization of the proton antiproton collider, in *Bruno Touschek Memorial Lectures, Frascati Physics Series*, vol. XXXIII, ed. by M. Greco, G. Pancheri (2004), pp. 57–60. http://www.lnf.infn.it/sis/frascatiseries/Volume33/volume33.pdf

22. N. Cabibbo, R. Gatto, Electron-positron colliding beam experiments. Phys. Rev. **124**(5), 1577–1595 (1961). https://doi.org/10.1103/PhysRev.124.1577

23. C. Bernardini, AdA: the first electron-positron collider. Phys. Persp. **6**, 156–183 (2004). https://doi.org/10.1007/s00016-003-0202-y

24. C. Bernardini, G. Corazza, G. Di Giugno, J. Haïssinski, P. Marin, B. Touschek, Measurements of the rate of interaction between stored electrons and positrons. Il Nuovo Cimento **34**(6), 1473–1493 (1964). https://doi.org/10.1007/BF02750550
25. C. Bernardini, G. Corazza, G. Ghigo, G. Di Giugno, J. Haïssinski, P. Marin, R. Querzoli, B. Touschek, Lifetime and beam size in a storage ring. Phys. Rev. Lett. **10**(9), 407–409 (1963). https://doi.org/10.1103/PhysRevLett.10.407
26. L. Bonolis, G. Pancheri, Bruno Touschek and AdA: from Frascati to Orsay. In memory of Bruno Touschek, who passed away 40 years ago, on May 25th, 1978 (2018). https://doi.org/10.48550/arXiv.1805.09434. arXiv:1805.09434 [physics.hist-ph]
27. G. Pancheri, L. Bonolis, Touschek with AdA in Orsay and the first direct observation of electron-positron collisions (2018). https://doi.org/10.48550/arXiv.1812.11847. arXiv:1812.11847 [physics.hist-ph]
28. J. Haïssinski, From AdA to ACO. Reminiscences of Bruno Touschek, in *Bruno Touschek and the Birth of e^+e^- Physics*. Frascati Physics Series, vol. XIII, ed. by G. Isidori (1998), pp. 17–31
29. F. Amman, C. Bernardini, R. Gatto, G. Ghigo, B. Touschek, Anello di Accumulazione per elettroni e positroni (ADONE), Frascati National Laboratories, Internal Report No. 68 (1961)
30. G. Salvini, From AdA to Tristan and Lep, in *Bruno Touschek Memorial Lectures*, Frascati Physics Series, vol. XXXIII, ed. by M. Greco, G. Pancheri (2004), pp. 61–68. http://www.lnf.infn.it/sis/frascatiseries/Volume33/volume33.pdf

Chapter 3
AdA at Orsay

Jacques Haïssinski

Abstract AdA, acronym for *Anello di Accumulazione* in Italian, was the first electron−positron storage ring ever built and operated. During a meeting held in the Frascati National Laboratories in February 1960, Bruno Touschek had proposed to equip these Laboratories with an accelerator of a new kind that would allow the investigation of matter−antimatter annihilation. Soon after, Touschek designed AdA, an e^+e^- storage ring prototype. Quickly built, AdA was first commissioned in Frascati where the behavior of the counter rotating beams was investigated in the low stored particle number regime. Then, in 1962, AdA was brought on a truck to the Orsay *Laboratoire de l'Accélérateur Linéaire* to benefit from the high intensity electron linear accelerator available there. The commissioning was then continued by a small Italo-French collaboration which pursued the study of the machine performances, collective effects included. Exploring this entirely new accelerator physics domain was a unique and exciting period for the team. It lasted close to two years during which the beam lifetime limitations, the size of the stored bunches and the particle-antiparticle collision rate were measured. By the end of that period, the basic underlying concept of e^+e^- colliders was established, showing that the road to future powerful colliders was open.

3.1 Introduction

The making of AdA can be set in a time sequence which illustrates the major steps which led to the proof-of-principle of electron−positron storage rings to be a major discovery tool for particle physics in the second half of last century:

> February 1960: Touschek proposes to turn the Frascati synchrotron into an electron-positron ring.

J. Haïssinski (✉)
Laboratoire IJCLab Orsay, Paris, France
e-mail: jhaiss@lal.in2p3.fr

© The Author(s) 2023
L. Bonolis et al. (eds.), *Bruno Touschek 100 Years*,
Springer Proceedings in Physics 287,
https://doi.org/10.1007/978-3-031-23042-4_3

Fig. 3.1 Bruno Touschek in
Catania in 1963. Credit:
Rome Sapienza University
Physics Department
Archives

March 1960: Decision to engage the Frascati Laboratory in an e^+e^- colliding beam experiment.

May 1961: First electrons/positrons stored in AdA [1, 2].

July 1962: AdA is brought to the Laboratoire de l'Accélérateur Linéaire (LAL) in Orsay.

1963: Discovery of the *Touschek effect* [3] and first evidence ever of collisions between counter-rotating stored particles [4].

Summer 1964: AdA goes back to Frascati.

In what follows, I shall outline why the AdA storage ring was brought to Orsay after having been commissioned in Frascati, the first commissioning period in France, AdA's operations at LAL, AdA beam lifetime and the Touschek effect, bunch size and luminosity measurements, to conclude with a summary of the main physics results obtained with AdA at Orsay [5].

I would start by showing a picture of Bruno Touschek, (Fig. 3.1), which I like particularly since it captures very nicely Bruno's personality which was characterized by a quick and imaginative mind and also by a permanent sense of humour.

To put the birth of AdA in perspective, I would like to recall what was the status of the conventional accelerators at the time in Europe. By conventional I mean accelerators which accelerate particles (projectiles) which are sent and hit some target at rest in the laboratory. Towards the end of the 1950s, several machines had been commissioned, as shown in Fig. 3.2.

There were three proton machines and two electron machines, in particular the Frascati electron synchrotron, which at the time offered the perspective for a fruitful experimental high-energy physics program.

Following the successful commissioning of the Frascati synchrotron, and Touschek's proposal, a decision by the Frascati Laboratories was taken in March 1960 to build AdA, Anello di Accumulazione, *Storage Ring* in English (Fig. 3.3), a small accelerator for an experiment to study electron−positron collisions, as suggested by Bruno Touschek [1].[1]

---

[1] See L. Bonolis and G. Pancheri's contributions to this conference.

Name	Country Laboratory	Particles	Energy	First beam
Synchrophasotron	USSR Dubna	protons & ions	10 GeV	1957
SATURNE	France Saclay	protons & ions	3 GeV	1958
Proton Synchrotron (PS)	CERN	protons	20 GeV	1959
Linac	France Orsay	electrons	1 GeV	1959
Electron Synchrotrone	Italty Frascati	electrons	1.1 GeV	1959

Fig. 3.2 Conventional accelerators commissioned in Europe between 1957–1959

Fig. 3.3 AdA in Frascati in 1961, before being moved to LAL. Credit: INFN-LNF

Less than one year later, in February 1961, the first electrons circulated in AdA. However, it happened that the capture rate of electrons and positrons in the ring was much lower than anticipated.

In July 1961, following Italy's success in constructing AdA and planning for a more powerful and bigger e^+e^- machine (ADONE), AdA was presented at CERN and discussions taking place during the conference inspired a visit to Frascati by Pierre Marin and Georges Charpak [6].

During this visit, Marin, a researcher from the *Laboratoire de l'Accélérateur Linéaire*, suggested to Carlo Bernardini and Bruno Touschek to move AdA to Orsay and use the newly built Linear Accelerator (LINAC) (Fig. 3.4, left panel) as injector and thus increase the e^+ or e^- capture rate achieved in Frascati from a few 10^2 particles per beam to a few 10^7 per beam. The LINAC provided a 500 MeV beam and this meant multiplying the number of stored particles in AdA by a factor of the order of 10^5.

After his visit to Frascati, Pierre Marin (Fig. 3.4, right panel) wrote a report where he already envisaged that AdA could be transported to Orsay (Fig. 3.5).

Official negotiations between André Blanc-Lapierre, LAL's director, on the one side and Edoardo Amaldi, INFN director in Rome, Giorgio Salvini and Italo Federico Quercia, first and second director of the Frascati Laboratories (Fig. 3.6) on the other,

Fig. 3.4 The Orsay linear accelerator wave guide, credits *Laboratoire de l'Accélérateur Linéaire d'Orsay*, and Pierre Marin (1927–2002) in a 1966 photograph. Credit: Yvette Haïssinski

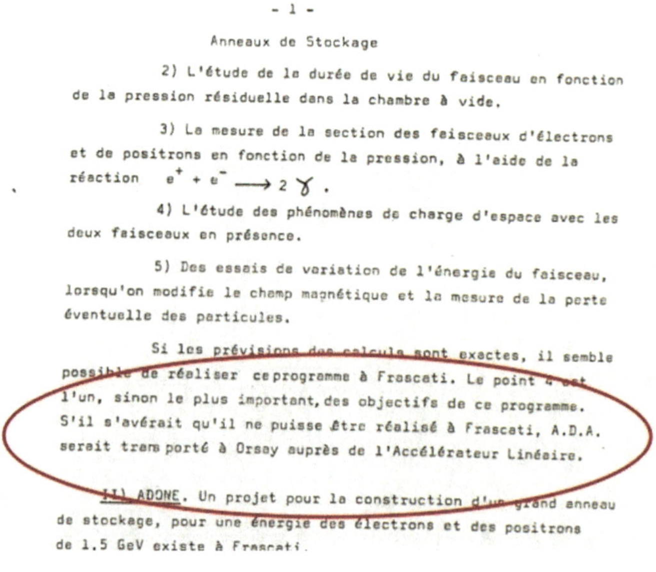

- 1 -

Anneaux de Stockage

2) L'étude de la durée de vie du faisceau en fonction de la pression résiduelle dans la chambre à vide.

3) La mesure de la section des faisceaux d'électrons et de positrons en fonction de la pression, à l'aide de la réaction $e^+ + e^- \longrightarrow 2\gamma$.

4) L'étude des phénomènes de charge d'espace avec les deux faisceaux en présence.

5) Des essais de variation de l'énergie du faisceau, lorsqu'on modifie le champ magnétique et la mesure de la perte éventuelle des particules.

Si les prévisions des calculs sont exactes, il semble possible de réaliser ce programme à Frascati. Le point 4 est l'un, sinon le plus important, des objectifs de ce programme. S'il s'avérait qu'il ne puisse être réalisé à Frascati, A.D.A. serait transporté à Orsay auprès de l'Accélérateur Linéaire.

III) ADONE. Un projet pour la construction d'un grand anneau de stockage, pour une énergie des électrons et des positrons de 1.5 GeV existe à Frascati.

Fig. 3.5 The report presented in September 1961 by Pierre Marin to LAL's director André Blanc-Lapierre after meeting with Ruggero Querzoli, at a conference in Aix-en-Provence. Credit: *Laboratoire de l'Accélérateur Linéaire d'Orsay*

led to AdA arriving in Orsay in July 1962 [7] and be installed at the experimental hall of the *Laboratoire de l'Accélérateur Linéaire* (Fig. 3.7).

The end part of the LINAC vacuum pipe—which was used to inject particles in AdA—is visible in the upper part of Fig. 3.7. The 500 MeV particles were going through the small scintillating screen on the right used to focus the beam, then the same beam was hitting the tantalum foil where they were producing high-energy bremsstrahlung photons which were entering the vacuum chamber of AdA. Within the vacuum chamber there was another tantalum plate where the high-energy photons were creating electrons and positrons, some of which were stored. AdA could in fact be rotated and translated and thus, depending on the geometric configuration of AdA with respect to the LINAC vacuum tube, one could move from electrons to positrons storing process.

A view of AdA's control room in Orsay is shown in Fig. 3.8, with Giuseppe Di Giugno. There was no computer at the time to control the injection process, nor the

Fig. 3.6 Left panel: Edoardo Amaldi, Giorgio Salvini (Frascati Director during approval of AdA's construction), Italo Federico Quercia (Frascati Director during AdA's construction). Credit: Rome Sapienza University Physics Department Archives, and INFN-LNF. Right panel: André Blanc-Lapierre with Pierre Marin in a later photograph. Credit: *Laboratoire de l'Accélérateur Linéaire*

Fig. 3.7 Ada in Salle 500 in Orsay. Credit: *Laboratoire de l'Accélérateur Linéaire*

Fig. 3.8 The youngest member of the collaboration, Giuseppe (Peppino) Di Giugno, in AdA's control room, in Orsay. Credit: Giuseppe Di Giugno

ring parameters, but a number of meters and screens and a number of knobs for the operator to turn.

In Fig. 3.9 I show the synchrotron radiation emitted by electrons and positrons circulating in AdA, which was in the visible region. The first picture shows that when the beam was unperturbed, the transverse profile of the bunch was flat and looked like the transverse view of an optical focussing lens. Its larger dimension was about 3 mm (FWHM). When some magnetic coupling is applied between the horizontal and the vertical betatron oscillations, the beam becomes slanted and rounder, and then a little wider.

In Fig. 3.10 I show a few AdA parameters. The overall size of AdA was about 1.2 m, the orbit length was 4.1 m long and we used to work at an energy of around 225 MeV. The value of the beam current in each beam was about 1/2 mA and the vacuum which was maintained in the ring was about 1 nTorr. At that time, it was a real challenge to maintain it at such a low level.

The AdA collaboration at Orsay was comprised of five physicists from Italy: Bruno Touschek, Carlo Bernardini, Gian Franco Corazza—who took care of much of the machine hardware—Ruggero Querzoli, see Fig. 3.11, and Giuseppe Di Giugno [8]. Together with Pierre Marin, François Lacoste was one of the two first French physicists who greeted AdA at Orsay, then I replaced him. I was a graduate student at the time and AdA became the subject of my Ph.D thesis [9]. Eventually, Bernardini and Touschek were members of the examination board.

3.2 The Orsay Scientific Program

The program envisioned for AdA in Orsay was based on measuring:

1. The beam life-time
2. The e^+ and e^- bunch size in the ring

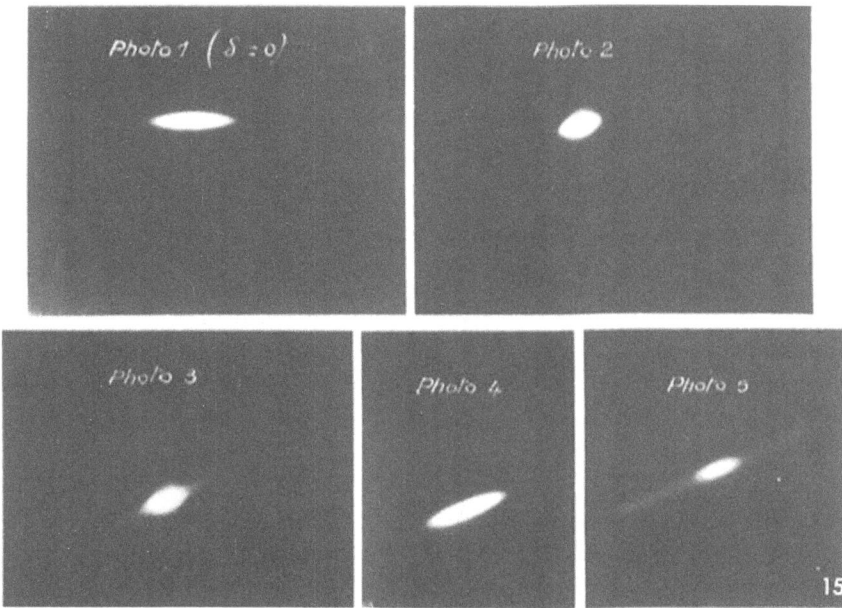

Fig. 3.9 AdA transverse bunch shape without and with an applied magnetic coupling [9]

Parameter	Value or typical operation value	Units
Orbit length	4.1	m
Energy per beam	225	MeV
Luminosity	$\sim 10^{25}$	$cm^{-2} s^{-1}$
Beam current, per beam	0.5	mA
Vacuum pressure	1	nTorr

Fig. 3.10 AdA parameters [9]

3. The collision rate (ring luminosity)

After checking the beam life-time and measuring the bunch size in the ring, the last step was the most important point of our program: measuring the ring luminosity.

In Fig. 3.12 the plot shows the current in nA *vs* time driven by a photomultiplier which could detect the synchrotron light produced by the stored particles. On the left of the graph one can see that at a certain point there were five electrons stored in the

Fig. 3.11 Members of AdA collaboration in Orsay, from left: Carlo Bernardini, Gianfranco Corazza working at the vacuum chamber of the Frascati synchrotron and Ruggero Querzoli. Credit: INFN-LNF

machine. Then, by playing with the radiofrequency (RF) power, one could shorten the lifetime of these particles. Each subsequent step in the graph corresponds to the loss of a single stored particle, until what remained is just the photomultiplier dark current. And so, one could observe even a few particles stored in the macroscopic AdA setup. I think this graph—which had been already obtained in Frascati—is quite remarkable and that it would have deserved much more advertisement.

Coming to the lifetime of the stored beams, one of the first measurements which were carried out was to check how this lifetime was affected when varying the power fed in the RF cavity which kept the electrons and positrons rotating despite the fact that they were continuously losing their energy in the form of the synchrotron radiation. If one increases the high voltage in the RF cavity, the lifetime of the bunch gets longer, as shown in the logarithmic graph at left in Fig. 3.13. What happens is the following: electrons and positrons stored in the beams were trapped in the potential well provided by the RF cavity and when such potential well was progressively increased by putting more power in the cavity, the lifetime increased very rapidly.

Fig. 3.12 The plot showing counting electrons one-by-one in AdA, from J. Haïssinski's dissertation (1965)

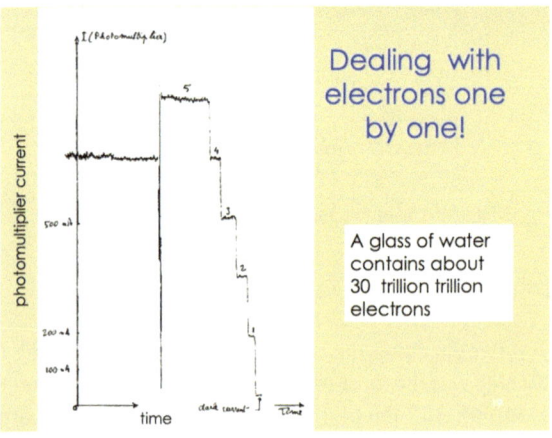

Dealing with electrons one by one!

A glass of water contains about 30 trillion trillion electrons

From this measurement one could infer the length of the stored bunch which, at the time of this observation, was about 10 cm.

To increase AdA's luminosity and take advantage of the superior electron current from the LINAC, various attempts were made with different injection techniques for the positrons, as well as looking for different final state processes.

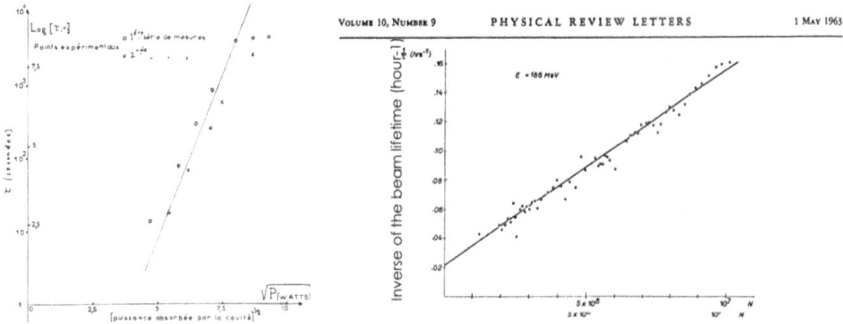

Fig. 3.13 At left, particle losses due to quantum fluctuations in AdA *versus* RF power [9], at right the plot showing the life-time decrease due to the Touschek effect. Figure reprinted with permission from [3], ©American Physical Society

Fig. 3.14 Synthesis of all experimental points with two beams, showing correlation between number of particles in beam 1 versus number of particles in beam 2, observed in coincidence with a bremsstrahlung photon. The point at $N_2 = 0$ is normalized to $p = 10^{-9}$torr [4]

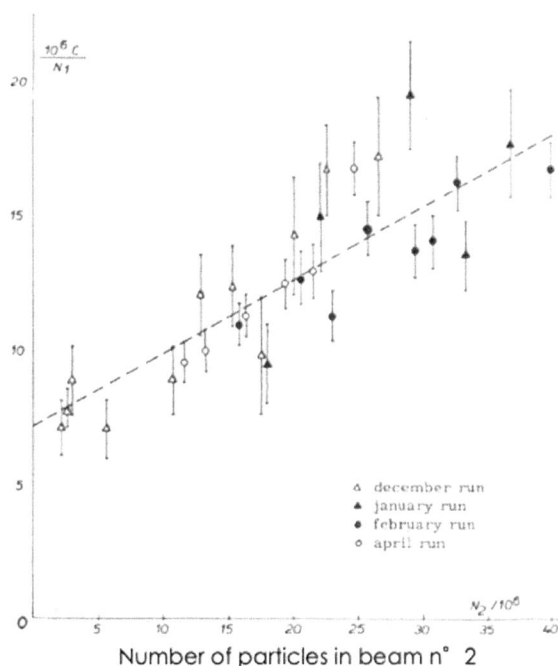

A breakthrough occurred when, one night, early in 1963, efforts to increase the beam current resulted in a decrease of the beam life-time. This was a totally unexpected effect which was right away interpreted by Bruno Touschek, and this is how it was later called the *Touschek effect*. This effect could not have been measured in Frascati, because of insufficient number of stored particles in a bunch. The observation of such effect was a major step towards understanding stored beam dynamics.

It was understood by Touschek to be a collective effect inside a single beam (intra-bunch effect) due to Møller scattering: his interpretation and calculation of the beam life-time τ from $1/\tau \approx N/V$ (particle number density within the bunch) led to infer the measurement of the bunch volume and its energy dependence. The effect was devastating for AdA's hope to see particle production through annihilation, but, decreasing with increasing beam energy, Touschek's calculation gave hopes for storage rings to work at higher energies: ADONE (at 3 GeV c.m. energy) [10] and ACO (at 1300 GeV) [11], respectively approved by INFN and CNRS, were safe.

The results were submitted to *The Physical Review Letters* [3] and gave confidence to the accelerator physics community that electron−positron storage rings could open the road to high energy physics.

The bunch size was crucial for measuring the collision rate. Its calculation depends on:

σ_{radial} was measured optically, giving $\sigma_{radial} = 0.5$ mm, taking a picture of the transverse dimensions of the bunches,

$\sigma_{longitudinal}$ was inferred from the measured lifetime due to quantum fluctuations $\sigma_l \approx 7$ cm (at $E_{beam} = 195$ MeV and $V_{RF} = 5.5$ kV),

$\sigma_{vertical}$ was the only big unknown.

From the Touschek effect $\sigma_{vertical} \approx 20\mu$, while $\sigma_{vertical}$ expected from synchrotron radiation recoil effects was only 2μ. When all this was understood, the team had a realistic estimate of which process could prove that collisions had taken place, namely $e^+e^- \rightarrow e^+e^- + \gamma$, whose theoretical calculation was being done in Rome by two students of Raoul Gatto, Guido Altarelli and Franco Buccella [8]. During fall 1963 and spring 1964 the team focused on gathering data and submitted an article about the first observation ever of electron−positron collisions in a storage ring [4].

3.3 Conclusion

The main storage ring physics results obtained with AdA were [5]:

- Check of the calculation of the beam scattering effects by the residual gas
- Confirmation of the theory of the RF lifetime due to synchrotron radiation quantum fluctuations
- Discovery and theory of the Touschek effect
- Evidence for the mechanism that determines the stored bunch height
- First evidence ever of collisions between opposite stored beams.

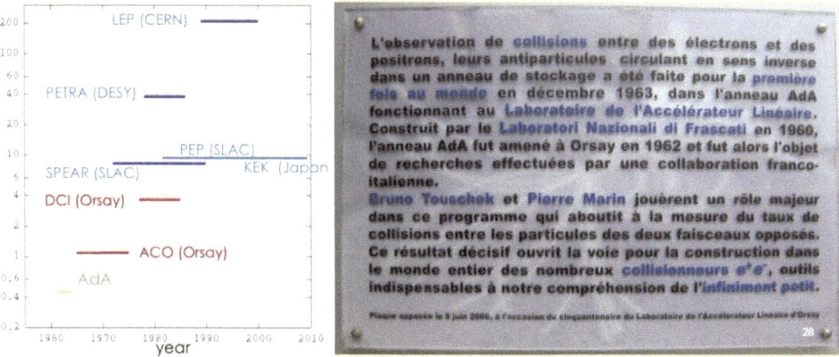

Fig. 3.15 The scientific impact of AdA (besides the ADONE ring in Frascati), eventually resulting in the building of LEP at CERN. Right: plaque placed at the entrance of the *Laboratoire de l'Accélérateur Linéaire* in Orsay, to commemorate AdA's pioneering contribution to the development of matter–antimatter storage rings

Thus, the basic underlying concept of e^+e^- colliders was established.

The scientific impact of the coming of AdA to Orsay resulted in the building of three storage rings, ACO, DCI and Super-ACO, and the opportunity for Ph.D students trained at LAL to pursue their research at other colliders (Fig. 3.15 left panel).

In 2006, when 50 years of the *Laboratoire* were celebrated, a commemorative plaque was placed at its entrance (Fig. 3.15 right panel).

Acknowledgements The text above is the transcription by Luisa Bonolis and Giulia Pancheri of the author's oral presentation during the Bruno Touschek Symposium. The author is extremely grateful to Luisa and Giulia for their excellent transcription.

References

1. C. Bernardini, G.F. Corazza, G. Ghigo, B. Touschek, The Frascati Storage Ring. Il Nuovo Cimento **18**(6), 1293–1295 (1960). https://doi.org/10.1007/BF02733192
2. C. Bernardini, U. Bizzarri, G. F. Corazza, G. Ghigo, R. Querzoli, B. Touschek, A 250-Mev Electron-Positron Storage Ring: The A. D. A. Contribution to HEACC, (Geneva, 1961), pp. 256–261
3. C. Bernardini, G.F. Corazza, G. Di Giugno, G. Ghigo, R. Querzoli, J. Haissinski, P. Marin, B. Touschek, Lifetime and Beam Size in a Storage Ring. Phys. Rev. Lett. **10**(9), 407–409 (1963). https://doi.org/10.1103/PhysRevLett.10.407
4. C. Bernardini, G. Corazza, G. Di Giugno, J. Haissinski, P. Marin, R. Querzoli, B. Touschek, Measurements of the Rate of Interaction between Stored Electrons and Positrons. Il Nuovo Cimento **34**(6), 1473–1493 (1964). https://doi.org/10.1007/BF02750550
5. J. Haïssinski, in *Bruno Touschek and the birth of* e^+e^- *physics*, ed. by G. Isidori. Frascati Physics Series Vol. 13 (1998) pp. 17–31
6. P. Marin, *Un demi-siècle d'accélérateurs de particules* (Editions du Dauphin, Paris, 2009)

7. L. Bonolis, G. Pancheri, Bruno Touschek and AdA: from Frascati to Orsay. In memory of Bruno Touschek, who passed away 40 years ago, on May 25th, 1978, arXiv:1805.09434 [physics.hist-ph](2018) doi: https://doi.org/10.48550/arXiv.1805.09434
8. G. Pancheri, L. Bonolis, Touschek with AdA in Orsay and the first direct observation of electron-positron collisions, arXiv:1812.11847 [physics.hist-ph] (2018) doi: https://doi.org/10.48550/arXiv.1812.11847
9. J. Haïssinski, Thèse d'État, 1965
10. F. Amman, C. Bernardini, R. Gatto, G. Ghigo and B. Touschek, *Storage Ring for Electrons and Positrons, ADONE*. Internal Report No. 68 (the Laboratori Nazionali di Frascati, 1961)
11. A. Blanc-Lapierre et al., The Orsay Project of a Storage Ring for Electrons and Positrons of 450 MeV Maximum Energy, in *Proceedings of 4th International Conference on High-Energy Accelerators*, (Dubna, 1963)

Chapter 4
Bruno Touschek and Statistical Mechanics

Giancarlo Rossi

Abstract In this talk I will describe the history of the birth of the "Meccanica Statistica" book, that Bruno Touschek and I wrote in 1970. The book was conceived and brought to conclusion in the broad context of the stimulating atmosphere that the Touschek scientific and teaching activity had created in the Physics Department of the University of Rome "La Sapienza". I will present a recollection of my memories of the years from 1965 to 1970 during which the book was imagined, written, rewritten, corrected and polished till its final version. I will also briefly describe the content of the book underlining the unmistakable footprint of Touschek unconventional way of thinking.

4.1 Introduction

Just like many of the people participating in this Memorial symposium, celebrating the 100th anniversary of Bruno Touschek's birth, I was one of his students. I graduated in 1966 under his supervision defending the thesis "$e^+e^- \to \mu^+\mu^- + \gamma$ annihilation and the Bloch–Nordsieck method". The calculations confirmed Touschek's conjecture [1] that soft-photon emission could be elegantly described within the coherent photon state formalism. The results presented in the thesis appeared in my first paper, written in collaboration with Greco [2], in which employing the formalism developed in [1] (and subsequently extended in [3]), we rigorously proved that, as expected, the resummation of soft photon emission results in an annihilation cross section proportional to the factor $(\Delta\omega/E)^\beta$, where $\Delta\omega$ is the energy resolution of the experimental apparatus, E is the center of mass energy of the beams and β is

G. Rossi (✉)
Dipartimento di Fisica, Università di Roma "Tor Vergata", Via della Ricerca Scientifica, 00133 Roma, Italy

Sezione di Roma "Tor Vergata", INFN, Via della Ricerca Scientifica, 00133 Roma, Italy

Museo Storico della Fisica e Centro Studi e Ricerche Enrico Fermi, Via Panisperna, 89a, 00184 Roma, Italy
e-mail: rossig@roma2.infn.it

© The Author(s) 2023
L. Bonolis et al. (eds.), *Bruno Touschek 100 Years*,
Springer Proceedings in Physics 287,
https://doi.org/10.1007/978-3-031-23042-4_4

the famous Bond-factor. Touschek named it this way because in the typical ADONE's kinematical conditions its value was just 0.07.

The coherent state formalism was subsequently extended to encompass the non-abelian case with applications to soft gluon emission corrections in parton processes in [4–6]. Although not explicitly acknowledged in the literature, after a moment of thought one recognizes that the kinematical kernel appearing in the description of the Maximally Helicity Violating amplitudes [7], today currently expressed in terms of spinor variables [8], is nothing else but the same basic kernel that describes soft photon emissions. It is amazing to realize how far Touschek's intuition has evolved.

4.1.1 Touschek's Teaching Activity

Many talks in this Conference are focused on the remarkable and at the time unexpected developments of the Touschek's simple and brilliant idea of having electrons and positrons running head-on in a ring, held on the same circular trajectory "by the CPT theorem". The construction of the "Anello di Accumulazione" (AdA) built in 1962 in the Frascati National Laboratories (LNF) proved the feasibility of workable e^+e^- colliders. The success of AdA prompted the construction of ADONE (big AdA, but also Adonis in English) which in the future years was followed by a number of e^+e^- machines all around the world with increasing center of mass energy and luminosity, culminating in the construction of the Large Electron-Positron Collider (LEP) at CERN.

In this contribution I'm not going to talk about these extraordinary developments, others in this Conference will do it. Instead, I want to focus on Touschek's large and varied teaching activity, a somewhat less known side of his scientific personality. In particular I shall describe the birth of the book "Meccanica Statistica" to which I had the privilege to contribute.

As a teacher Touschek carried out a wide, dedicated and highly valued activity. The list of courses below is just what I could reconstruct from the information I was able to collect from colleagues and former students, but it is certainly not complete

1. "Meccanica Statistica" (IV year)
2. "Metodi Matematici della Fisica" (III year)
3. "Renormalization" (Scuola di Perfezionamento)
4. "Quantum Electrodynamics" (Scuola di Perfezionamento)
5. "Sull'insegnamento della teoria dei quanti" (Lincei)
6. "The LASER effect" (private notes[1])
7. ... and maybe more

---

[1] These notes are in my possession. In Touschek's view they should have been the basis of my "Tesi di laurea" which originally was supposed to be about the "LASER effect". I must confess, however, that the argument looked too difficult to me. A bit desperate, I asked Touschek to allow me to change the subject of my "Tesi", which he kindly did by associating me to the group of people (P. Di Vecchia, F. Drago, E. Etim, M. Greco, L. Pancheri, Y. Srivastava, ...) already intensely working on the many theoretical aspects of relevance for the forthcoming of e^+e^- machines.

Fig. 4.1 Touschek at the blackboard with cigarette on the left hand and chalk on the right (photo from [9])

This list better that anything else may give an idea of the wide range of subjects on which Touschek had been lecturing in the years he spent in the Department of Physics of "La Sapienza" and of the amazingly large diversity of interests his scientific activity was covering (Fig. 4.1).

Between 1959 and 1968 Touschek was particularly keen to lecture on Statistical Mechanics. One of the reasons for this was that he thought it was necessary to update the program of the course and fill a number of holes in the curriculum followed by Physics students. One aspect of this was the fact that the teaching of Statistical Mechanics was commonly mainly focused on equilibrium physics and the theory of thermodynamic *ensembles*. In the standard courses there was nothing or very little about stationary but open systems and the many dynamical problems that the "Master Equation" could deal with. In other words, not much was usually taught about "statistical dynamics". The enlargement of the scope of the course with the inclusion of statistical dynamics as well as some unconventional and rarely discussed applications was highly appreciated by the students. Despite the difficulty of following Touschek lectures, "il corso di Meccanica Statistica" quickly became a "must" for many students, including myself.

4.2 The Birth of the Book "Meccanica Statistica"

The book "Meccanica Statistica" was published by Boringhieri in 1970 in the Series "Programma di Matematica Fisica Elettronica" [10]. It was the result of a rather long and elaborated journey. The first draft of the manuscript dates back to the winter of 1967. It was based on the notes that, as a fourth year student in Physics, I had taken two years before (i.e. in the academic year 1964–65), when I followed Touschek's course on Statistical Mechanics. This manuscript, carefully revised by Touschek himself, was then published by a local Editor, "La Goliardica", as "dispense" for the students in 1969 [11].[2]

Almost contemporarily, with the idea of publishing a text-book on Statistical Mechanics, in the academic year 1967–68, Touschek started to put on paper with the help of his beloved Olivetti Lettera 22 (no PC's were available at that time!) an English version of the lectures he was delivering week by week.

The final text of the published book resulted from the intersection of my translation of the English notes written by Touschek with the text I had drafted in Italian while attending (again) his lectures. As mentioned before, the manuscript was published in Italian by Boringhieri in 1970 in the Series "Programma di Matematica Fisica Elettronica" with the title "Meccanica Statistica" [10]. Touschek' hope was that the book could also appear in English. Unfortunately, despite some initial interest from Wiley and Academic Press to his great regret this project never materialized.

In the long process of deciding the content of the book and the style of the text, Touschek's guiding principle and his main concern was always clarity, as the book was supposed to be addressed to undergraduate students. For this reason we chose to use as plain and simple language as possible, and decided to end each chapter with a summary of the results and the main ideas that would be developed in the following chapter. Simplicity did not mean that all the subtleties inherent in the conceptual construction of the methods of Statistical Mechanics methods were ignored or overlooked. Quite the contrary! In the book not only standard subjects, like the construction of the various statistical *ensembles*, the proof of their equivalence and the derivation of the "Master Equation", were inserted and discussed in a somewhat original, yet elementary way. A few unconventional problems were also addressed.

4.3   The Content of the Book

It is illuminating to look at the content of the book because it shows how modern and original was Touschek's point of view even when covering standard topics in Statistical Mechanics. More than many words, Fig. 4.2, which was used to illustrate in an intuitive way the mathematics behind the saddle point method, is emblematic of the unmistakable footprint that Touschek had left in the book.

Here is the Table of content of the book.

- PARTE PRIMA: STATICA STATISTICA

    1. Meccanica statistica e termodinamica dell'oscillatore armonico
    2. Teoria dell'*ensemble* di Gibbs
    3. Termodinamica covariante
    4. Termodinamica di un gas ideale di particelle identiche

---

[2] I must add that a little earlier notes collected by another student, M. Gambarelli, where available to the students.

Fig. 4.2 The saddle point
(method), from [10]

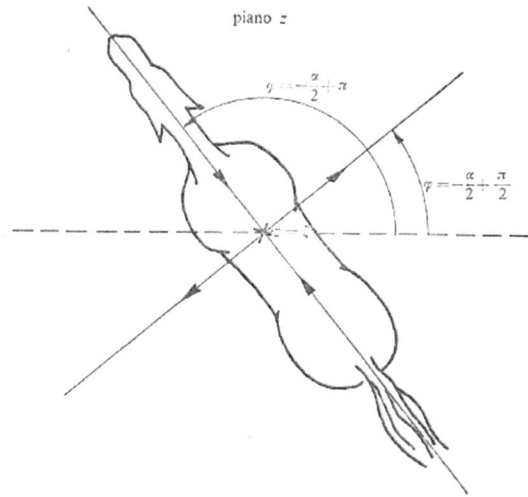

5. Gas degenere e imperfetto
6. Sistemi in cui il numero di particelle non è costante

• PARTE SECONDA: DINAMICA STATISTICA

1. Proprietà degli stati di non-equilibrio
2. I fondamenti microscopici della *master equation*
3. Applicazioni della *master equation*
4. Teoria del trasporto

• APPENDICI

1. A. Teoria quantistica del conteggio
2. B. Teorema adiabatico
3. C. Un esperimento statistico

Among the topics discussed in the book, it is worth to highlight three arguments that attracted the attention of the physics community as witnessed by the significant interest they stimulated in the specialized literature. The first topic is the solution of the problem posed by the definition of Temperature of a moving body in Special Relativity. The second is the application of the Master Equation to the study of the hourglass with an infinite or a finite number of grains. The third is a proposal to understand the apparent antinomy between microscopic reversibility and macroscopic irreversibility. Because of the general relevance of these questions in Physics and the originality of the arguments developed in the book, I would like to devote a few words to illustrate the physical and conceptual content of these problems and the proposed solutions. I will say something about each of these three items in the next three subsections.

4.3.1 Covariant Thermodynamics and the Lorentz Transformation Properties of Temperature

A rather non-standard topic which Touschek decided to include in the book (see chapter 3 PARTE PRIMA) was the discussion of the question of how to define the temperature of a moving body in Special Relativity. The problem, first analyzed in [12], is that it looks like we have a contradiction between two apparently equally acceptable physical definitions of temperature for a system in motion.

In fact, if one decides to make reference to the ideal gas law, $PV = RT$, in order to define the temperature of a moving system, one is led to conclude that the temperature should transform like a length under a Lorentz transformation, because P is a relativistic invariant (it is the trace of the energy-momentum tensor). As a result, the relation between the temperature of the system at rest and the temperature of the system in motion with (uniform) velocity v would be (we set $c = 1$)

$$T(v) = \frac{T(0)}{\gamma}, \qquad \gamma = \frac{1}{\sqrt{1 - v^2}}. \tag{4.1}$$

On the other hand, if one decides to look at the second Law of Thermodynamics, $dS = \delta Q / T$, one is led to conclude that the temperature must transform like an energy, as entropy is just a number (it is the logarithm of the number of micro-states of the system) leading to the formula

$$T(v) = T(0)\gamma, \qquad \gamma = \frac{1}{\sqrt{1 - v^2}}. \tag{4.2}$$

The way-out of this paradox lies in the observation that the usual operative definition of temperature actually refers to a measurement performed in the rest frame of the system (the one in which the thermometer is at rest with respect to the system). What one might call the temperature of a moving system is actually a matter of conventions, in the sense that different measurement procedures give raise to different definitions of the temperature of a body in motion, hence to apparently different Lorentz transformation properties.

This argument can be made rigorous by constructing a covariant formulation of Thermodynamics and Statistical Mechanics which can be done by making fully covariant the construction of statistical *ensembles*. For instance, with reference to the Gibbs *ensemble*, in addition to including the constraint energy conservation, one needs to enforce the conservation of three-momentum. The occupation numbers will then satisfy the equations

$$\ln a_n + \lambda + \beta_\mu p_n^\mu = 0, \tag{4.3}$$

which represent the solution of the covariant extension of the standard variational problem of classical Statistical Mechanics. In (4.3) λ is the Lagrange multiplier associated with particle number conservation, $\sum_n a_n = N$, and β_μ the time-like four-vector associated with the four-momentum conservation, $\sum_n a_n p_n^\mu = P^\mu$. The covariant formulation of Statistical Mechanics will then be expressed by the equations

$$a_n = \frac{N}{Z} e^{-\beta_\mu p_n^\mu} \tag{4.4}$$

$$Z = \sum_n e^{-\beta_\mu p_n^\mu} \qquad P_\mu = -N \frac{\partial \ln Z}{\partial \beta^\mu}. \tag{4.5}$$

One can prove that the relation between β_μ and the temperature at rest, $T(0)$, is given by (k_B is the Boltzmann constant)[3]

$$\beta_\mu = \frac{u_\mu}{k_B T(0)}, \qquad u_\mu = \gamma(1, \mathbf{v}). \tag{4.6}$$

This formula shows that the Lorentz transformation properties of the temperature depend on how the latter is measured, i.e. which component of β_μ is employed to define T for a body in motion with four-velocity, u_μ.

4.3.2   The Hourglass and the Periodic Statistical Clock

In this section I want to present an amusing application of the Master Equation to the time evolution of the hourglass (see Fig. 4.3) and the periodic statistical clock, discussed in chapter 3 of the PARTE SECONDA of the book.

4.3.2.1   The Hourglass

The statistical description of the time behaviour of the hourglass can be described by the equations

$$\dot{p}_0(t) = -\lambda p_0(t) \qquad\qquad p_0(0) = 1 \tag{4.7}$$

$$\dot{p}_{s+1}(t) = \lambda[p_s(t) - p_{s+1}(t)], \qquad p_{s+1}(0) = 0, \quad s = 0, 1, \ldots \tag{4.8}$$

where $p_s(t)$ is the probability of having s grains in the lower part of the hourglass at time t and λ is the transition rate from the state s to the state $s + 1$. Naturally the boundary conditions we are interest is are such that the initial time ($t_0 = 0$) state is the one where the lower part of the hourglass is empty as indicated in the right part of the Eqs. (4.7) and (4.8). Dots in (4.8) mean that we are considering the case is which N, the number of grains, is infinitely large. In this situation the system (4.8)

---

[3] Technically (4.6) holds under the rather mild assumption that the partition function is a relativistic invariant.

Fig. 4.3 The hourglass
(figure by G. C. Rossi)

can be solved exactly, yielding

$$p_s(t) = \frac{(\lambda t)^s}{s!} e^{-\lambda t}, \quad s = 0, 1, \ldots \tag{4.9}$$

The probability distribution is therefore Poissonian with

$$\langle s \rangle = \lambda t, \qquad \frac{\sigma}{\langle s \rangle} = \frac{1}{\sqrt{\langle s \rangle}}. \tag{4.10}$$

From these equations we see that we can actually use the hourglass as a clock as the average number of grains in the lower part grows proportionally with time and the relative fluctuations in the number of grains dies out like $1/\sqrt{t}$.

4.3.2.2 The Periodic Statistical Clock

We now consider the case in which N is finite. In this case the hourglass Master Equations cannot be solved exactly. A related soluble problem is, however, the "periodic clock", i.e. a system in which, defining $p_s(t)$ the probability for the system to be in the state s at time t, we have

$$p_s(t) = p_{s+N}(t), \quad s = 0, 1, \ldots, N - 1. \tag{4.11}$$

The Master Equation for the periodic statistical clock reads

$$\dot{p}_{s+1}(t) = \lambda[p_s(t) - p_{s+1}(t)]. \tag{4.12}$$

This set of linear first order differential equations can be solved by normal modes decomposition. If we start with the initial condition $p_s(0) = \delta_{s,0}$, one finds that the system visits it periodically, every $T = N/\lambda$. In principle for appropriate choices of N and λ this system can then be used as a clock.

4.3.3 Micro-reversibility Versus Macro-irreversibility

In this subsection I want to illustrate the proposal, discussed in the APPENDICE C of the book, for a solution (or an understanding) of the problem of reconciling the invariance under time reversal of the fundamental laws of Physics (both Newton's equations in classical physics and the Schrödinger equation in quantum physics), and the irreversibility we observe in macroscopic processes.

The question has not only drawn an enormous attention in the specialized literature, but it has also generated a large philosophical debate because of its epistemological implications for the very definition of the notion of time. The solution of the reversibility versus irreversibility paradox proposed in the book is actually very simple and can be briefly summarized as follows.

(1) The fundamental laws of micro-physics are invariant under time reversal

(2) The lack of symmetry under time inversion that we routinely experience is a consequence of the very peculiar initial conditions from which the macroscopic systems we usually observe, evolve.

Indeed, the unavoidable mixing process between wine and water taking place in a glass ensues from the sharp separation of the two liquids at the initial time. Starting from this (largely out of equilibrium, hence statistically highly improbable) initial condition, the time evolution of the system molecules is then completely controlled by perfectly time reversible equations. From a thermodynamic point of view the system evolves towards its equilibrium state.[4] This means that repeating the experiments many times the final state of the system (i.e. the state reached after a very long time lapse) is within statistical fluctuations always the same.

A simple and direct check of the validity of this point of view is provided by the following simple but paradigmatic numerical example. Let us consider the "time" evolution of the stochastic variable represented by the mean value of the sum of $N = 100$ numbers, s_n, between $-1/2$ and $1/2$, randomly extracted from a uniform

---

[4] We are excluding here mathematical instances of systems with unrealistic initial conditions which would prevent the system to ever reach equilibrium, for instance a system of perfectly elastically interacting balls in a cubic box with perfectly reflecting walls, having initial velocities all equal and oriented perpendicularly to the box walls.

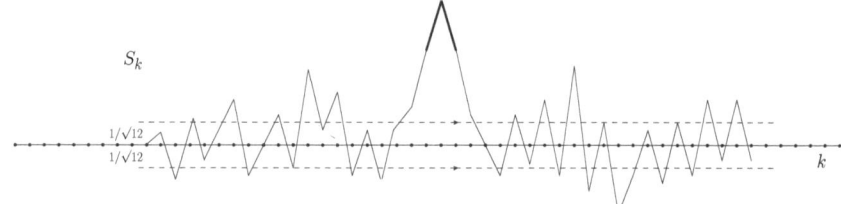

Fig. 4.4 The typical behaviour with "$k \sim$ time" of the stochastic variable defined in (4.13)

distribution, when successively a new term is added to replace the first one. In formulae we write

$$S_k = \sum_{n=k}^{99+k} s_n , \qquad k = 1, 2, \ldots, N_{config} . \tag{4.13}$$

The "time" history of this stochastic variable was followed for $N_{config} = O(10^7)$ steps (the actual calculation was carried out with the historic UNIVAC machine located in the "Centro di Calcolo" of Sapienza University of Rome, at a time when codes were still written on punched cards). The "$k \sim$ time" evolution of S_k is very illuminating and it is sketched in Fig. 4.4.

For a perfectly flat distribution one expects the value of S_k to fluctuate around $\langle S \rangle = 0$ with $\langle S^2 \rangle = 1/12$. Looking at Fig. 4.4, one notices that there are many small fluctuations where S_k slightly differs from zero and a very few large fluctuations in which S_k is significantly different from zero. But the key aspect of the time behaviour of the system is that the latter is symmetric around any point where S_k is at a local maximum or minimum. As it follows from the second Law of Thermodynamics, the left and right slopes around those points can be shown to be equal giving rise there to cusps. This means that, looking at the behaviour of S_k, it is impossible to tell which way time is running. In other words the actual history of the system and the one obtained by time reversal are indistinguishable.

On the other hand, macro-physics looks irreversible because most often the system whose time behaviour we are observing is (almost) always largely out of equilibrium. Starting from a large fluctuation, the system, if left alone, (almost) inevitably moves away from this very low probability state, since the probability of moving towards an even larger fluctuation is totally negligible. As a result, a system finds itself (almost) always at a maximum of a fluctuation.

We conclude with an amusing observation. It appears that these considerations provide a practical way to check the degree of randomness of random numbers. In fact, since the fluctuation probability distribution for random number generation can be computed exactly, one can measure the probability $P[\langle S \rangle > \overline{S}]$ of finding $\langle S \rangle$ greater than a given threshold, say $\overline{S}$. In our 1969 simulation it turned out that the measured $P_{exp}[\langle S \rangle > \overline{S}]$ didn't agree with theoretical expectation, meaning that the random numbers generated by our computer were not so random! Touschek was very upset by this situation to the point that he came up with a new algorithm aimed at "randomizing random numbers" [13].

4.4 Some Personal Recollections

The revision of the draft of the Italian version of the manuscript usually took place in Touschek's apartment situated in Via Pola either in the morning from 10 to lunch time or in the afternoon from 3 pm onwards. The reason was that too often we could not access the Physics Department because it was "occupato dal movimento studentesco". We are in 1968, in the hottest moment of the turmoil.

When I arrived, on Touschek's desk submerged by his typewritten notes with ashtrays full of butts, there was always a bottle of Chianti, open and ready, together with two glasses. According to Touschek "Chianti was the ideal magic potion making the brain work smoothly and brilliantly". Actually not mine! After a couple of hours my brain wasn't so much focused and "brilliant" any more. But it was very difficult to refuse to drink the wine repeatedly poured into my glass! Until, after a week or so, I discovered that the only working strategy I could put in place was to never empty the glass after the first shot.

These meetings were for me extraordinary occasions in which we didn't only talk about the book we were preparing and Physics in general, but also about every day life, philosophy, politics (remember this was in 1968). An argument which came up frequently was the issue of the famous "pompe di m...", a commercial activity in Rome (in Piazza Indipendenza) initially run by his aunt. With polemical irony those were indicated by Touschek as the real source of the income of his family, because he considered the meagre salary he was getting as "Professore Aggregato" like a kind of charity graciously handed by the Ministry of Education. One day, particularly angry because the procedure initiated to get him a tenured position as "Professore Ordinario" was being delayed, he put up on the door of his office his pay slip with the provocative purpose of denouncing the total failure of his fight against the dull Italian bureaucracy. Actually Touschek wanted also to complain about the low funding of research and the enormous delay in the construction of ADONE. The latter was blamed by Touschek mainly on the inability of the LNF (Laboratori Nazionali di Frascati) management to deal with the repeated strikes and "occupations" going on in those years.

In everyday relations Touschek was an exquisite person with an extraordinary sense of humor and a sort of disenchanted cynicism making people around him fascinated and disoriented. But at the same time he was always keen to talk and ready to patiently explain things. The time I spent working on the book not only served as a guide for my career as a Physicist, but was also a school of formation as a person.

4.5 Conclusions

Touschek was a great scientist, a brilliant teacher and an amazing person, and for me a source of invaluable inspiration. Interacting with him was a fantastic human and scientific adventure.

I want to conclude these considerations on the birth and the content of the book "Meccanica Statistica" with a few remarks that I believe are the key to explain the forthcoming developments of theoretical research in Italy.

The point I want to make is that, despite the fact that Touschek had been lecturing on the subject of Statistical Mechanics for only 4 or 5 years at Sapienza University of Rome (at a certain point he moved to the course of "Metodi Matematici della Fisica"), his cultural legacy had an enormous impact on the development of Theoretical Physics in Rome. One cannot consider it to be just a mere coincidence the fact that in the following years Statistical Mechanics has grown to be one of the major areas of investigation in Rome and in Italy, culminating in the award of this year's Nobel Prize for Physics to Giorgio Parisi. Indeed, the root of the many important contributions that Italian physicists have given to a number of research fields related to Statistical Mechanics (among which, besides the theory of Spin Glasses and Complex Systems [14, 15], I want to mention Turbulence [16], Lattice QCD [17, 18] and the emerging field of Biophysics [19]) can be traced back to the crucial influence that Touschek scientific and teaching activity had on a whole generation of physicists.

Finally, just allow me to repeat that a great regret for Touschek (as well as for myself) was that he didn't manage to publish an English version of the book. Is it too late today?

Acknowledgements I wish to thank the Organizers of this Memorial. It was not only a moving and intense occasion but also a stimulating overview of the recent developments in the theory of fundamental interactions and the prospects for the construction of future particle accelerators brought about by Touschek's far-reaching vision.

References

1. G.E. Etim, G. Pancheri, B. Touschek, The infra-red radiative corrections to colliding beam (Electrons and Positrons) experiments. Nuovo Cim. B **51**, 276 (1967)
2. M. Greco, G.C. Rossi, A note on the infrared divergence. Nuovo Cim. **50**, 168 (1967)
3. M. Greco, F. Palumbo, G. Pancheri-Srivastava, Y. Srivastava, Coherent state approach to the infrared behavior of non-abelian Gauge theories. Phys. Lett. B **77**, 282 (1978)
4. G. Curci, M. Greco, Y. Srivastava, QCD jets from coherent states. Nucl. Phys. B **159**, 451 (1979)
5. G. Curci, M. Greco, Y. Srivastava, Coherent Quark-Gluon jets. Phys. Rev. Lett. **43**, 834 (1979)
6. S. Catani, M. Ciafaloni, G. Marchesini, Non-cancelling infrared divergences in QCD coherent states. Nucl. Phys. B **264**, 588 (1986)

7. S.J. Parke, T.R. Taylor, An amplitude for n Gluon scattering. Phys. Rev. Lett. **56**, 2459 (1986)
8. M.L. Mangano, S.J. Parke, Z. Xu, Duality and multi-Gluon scattering. Nucl. Phys. B **298**, 653 (1988)
9. C. Bernardini, G. Pancheri, C. Pellegrini, Rev. Accel. Sci. Tech. **08**, 269–290 (2015). https:// doi.org/10.1142/S1793626815300133. arXiv:1510.00933 [physics.hist-ph]
10. B. Touschek, G.C. Rossi, *Meccanica Statistica (Programma di Matematica, Fisica, Elettronica)* (Boringhieri, Torino, 1970)
11. B. Touschek, G.C. Rossi, *Appunti di Meccanica Statistica*, Quaderni di Fisica 13 (La Goliardica, Roma, 1969)
12. H. Ott, Lorentz-Transformation der Wärme und der Temperatur. Z. Phys. **175**, 70 (1963)
13. B. Touschek, A note on random numbers. (unpublished)
14. M. Mézard, G. Parisi, M. Virasoro, *Spin Glass Theory and Beyond*. Lectures Notes in Physics
15. G. Parisi, *Statistical Field Theory*. Frontiers in Physics (Addison-Wesley Pub. Co., Redwood City, 1988)
16. R. Benzi, L. Biferale, S. Ciliberto, M.V. Struglia, R. Tripiccione, Generalized scaling in fully developed turbulence. Phys. D: Nonlinear Phen. **96**, 162 (1996)
17. K.G. Wilson, Confinement of quarks. Phys. Rev. D **10**, 2445 (1974), in *New Phenomena in Subnuclear Physics*, ed. by A. Zichichi (Plenum, New York, 1977)
18. M. Bochicchio, L. Maiani, G. Martinelli, G.C. Rossi, M. Testa, Chiral symmetry on the lattice with Wilson Fermions. Nucl. Phys. B **262**, 331 (1985)
19. S. Morante, G.C. Rossi, The notion of scientific knowledge in biology. Sci. & Educ. **25**, 165–197 (2016)

Chapter 5
Role of Bruno Touschek in the Realization of the Particle-Antiparticle Colliders

Carlo Rubbia

Abstract Recollections about meeting and collaborating with Bruno Touschek.

I have met for the first time Bruno when I was a student at the Scuola Normale di Pisa. Luigi Radicati had succeeded in convincing Bruno to come periodically to Pisa by train from Roma and to give some lectures on subjects of his choice.

Parity violation had just been discovered and the question of the true nature of the neutrino fascinated and obsessed Bruno. But he was even more fascinated by the role in Nature of fundamental symmetries like C, P, and T. The originality and uniqueness of his personality and of his ideas, even his strange accent, and, most of all, the enthusiasm and the drive with which he was literally aggressing subjects in his lectures and in the subsequent long discussions, made a deep impression on all of us, then young students.

On my return to Italy after a period spent in the United States, I moved to the University of Roma, where I met then again Bruno. He had not changed, not even a bit. At that time he was in his full creative effort on electron positron colliding beams. I was extremely surprised that he could be talking about such "practical" devices, like those needed to accumulate particles, since I had known him only as a "champion of the Majorana neutrino".

Then I understood that in his mind electron–positron collisions were nothing else than the way of realizing in practice the idea of symmetry between matter and antimatter, in the deep sense of the Dirac equation.

I still remember him saying with a very loud voice, resonating in the corridors, "the positron and the electron must collide because of the CPT theorem". His boundless enthusiasm for particle-antiparticle collisions was dominated by a sense of perfect and intellectual aesthetics, rivalled only by his contempt for the other and more

C. Rubbia (✉)
CERN, Geneva, Switzerland
e-mail: Carlo.Rubbia@cern.ch

© The Author(s) 2023
L. Bonolis et al. (eds.), *Bruno Touschek 100 Years*,
Springer Proceedings in Physics 287,
https://doi.org/10.1007/978-3-031-23042-4_5

mundane alternatives of collisions of electrons or of protons, being explored at that time for instance by Jerry O'Neill, Andy Sessler and others.

One must recognize that talking about practical collisions between articles and antiparticles was at that time perfectly and totally crazy in the views of most of the so-called "reasonable" scientists, since neither the accelerator technology, nor the vacuum, —without mentioning the problem of accumulating realistic amounts of positron current—were known at the time.

Norman Ramsey told me later that returning in those days from a trip to Europe and the Soviet Union he got as an answer: "there will never be enough luminosity to do any physics".

It was however evident that all these concerns had absolutely no influence on Bruno and that he was only attracted by the perfection and the beauty of a machine capable of producing "an excited vacuum". I remember him explaining that in this way "all possible (charged) particle states must be produced, the "ultimate and definitive spectroscopy".

Later, when I met Budker, I realized how similar his and Bruno mental attitudes were. I met Budker in the United States, where he had come for a short visit in California, at a dinner with Wheeler at the O'Neill home.

At that time proton-antiproton collisions had become the next "unthinkable idea". Shortly afterwards, Budker visited CERN with Skrijnsky, since he was very curious to see the progress on the ISR, which was being started at that time. However he was not very respected by the CERN accelerator community, much too conservative and attached to formalisms to fully appreciate the genius of the man. So I had to take personally a significant role in the visit, showing him around CERN.

In order to smooth further the harshness of the reception at CERN and also in order to have a further chance "to pick at his brain", I decided to accompany both Budker and Skrinsky in their visit to Roma and to Frascati, where instead he was received very warmly and with an immense enthusiasm.

On the next day—which was some kind of a holiday—we were all invited for lunch in Bruno's home. Of course, the "lunch" lasted a major fraction of the afternoon. This has been for me the occasion of witnessing the interesting interactions between Bruno and Budker, at the same time so similar and so different. While Budker tended to jump constantly from one subject to the other in a continuous firework of ideas, Touschek was saying much less and concentrating stubbornly on the same idea.

It is usually believed that the idea of transforming a conventional accelerator into a proton-antiproton collider was developed by me and collaborators in the late seventies and in order to observe the production of intermediate vector bosons. Actually the idea is to be traced far back in time and to Italy. About ten years before, Giorgio Salvini—at that time President of INFN—had asked a number of physicists, including myself, to meet in Pisa under Stoppini as a "coach" in order to come up with a recommendation for the next step in accelerators in Italy. At that time the SPS was not yet accepted and many people thought that one should have launched the "next step" on a national basis, and why not, also in Italy. I must say that hopes were not riding very high, if one considers that the name with which the project was unofficially labeled was

Macchina Acceleratrice Italiana Protosincrotrone Inter Universitario—MAI-PIU' (Never Again).

At that time, we had two alternatives: one was a conventional 80 GeV proton synchrotron, the other a proton-antiproton collider, based on Budker's electron cooling, in the same tunnel and 160 GeV in the centre of mass.

I remember I had a long discussion with Bruno on what one should do next. He had no doubt that the colliding beam solution was the correct line to follow. Clearly in his and in our mind at the time the proton antiproton option was the logical continuation of the ADA-Adone line.

Bruno's enthusiasm was—as usual—very contagious and Ghigo and myself started to work out in detail a possible and "least unrealistic" scheme. We concluded that the first step was the one of testing the idea of electron cooling experimentally. To that effect, we had planned to borrow from CERN the "electron analogue" of the ISR, at that time left unused in the Adam's Hall at CERN. We spent in fact several days at CERN and found that all components for cooling experiment were easily at hand at that time. What was lacking—and that we were prepared to provide—was the real interest in proceeding with the studies and the courage to take these things seriously.

Unfortunately, the end of that summer coincided with the end of our dreams, shortly followed by the tragic and sudden death of Ghigo. The whole matter was set to rest, since it was decided by the scientific community at large to concentrate all European efforts toward the political consensus needed for construction of the SPS.

The Italian initiative for a collider-accelerator, as well as the projects in France and Germany for conventional medium energy accelerators were in the way of the larger CERN machine and had to be sacrificed. In a way this has not been all bad, since the MAI-PIU' option would never had the energy to reach W and Z thresholds!

Ten years later the fire of the proton-antiproton collision was still burning in the back of my mind, and I must say that so it was in the mind of Bruno. (The third person would—no doubt—have been Ghigo, if still alive!). As soon as he knew that the proton-antiproton collision adventure at last was actually going to start—although already terribly affected by his illness—Bruno decided to move immediately to CERN.

I remember having long discussions with him first at CERN and then, toward the end, at the nearby Hospital de La Tour, where he was periodically admitted for intensive care. Although the body was clearly weakening, his mind was as sharp and lucid as ever.

He was trying to assess for his own mind the relative merits between the electron cooling of his old friend Budker and the more modern stochastic cooling being worked by SimonVan Der Meer, Lars Thorndhal and Frank Sacherer (also tragically deceased soon after).

His approach was very indicative of the way in which his mind worked, totally polarized and almost uninterested of the way in which the problem was being tackled by others. His last paper—posthumously published by one of his then young disciples at Frascati—has been on stochastic cooling. Although it is clearly an unfinished job and it does not contribute significantly to the practical realization of the new device, it has all the flavours of his unique way of observing the world through the eyes of a true theoretical physicist.

It has been often pointed out that the contributions of Bruno in the field of antiproton cooling have been negligible. It is very likely so especially if one looks at the impact of such a last, notebook paper.

However there are ways of contributing to a field of science which cannot be quantized in terms of published papers and identifiable contributions.

So it has been for instance the case of Niels Bohr who, in comparison with other top scientists of his time, has produced almost nothing—there is no Bohr equation, no Bohr effect, no Bohr constant, no Bohr discovery. As yet, without Bohr, today there will be no quantum mechanics. Likewise without Touschek's and Budker's contributions today there will be no colliding beams of matter−antimatter. At the end, Bruno returned to his native Austria accompanied all along by my CERN driver Willy Aigner also from Austria. He died at age 57 in Innsbruck on 25 May 1978.

I have learned from Bruno to love matter−antimatter reactions. Without this fact, my own scientific career would certainly have been very different. So I believe it has been the case also for many other of us.

Chapter 6
Adone, Asymptotic Freedom and QCD

Giorgio Parisi

Abstract I am collecting some personal remembrances of the construction of the Adone intersecting storage ring and of the very first years of operations. I will discuss mainly particle theory at the beginning of the sixties and its impact on the design of experimental apparatus and the contribution of Adone to the emergence of QCD.

6.1 Introduction

This chapter is mainly based on personal remembrances. I started to be interested in Adone physics in 1970 [1] and I was a researcher in the Frascati National Laboratories starting from January 1971 for 10 years. The construction of the Adone was quite a recent effort and the exploitation of the machine was just starting, I have been a witness to those events: this paper can be considered mostly a primary historical source.

In this chapter, I discuss in sequence the following arguments.

- The Adone intersecting storage ring.
- Particle theory at the beginning of the sixties.
- The impact of the theory on the design of experimental apparatus.
- Partons: back to field theory.
- A new paradigm: QCD, Asymptotic freedom and quark confinement.

Contrary to many other speakers I had not too many interactions with Bruno Touschek. For various reasons (that at the time I had not fully investigated) Bruno was not coming too often to Frascati. I remember some discussions of physics with him at the University. A particular episode stuck in my mind.

Around 1971 a preprint arrived where it was claimed that the cross-section for the process

$$e^+e^- \rightarrow 2e^+2e^- \tag{6.1}$$

G. Parisi (✉)
Department of Physics, Rome University La Sapienza, Piazzale Aldo Moro 5, 00199 Roma, Italy
e-mail: giorgio.parisi@gmail.com

© The Author(s) 2023
L. Bonolis et al. (eds.), *Bruno Touschek 100 Years*,
Springer Proceedings in Physics 287,
https://doi.org/10.1007/978-3-031-23042-4_6

was of order $1/m_e^2$ and not $1/E^2$ as expected. The difference in the prefactor produces an enhancement of the order 10^6. It was very clear to Nicola Cabibbo that the paper was wrong. A factor $1/m_e^2$ is present in individual Feynman diagrams, but it cancels out if you sum different diagrams without doing algebraic mistakes. If you do the right computation, you find an enhancement factor of only $\log(E/m_e)^2$. I remember that I and someone else were in Nicola's room where Nicola was presenting his ideas on the subject showing that one can do a simple *Blitz* computation.[1] Bruno entered in the room and he started the conversation by immediately declaring: *I do not understand how a grown-up man can write such a nonsense.* He was right!

Of course, at Frascati, there were many circulating anecdotes about him. In the summertime after lunch, he was going to the nearby lake of Castel Gandolfo for the *tocca tinca*, i.e. while snorkeling, trying to touch the tail of a "tinca" (a tench, a freshwater fish); he also asked friends to come with him claiming that this was a very good occupation for theoretical physics. But without too much success.

In this talk, I will also not cover in detail the Physics at Frascati with AdA and Adone that is the subject of Mario Greco's talk in these *Proceedings*.

6.2 The Adone Intersecting Storage Ring

The idea of constructing an e^+e^- intersecting storage ring was presented in a historical seminar at Frascati in March 1960. After the seminar things moved quite fast. In the proceedings of a conference held at CERN in June 1961 [5] Touschek describes the situation at the beginning of his contribution (Fig. 6.1).

Frascati is developing two storage rings. The first (code name AdA = anello d'accumulazione = storage ring) designed for storing electrons and positrons of up to 250 MeV is actually undergoing the first tests, the second (code name Adone) a storage ring for electrons and positrons of up to 1.5 GeV, is still being planned.

The AdA team consists of C. Bernardini, G. F. Corazza, G. Ghigo, B. Querzoli and myself. The magnet was planned by Dr. Sacerdoti and built in Terni, the radiofrequency by Dr. Puglisi.

Adone is a national effort. A design team headed by Dr. Amman has the task of arriving at a specific design proposal by the beginning of 1962. Simultaneously a committee is preparing the experiments to be carried out with the machine. If by the beginning of 1962 it is found that the project has a reasonable chance of success from a technical viewpoint, it is expected that the machine should be working in late 1964.

Let me be very brief on the Adone project: electrons and positrons circulate in a magnetic ring designed to contain particles up to 1.5 GeV energy. The energy losses

---

[1] This episode was important in a historical perspective. Nicola's computations were published much later in a joint preprint with Rocca. They computed the probability of finding inside a γ a triplet γ, e^+, e^- [2]. I had many discussions with Nicola on these generalized Weizsäcker–Williams relations and his work had deeply influenced my paper with Guido Altarelli on *Altarelli Parisi equations* [3].

Fig. 6.1 A photograph of the Adone Intersecting ring with the experimental apparata. Copyright INFN—LNF

are replaced by R.F. Injection is effected by means of a low energy (of between 50 and 200 MeV). The final energy is reached by raising the magnetic field to the desired value.

AdA was a great success [4]. After the first positive results the construction of Adone started immediately and in 1968 Adone was completed [6, 7]: the beams of electrons and positrons circulated in a ring 105 m long, divided into 12 equal sectors composed of a bending dipole, followed by a pair of quadrupoles and a straight section of 2.5 m.

Four of the 12 straight sections available allowed the intersection of the beams, resulting in many points of interaction for the installation of the experimental equipment. Four experimental apparatus have been constructed. The remaining 8 sections housed the radiofrequency cavities and the equipment for beam injection and diagnostics.

The maximum project energy of Adone was 1.5 GeV per beam, but after November 1974 it operated also at 1.55 GeV. The minimum energy was around 0.7 GeV.

I remember that someone told me that during the construction of Adone, Bruno and the others had to take irreversible decisions without no hints. In all the cases the decisions were the right ones, otherwise, Adone could have had years of delay. This was a nice example of serendipity. The project was not perfect and there were some unforeseen phenomena: fortunately in all the cases the cure, often found by Bruno, was not too complex.

6.3   Particle Theory in the Beginning of the Sixties and the Design of the Experimental Apparatus

Around 1960 Geoffrey Chew proposed the *boostrap* philosophy [8]. There were no elementary constituents, all particles were supposed to be on the same footing. This approach became extremely popular and also Murray Gell-Mann was not taking quarks seriously: he considered them as a mathematical model to implement $SU(3)$ symmetry. Field theory was discredited and some authors suggested that a good knowledge of quantum field theory was detrimental to the comprehension of bootstrap. This was a somewhat strange position because dispersion relations were a cornerstone of bootstrap and they were firstly derived in the context of local field theory. Ironically the bootstrap approach was instrumental to the birth of string theories that are among the most sophisticated quantum field theories.

The absence of elementary point-like objects was suggesting that hadrons were extremely soft. This viewpoint was confirmed by the very fast decay of the proton form factor, by the exponential suppression of particle productions at large momentum transfer, and by Hagedorn theory where a maximum temperature somewhat less than 200 MeV was supposed to be present in nature: also a very energetic hadronic fireball would emit hadrons of no more than few hundredth MeV.

There was also a different viewpoint: electromagnetism, Fermi interactions, and the V-A theories for weak interaction were based on local currents and quantum field theory. The purely leptonic electro-weak was described by a field theory that was non-renormalizable, however, there was some hope that the introduction of heavy vector Bosons could make the theory renormalizable.

The semi-leptonic weak interactions were mediated by a hadronic current, and the resulting current algebra (based on local commutators) was crucial to normalizing the weak interaction vertices, which played a fundamental role in Cabibbo's theory of weak interaction. This quantum field theory approach to physics was strongly pushed in Europe: there were strong collaborations among different scientific institutions that were later formalized in the Triangular Meetings (Paris-Rome-Utrecht). The impact of the theory on the design of experimental apparatus was obviously strong. The bootstrap approach was the most popular and the alternative approach based on quantum field theory was not mature to propose a credible alternative.

There was a widespread belief that the dominant process should be the production of 2 particles (like $\pi^+ + \pi^-$) or 3 particles (like $\pi^+ + \pi^- + \pi^0$). The cross section was supposed to be small at energies greater than 1 GeV in the center of mass: indeed the 2 body processes (and in a similar way the three body processes) were controlled by form factors that were known to decrease very fast (at least in the case of the proton) at large momentum transfer (q^2).

Most of the interesting physics was supposed to be the discovery of new resonances like ρ', or of exotic objects like a heavy electron (τ in modern language) that unfortunately had slightly too high energy to be observed at Adone.[2]

Consequently, the four experimental apparatus had a small fraction of solid angle that was enough for collinear and planar events: in these cases, the distribution of the produced particles was easy to predict. For two-body collinear events the angular dependence is well known. For three-body events one has to study the Dalitz plot of the events: this is a relatively simple job as far as the Dalitz plot could be easily parametrized in terms of a few parameters.

6.4 Partons: Back to Field Theory

6.4.1 Partons in Electron Collisions

In 1968 experiments at SLAC show a large cross section for the process (deep inelastic scattering)

$$e^+ + \text{proton} \rightarrow e^+ + \text{hadrons}. \tag{6.2}$$

Bjorken shows that these large cross sections are consequences of current algebra sum rules. Feynman introduces the idea of partons (field theory under disguised form) suggesting that these events were the signature of the presence of hard, point-like components inside the hadrons.

Adone started to work in 1969 and from 1970 it was clear that the total hadron production was quite high, more than the $\mu^+\mu^-$ cross sections and many events had a charged multiplicity greater than 2 (multi prong events). It was a mess to be sure of the value of the cross section for hadron production as far as the efficiency for detection depends on the production model. After some time it became clear that the ratio

$$R = \frac{e^+e^- \rightarrow hadrons}{e^+e^- \rightarrow \mu^+\mu^-} \tag{6.3}$$

was around 2 without a great dependence on the energy [9–11, 18].

In 1970 Nicola Cabibbo, Massimo Testa, and I derived the formula

$$R = \frac{e^+e^- \rightarrow hadrons}{e^+e^- \rightarrow \mu^+\mu^-} = \sum_i Q_i^2 + \frac{1}{4}\sum_i B_i^2, \tag{6.4}$$

---

[2] I remember that around 1972 I was discussing with Luciano Maiani about the signature of charmed mesons or heavy electrons and we noticed that direct production of $e - \mu$ and hadrons would be present in both cases, so we could not discriminate directly the two cases and some confusion could arise if the two channels would open at similar energy. After some reflections, we concluded that it was extremely unlike that the threshold should be similar. Amazingly the lowest mass of a charmed charged meson is 1870 MeV and the mass of the τ 1777 MeV.

where Q_i are the charges of the spin 1/2 partons and B_i are the charges of the spin 0 partons. In the naive quark model

$$R = \frac{2}{3} \tag{6.5}$$

For partons that are $SU(3) \times SU(3)$ mesons $R = 1$.

Around 1971 Fritzsch, Gell-Mann, and Leutwyler suggested that the correct theory is given by colored quarks interacting via an octet of colored gluons. I remember a seminar given by Gell-Mann (1972) in "Aula Fisica Superiore",[3] where Gell-Mann proposed QCD. During that seminar, he said that in QCD

$$R = 2 = 3\frac{2}{3}, \tag{6.6}$$

where the extra factor 3 comes from the color of quarks.

After the seminar Conversi remarked that this was in agreement with Frascati data. Gell-Mann was quite impressed.

6.4.2 Asymptotic Freedom

Landau noticed in 1955 that all known renormalizable theories had a divergent effective coupling at high energy, but for gauge Yang–Mills theory the computation was not done. Around 1966 the Russian physicist Iosif Benzionovich Khriplovich [15, 16] computed the beta function for Yang–Mills theory in a very elegant way: I do remember the argument very well.[4] The paper started to be quoted after 1977, unfortunately it did not affect history, most of the citations started nearly ten years later. I noticed it at the end of the seventies looking for something else on.

In 1972 things started to move fast: Symanzik noticed that the $g\phi^4$ theory is asymptotically free for negative g: in this way, one obtains *a field theory with computable large-momenta behaviour.*

In the same year t'Hooft presented at the Marseille conference his result on the negative sign of the beta function for small coupling constant in Yang–Mills theory (a five-minute remark) that received not much attention apart from Symanzik: no preprint or paper was ever written.

---

[3] The classroom is currently dedicated to Marcello Conversi at the Physics Department of Sapienza University in Rome.

[4] The idea was to compute the logarithmic shift in the vacuum energy in presence of a constant *magnetic field.* In electrodynamics the computation was done by Heisenberg and Euler in 1936 [17]. In general there are two contributions that are well known for electromagnetism: the effects of Landau levels for free electron of the sea and the polarization of the corresponding spins. Generalizing these results to particles of spin 1, it was very simple to get the results without doing a loop computation.

In 1973 Politzer and Gross & Wilczek computed the beta function, and published their results showing that QCD is asymptotically free. This opened the way to a detailed comparison of the theory with experiments. At the end of 1973, I presented the first evaluation of the running coupling constant from violations of Bjorken scaling, so it was possible to start to compare the theory with experiments.

In 1974 the discovery of J/ψ at Brookhaven and SLAC, immediately reconfirmed at Frascati proved the existence of charm. The value of the widths of the ψ particle in hadrons or $\mu^+\mu^-$ were measured both at SLAC and Adone: They were in good agreement with the theoretical prediction using the charmonium model and provided an independent estimate of the running strong coupling constant.

The success of the charmonium model was undisputed when the first excited state was found at SLAC (ψ') and its pseudo scalar partner η_c was found at DESY. The same model gave good predictions for the mass splitting $\psi - \eta_c$, for the width for the process $\psi \to \eta_c + \gamma$ and the width of η_c.

At the end of the story there were two crucial contributions of Adone to the development of QCD:

- The measurement of R that was compatible with the colored quark model.
- The measurement of the decay widths of the ψ that were crucial in establishing directly the correctness of the colored gluon model.

References

1. N. Cabibbo, G. Parisi, M. Testa, Hadron production in $e^+ e^-$- collisions. Lettere al Nuovo Cimento (1969–1970) **4**(1), 35–39 (1970)
2. N. Cabibbo, M. Rocca, The ρ Bremsstrahlung, Ref TH. 1872-CERN (1974)
3. G. Altarelli, G. Parisi, Asymptotic freedom in parton language. Nucl. Phys. B **126**(2), 298–318 (1977)
4. C. Bernardini, AdA: the first electron-positron collider. Phys. Perspect. **6**(2), 156–183 (2004)
5. B. Touschek, The Frascati storage rings, in *Proceedings of International Conference on Theoretical Aspects of Very High-Energy Phenomena, CERN, Geneva, 5–9 June 1961*. CERN Report 61-22 (1961)
6. F. Amman, et al., Status report on the 1.5 GeV electron positron storage ring - Adone, in *Proceedings of the 4th International Conference on High-Energy Accelerators*, 21–27 August 1963 (1965), pp. 309–327
7. F. Amman, F. Andreani, M. Bassetti, Adone: the frascati 1.5 GeV electron positron storage ring, in *International Conference on High Energy Accelerators. Rome, Comitato Nazionale per l'Energia Nucleare, 1966*. CNEN, Frascati, Italy (1967), p. 703
8. G.F. Chew, The S-matrix theory of strong interactions. Preprint UCRL 9701 (1961)
9. B. Borgia, Results from storage rings. J. Phys. A: General Phys. **5**(2), 179 (1972)
10. F. Ceradini et al., Multiplicity in hadron production by e^+e^- colliding beams. Phys. Lett. B **42**(4), 501–503 (1972)
11. V. Alles-Borelli et al., e^+e^- Annihilation into two hadrons in the energy interval 1400–2400 MeV. Phys. Lett. B **40**(3), 433–436 (1972)

12. M. Grilli, et al., Multihadron production in $e^+ e^-$ collisions at high energy. Il Nuovo Cimento A (1965–1970) **13**(3), 593–644 (1973)
13. K. Symanzik, A field theory with computable large-momenta behaviour. Lettere al Nuovo Cimento (1971–1985) **6**(2), 77–80 (1973)
14. H. Fritzsch, M. Gell-Mann, H. Leutwyler, Advantages of the color octet gluon picture. Phys. Lett. B **47**(4), 365–368 (1973)
15. I.B. Khriplovich, Green's functions in theories with non-abelian gauge group. Yad. Fiz. **10**, 409 (1969). [Sov. J. Nucl. Phys. **10**, 235 (1970)]
16. M. Shifman, Historical curiosity: how asymptotic freedom of the Yang-Mills theory could have been discovered three times before Gross, Wilczek, and Politzer, but was not (2022). arXiv:2203.12030
17. W. Heisenberg, H. Euler, Consequences of the Dirac theory of positrons. Z. Phys. **98**, 714–732 (1936)
18. D.J. Gross, F. Wilczek, Ultraviolet behavior of non-abelian gauge theories. Phys. Rev. Lett. **30**(26), 1343–1346 (1973)
19. H.D. Politzer, Reliable perturbative results for strong interactions? Phys. Rev. Lett. **30**(26), 1346–1349 (1973)
20. C. Bacci et al., Preliminary result of frascati (ADONE) on the nature of a new 3.1-GeV particle produced in e^+e^- annihilation. Phys. Rev. Lett. **33**(23), 1408–1410 (1974)

Chapter 7
The Standard Theory and Theoretical Physics in Roma

Luciano Maiani

Abstract After a brief description of the rise of the Constituent Quark Model for hadrons, I illustrate the contributions of the Roman Theoretical School to the formation and exploration of the Standard Theory of fundamental particles, in the years 1970 to 1990.

7.1 Introduction

In 1937, C. Anderson and S. Neddermeyer discovered a new particle produced in the upper atmosphere by the collisions of Cosmic Rays. In 1946, in Roma, M. Conversi, E. Pancini and O. Piccioni proved that the mesotron (μ particle, today) is not the particle responsible for the nuclear forces, proposed by H. Yukawa. Many consider this discovery to be the birth of modern Elementary Particle physics.

Everybody (Fermi, Marshak, etc.) was worried: where is the pion?

Pontecorvo asked an apparently simpler question: what is the mesotron? and proposed a surprising answer: it is a second generation electron. I. Rabi commented: who ordered that?

What is the role of μ particle in the fundamental forces? A provisional answer was the concept of Universality of the Weak Interactions (Puppi, 1950). This line of research, after Feynman and Gell-Mann, Marshak and coll. and, later, Cabibbo, led eventually to electroweak unification (Glashow, Weinberg, Salam, Higgs and Englert). But we still do not have a plausible explanation of why are there different generations.

L. Maiani (✉)
Dipartimento di Fisica, Università di Roma La Sapienza and INFN, Sezione di Roma1, P.zza A. Moro 2, 00185 Roma, Italy

TH Division, CERN, 1211 Geneve 23, Switzerland
e-mail: luciano.maiani@roma1.infn.it

© The Author(s) 2023
L. Bonolis et al. (eds.), *Bruno Touschek 100 Years*,
Springer Proceedings in Physics 287,
https://doi.org/10.1007/978-3-031-23042-4_7

7.2   Elementary Constituents Versus Nuclear Democracy

In 1940–1950, a particle zoo emerged from the study of cosmic ray interactions. The new particles do not arise from further subdivision of normal matter (atoms, nuclei, nucleons, atomic and nuclear forces) and *the probability that all such particles should be really elementary becomes less and less as their number increases* (Fermi and Yang [1]). The proposal by Fermi and Yang was that not all the observed particles are elementary. They proposed the Yukawa particles, the pions, to be nucleon-antinucleon bound states, e.g. $\pi^+ = (p\bar{n})$. The natural symmetry of the Fermi–Yang scheme was the Isospin symmetry displayed by the nucleon doublet, the $SU(2)$ symmetry, which propagates to their mesonic bound states, as indeed is observed.

The only, but very startling consequence one could derive from the Fermi–Yang hypothesis was that the pion has to have a negative parity, as indeed was indicated in these years by pion photo-production experiments.

With the discovery of the strange particles, in 1956 Sakata [2] proposed the Λ baryon as the additional elementary constituent

$$\text{elementary} : S = (p, n, \Lambda)$$
$$\text{mesons} = S\bar{S}; \quad \text{baryons} = SS\bar{S} \tag{7.1}$$

The natural symmetry was now promoted from $SU(2)$ to the unitary transformations in a three-dimensional complex space, $SU(3)$. The Sakata model reproduces well the meson spectrum, and it makes a clear prediction: there must exist baryons with strangeness $S = +1$. Unfortunately it is a wrong prediction, no such particle has been seen until today!

On a different line of thinking, one can argue that, in the presence of very strong (unitarity saturated) interactions and using crossing symmetry, there is no clear distinction between composites and constituents:

$$\pi^+ = p\bar{n}, \text{ or rather } n = \bar{p}\pi^+ ??? \tag{7.2}$$

The most natural principle to start with, in the Sixties, was considered to be **Nuclear Democracy**: all strongly interacting particles, collectively called *hadrons*, are to be treated on the same footing (G. Chew and S. Frautschi).

It was also believed that unitarity and maximal analyticity of scattering amplitudes could provide the dynamical equations needed to determine the hadron spectrum. Recalling the story of the Baron of Munchausen, this approach received the suggestive name of *bootstrap*: to lift up himself by pulling the boots of his own shoes (or his pigtail as the Baron did to get out of the swamp).

Nuclear Democray and bootstrap had no real, recognized success. However, these principles inspired the dual model of Gabriele Veneziano, which, later, gave rise to *String Theory*, the basis of many theories of Quantum Gravity.

For sub-nuclear particles, the meaning of constituents was understood only in 1973.

Nuclear democracy holds. Subnuclear particles are on the same footing: they are all composite and the elementary constituents are quarks and gluons, endowed with a new quantum number, colour

elementary : $q = (u, d, s)$, *gluons* $\leftrightarrow$ generators of the colour group $SU(3)$.

and, after Gell-Mann [3] and Zweig [4]:

$$\text{mesons} = q\bar{q}, \; qq\bar{q}\bar{q}, \; \ldots$$
$$\text{baryons} = qqq, \; qqqq\bar{q}, \ldots \tag{7.3}$$

The fundamental strong interaction is the gauge theory associated to colour, Quantum Chromo Dynamics. It becomes weak at short distance, as shown by Gross and Wilczek [5] and by Politzer [6], so as to allow the unambiguous identification of the constituents.

7.3 Gatto, Cabibbo, Touschek and Cini in Roma and Frascati

In 1951, after the Diploma at Scuola Normale di Pisa, Raoul Gatto went to Roma as assistant to Bruno Ferretti.

In 1956, he left for the United States, to become a staff member of the Lawrence Radiation Laboratory in Berkley. The group of Luis Alvarez was in full production, discovering new hadrons with the hydrogen bubble chamber. Gatto absorbed quickly the exciting atmosphere of the laboratory. In close contact with the Alvarez group, he produced works on the symmetries of the weak interactions (Fermi's imprint on Italian theoretical physics) and the phenomenology of weak decays of hyperons.

Coming back in 1960, Gatto became the director of the newly formed theory group at Frascati, bringing to Italy the new ideas flourishing at the time in the US, concerning the application of symmetry and group theory to particle physics. He found, as junior partner, Nicola Cabibbo.

Nicola had graduated in 1958, tutor Bruno Touschek, and was the first theoretical physicist hired in Frascati by Giorgio Salvini, then director of the Laboratory.

These were exciting times in Frascati. Touschek and collaborators were building the first collider (AdA), to be followed by Adone (1969), a new particle (the eta meson) was discovered, checking the freshly introduced SU(3) symmetry.

Cabibbo and Gatto authored an important article on e^+e^- physics in 1961, promptly named *the Bible*. Among other works, they investigated the weak interactions of hadrons in the framework of the SU(3) symmetry (a precursor of Cabibbo theory of the weak interaction angle [7], made at CERN two years later).

Brilliant younger collaborators joined in Frascati: G. Da Prato, U. Mosco, G. De Franceschi and, later, G. Altarelli and F. Buccella (graduated with Gatto), G.

Gallavotti (with Touschek). The preparation of ADONE experiments prompted the renewal of QED studies.

In year 1962–63, Touschek was teaching Statistical Mechanics in Roma. He would present his lecture consulting personal notes that he had probably prepared the night before. In the presentation, it seemed as if he had just discovered what he was explaining.

Extremely clear and precise, Touschek spoke a perfect Italian with a fascinating Austrian accent and sometime old fashioned expressions. He referred to the heat bath to reach thermal equilibrium as "vasca di bagno" and described his revolutionary idea of making head-on electron-positron collisions as "treno-contro-treno". One could see the perfect image of a scientist and a perfect illustration of how rewarding research in physics might be.

Most importantly, in Roma Touschek kept alive the interest in field theory, given as dead in most countries, actively cultivating the study of QED, Fermi theory and of fundamental symmetries (Majorana and the two components neutrino theory).

During the late Fifties and early Sixties, theoretical alternatives to field theory were explored in Roma by Marcello Cini, with many young collaborators. Among them, M. Cassandro, L. Sertorio, M. Restignoli and, later L. Violini, M. Lusignoli, M. Toller, D. De Maria. Subjects went from the Fundamentals of Quantum Mechanics to Dispersion Relations, Regge Poles, Relativistic Thermodynamics.

7.4 Particle Physics in the Late Sixties

Hopes to reach a basic theory for strong, e.m. and weak interactions flourished in the late Sixties, based on several important results obtained in the first part of the decade

- quarks in three flavours, introduced by M. Gell-Mann and George Zweig, gave an excellent explanation of the observed meson and baryon spectrum;
- the Cabibbo theory of semileptonic $\Delta S = 0, 1$, hadron decays, marked a substantial progress with respect to the Fermi theory, enlarging universality to strange particle decays, via $d - s$ quark mixing.

There were clouds as well, indicating that something was still missing

- the clash with Fermi statistics of quarks in the baryon wave functions, with the first ideas about colour by Han and Nambu [8];
- the unexplored form of quark strong interactions inside hadrons; the exchange of an abelian gluon was often used as a toy model;
- The Fermi interaction was known to be non-renormalizable. Could W boson intermediary help?

Schwinger ideas about Yang-Mills theory and ElectroWeak unifications had been substantiated by Glashow [9] in 1961, with the introduction of the $SU(2)_L \otimes U(1)_R$

gauge group, with interactions mediated by the photon and by the massive intermediate bosons $W^\pm$, Z. The Brout–Englert–Higgs Mechanism had been worked out in 1965, leading to the Weinberg-Salam electroweak theory for leptons [10], in 1967.

It was known that embedding Cabibbo theory with three quarks in $SU(2)_L \otimes U(1)_R$ led to unobserved Flavor Changing Neutral Currents. Open questions: does Unification work for leptons only? may form factors suppress these processes?

In the late Sixties, few people believed that the basic strong interactions between quarks could be described by field theory. In the more established framework, Bootstrap, Regge Poles etc., a very promising new idea came out, the Dolen–Horn–Schmid Duality [11]: *the sum of baryon, s-channel resonance amplitudes reconstructs the (is dual to) Regge behavior in the t-channel and vice-versa.*

Duality was a new kind of bootstrap condition, and the result raised a lot of interest, which reached its maximum in september 1968 when Veneziano proposed a Dual Model of pion-pion scattering [12]. *Everybody went Dual.* Field theory for particle physics became an exoteric discipline, with few practitioners worldwide.

Nonetheless, few authors addressed the problem of higher order weak interactions in a bottom-up fashion, using the simplest theory with one charged vector boson coupled to the Cabibbo currents. At one-loop level, they found a startling, unexpected result [13–16]. The Vienna HEP Conference, August-September 1968, marked the real turning point.

Ideas about Duality were presented and widely discussed. Higher order weak interaction results about flavour changing neutral currents were presented and discussed (Cabibbo was the convenor of the weak interaction session). SLAC data on deep inelastic electron scattering were presented for the first time, indicating the onset of Bjorken scaling.

7.5 The First Weak Interaction Loop

At one loop, with one charged vector boson coupled *à la* Cabibbo, $K_L \to \mu^+\mu^-$ and $K^0 \bar{K}^0$ mixing are generated, with amplitudes of order: $\sin\theta \cos\theta\ G(G\Lambda^2)$, where G is the Fermi constant and Λ an ultraviolet cutoff.

The strict experimental limits existing at the time implied the surprisingly small value: $\Lambda \sim 2 - 3$ GeV, to be compared with the naturally expected value: $\Lambda = G^{-1/2} \sim 300$ GeV. The result was obtained by using current algebra commutators and showed that the ultraviolet divergence, in the theory with three quark flavours, *is not damped* by form factors, as a consequence of the intrinsically point-like, current algebra commutators.

The result eventually[1] led, in 1970, to the GIM Mechanism [18]: the introduction of a charm quark to cancel the quadratic divergence and the related interpretation of the Ioffe–Shabalin cutoff as a prediction of the charm quark mass, $m_c \sim 1.5$ GeV.

---

[1] The road to the GIM mechanism is described in [17].

GIM gives the possibility to include quarks in the Glashow-Weinberg-Salam gauge theory based on $SU(2)_L \otimes U(1)_R$. It was the first instance in which quark and W loops were taken seriously and led to startling predictions that indeed have been verified a few years later.

By the end of January 1970, in Harvard I think we had understood all the essentials. I remember one day going to the Legal Sea Food restaurant for lunch, where my wife Pucci joined us. Pucci told Shelly (Glashow) how happy I was about the new result and the work we were doing. He replied: *He is right, this paper is going to be in all school books.* Shelly was fantastic. In another occasion, a seminar given by him to the experimentalists of Harvard working at the CEA (Cambridge Electron Accelerator), Shelly introduced his talk by saying: *Look, with charm we have essentially solved particle physics. Except*, he added, *for CP violation.* Something that had to be reconsidered three years later by Makoto Kobayashi and Toshihide Maskawa [19], with the introduction of a third generation.

7.6 May 1970: Back in Roma

ADONE started operating at the end of 1969 and produced its first results in 1970.

All detectors observed an unexpected abundant production of hadrons. Beyond the ρ, ω, ϕ and ρ' resonances, the ratio of the hadronic to the $\mu^+\mu^-$ cross section was nearly constant and of order unity, as if the process went via the production of point-like constituents.

Coming back to Roma from the US, in may 1970, I found Nicola Cabibbo and his present and former students (Giorgio Parisi and Massimo Testa) very excited by the ADONE results. Needless to say, Touschek was greatly excited as well: the unexpected result indicated the crucial importance of e^+e^- collisions to investigate the deep structure of matter. In analogy with the formulae found by Drell and Yan for deep-inelastic muon pair production, they were playing with the formula [20]

$$R = \frac{\sigma(h)}{\sigma(\mu^+\mu^-)} = \frac{1}{4}\sum (Q_i^0)^2 + \sum (Q_i^{1/2})^2 \tag{7.4}$$

with $Q_i^{0,1/2}$ the electric charge of elementary constituents with spin 0 or 1/2. But: which constituents? Cabibbo, Parisi and Testa made different hypotheses, including spin 0 constituents, but it was only with more precise data that a definite conclusion has been possible.

Figure 7.1 gives a recent picture of the experimental determination of R [21]. The virtual photon created by the e^+e^- annihilation is a universal probe for any form of electrically charged matter. Different hypotheses about the quarks present in the final state give the values:

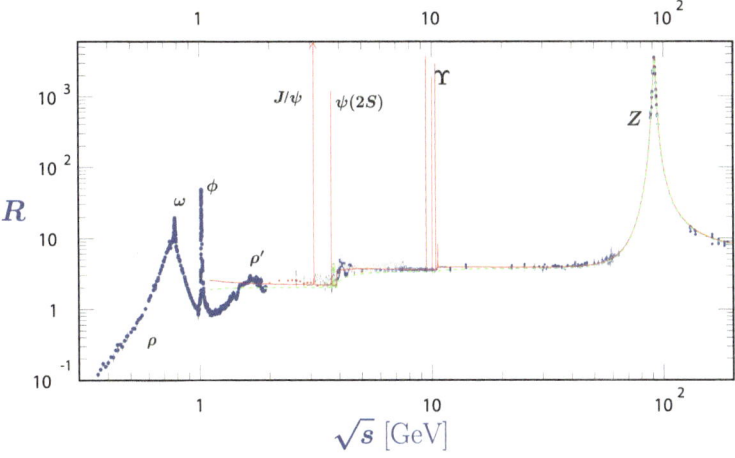

Fig. 7.1 The ratio R, (7.4), in the energy regions of Adone, SLAC and LEP. Particle Data Group [21], courtesy of the COMPAS (Protvino) and HEPDATA (Durham) Groups, May 2010

$$R =$$
$$= 2/3 \text{ (u, d, s, no colour)}$$
$$= 2 \text{ (u, d, s, each in 3 colours)}$$
$$= 2 + 4/3 \text{ (u, d, s, c in 3 colours)}$$
$$= 5 \text{ (three generations in 3 colours)} \quad\quad\quad (7.5)$$

In 1972, Gell-Mann was visiting CERN. In May he participated in a Conference in Frascati organised by Gatto. In a talk to the conference, Gell-Mann reported about work done with Bardeen and Fritzsch, where the idea of QCD was proposed [22, 23], based on a colour gauge group commuting with the electroweak group. This was before the discovery of Asymptotic Freedom of non-abelian gauge theories [5, 6].

In that occasion, Gell-Mann visited our Department in Roma and gave a seminar on QCD. He remarked that the hypothesis of fractionally charged quarks with three flavours and three colours gives $R = 2$. Conversi, present in the seminar, stated that the value observed by Adone was in fact converging to a ratio ~ 2, see Fig. 7.1.

After the J/Ψ peak, the ratio has a small increase, consistent with the $+4/3$ difference expected for the additional production of charm quark pairs. The association of J/Ψ with the $c\bar{c}$ threshold was first done in [24].

At even higher energies, $Y = b\bar{b}$ resonances appear, associated to a much smaller increase $\Delta R \sim +1/3$, corresponding to the creation of b-quark pairs. The very large Z^0 peak appears in the LEP region, associated with the neutral intermediary of the weak interactions. A further increase is expected, but not yet observed, after the $t\bar{t}$ threshold, where R should saturate the three-generation level, $R = 5$.

No signal of further structures, associated with new kinds of constituents or intermediary vector bosons has been observed so far in the single, virtual photon channel. It is left to future electron-positron, circular or linear colliders [25] to explore the region above the $t\bar{t}$ threshold, in search of further constituent matter.

7.7   Going Electroweak

Guido Altarelli was back in Roma in 1970, as Assistant Professor.[2] In 1971, Veltman and 't-Hooft proved that the Weinberg Salam theory is renormalizable: *everybody became electroweak.*

Guido and I began discussing with Nicola how to compute Electroweak corrections to the muon $g - 2$, due to the exchange of vector bosons and the Higgs boson. It was a new territory, at least for us, a lot of calculations and a lot of fun. Also many difficulties with inconsistent calculations: we called it *the rebellion of the matrices!* but we got it [26], at about the same time as other distinguished people [27].

The Adler, Bell–Jackiw anomalies in $SU(2)_L \otimes U(1)_R$ were the last obstacle towards a renormalizable electroweak theory.

Anomalies affect both quark and lepton currents. Bouchiat, Iliopoulos and Meyer [28] worked out the conditions for a cancellation to be operative. John's description of this work, in a short letter he sent me, was: *there must be charm, quarks have color and are fractionally charged.*

Asymptotic Freedom in QCD was found shortly after [5, 6]: the era of the Standard Theory had started.

7.8   Working in Roma in the Seventies

In Roma, Pucci and I used to see Guido and Nicola out of work, with wives and small kids.

Sometime we would go to Fregene, in the nice seaside house of the Altarellis, and to Grottaferrata, in the country house of the Cabibbos. We saw also other Roma professors, Salvini, Conversi, Bernardini, Careri and families.

New younger people joined in: Massimo Testa, Giorgio Parisi, Keith Ellis (a young Scottish, Italian-speaking student, attracted to Roma by Preparata and recruited in our group by Guido), Roberto Petronzio and, later, Guido Martinelli (also recruited by Guido). You will find their names appearing in the literature at first in association with Nicola, with Guido and sometimes with me.

From time to time the Physics Department was occupied by students, but we could always find a quiet office in Istituto Superiore di Sanità, across the road, where I worked. Roma and Italy were hit by social turmoil and terrorism, but our was a

---

[2] This and the following Sections are taken almost literally from [17].

quiet, intellectually stimulating, academic life that I remember with pleasure and that did not come back.

I moved in the University as full professor in 1976 and Guido took the chair shortly after, in 1980.

With John Iliopoulos in Paris, close relations were established between Roma and the group of Phil Meyer in Orsay. When Meyer's group moved from Orsay to École Normale Supérieure, in 1974, Guido Altarelli and I were living in rue d'Ulm (Keith Ellis was also around).

The discovery of the J/Ψ raised a lot of questions and we (Roma + Paris) offered to go to Utrecht to discuss with Tini Veltman and Gerardt 't-Hooft, a meeting which became the annual Triangular Meeting Paris-Roma-Utrecht, rotating among the three towns.

Guido took a crucial sabbatical in ENS in 1976–1977. Later, Giorgio Parisi came in and so Nicola Cabibbo, during my sabbatical in ENS, 1977–1978. It was remarked, at that time, that Roma people saw CERN only from the airplane, flying to Paris …and we all lived under the surveillance of Claude Bouchiat and the quiet but firm protection of Philippe Meyer.

7.9 The Altarelli–Parisi Equations

J. Iliopoulos, in his Plenary Report at ICHEP London, 1974, speaking about Asymptotic Freedom, observed [29]

as it is often the case, whenever someone talks about freedom, it invariably turns out that he really means something else. The same is true here.

Quarks are not really free in deep inelastic reactions. Deviations from exact Bjorken-Feynman scaling must be expected. Asymptotic freedom in QCD makes them calculable.

Parisi was after scaling violations very early, but all seemed very complicated and unintuitive. Then came the paper by Altarelli and Parisi [30], 1977, with a similar contribution from Dokshitzer [31] in the same year, anticipated by Gribov and Lipatov [32] in 1972: this became known as DGLAP.

The AP paper has had an enormous impact, it made easier to understand the physics and simpler to compare experimental data with theory. Guido was amused to see that their paper was rated at that time as the most quoted French theoretical paper in particle physics.

7.10   New Research Lines and New Younger Generations in Roma

Few results obtained in Roma during the wonderful years from 1974 to 1991 opened new research lines in the Standard Theory. They also saw the emergence of new generations of theorists. A, possibly incomplete, list goes as follows

- Enhancement of non-leptonic weak interactions due to QCD renormalization effects [33];
- Calculations of parton densities in the hadrons [34]
- Parton calculations of the electron beta decay spectrum of heavy quarks [35].
- QCD prediction of a phase transition from hadrons into deconfined quarks and gluons [36]
- Bounds to the Higgs boson and heavy fermion masses in grand unified theories [37]
- Lattice QCD calculation of weak parameters with lattice QCD [38].
- Proposal and realisation of the APE parallel supercomputer for lattice QCD calculations [39].
- Lattice QCD calculation of the decay constants of pseudoscalar charmed mesons, f_D and f_{D_s}, and beauty mesons, f_B [40].

7.11   Conclusions

The discovery of W [41] and Z [42] in 1983 concluded the heroic phase of the Standard Theory. Since then, up to the observation of the Higgs Boson in 2012 [43], we have had only confirmations.

The Standard Theory may not be the final word. There are many hypotheses advanced, only future experiments will tell what is really beyond ST.

Field theory has come back: will it be superseded by more subtle concepts? supersymmetry? strings?

Touschek idea of colliders has been essential for discovery and verification of the Standard Theory and it plays now a fundamental role in particle physics.

Our generation has been very lucky to be there, at the right place and the right time, and it all has been, indeed, a great fun.

References

1. E. Fermi, C.N. Yang, Phys. Rev. **76**, 1739 (1949)
2. S. Sakata, Progress Theor. Phys. **16**, 686 (1956)
3. M. Gell-Mann, Phys. Lett. **8**, 214–215 (1964)

4. G. Zweig, *An SU(3) Model For Strong Interaction Symmetry and Its Breaking. 2*, CERN-TH-412
5. D.J. Gross, F. Wilczek, Phys. Rev. D **8**, 3633 (1973); Phys. Rev. D **9**, 980 (1974)
6. H.D. Politzer, Phys. Rev. Lett. **30**, 1346 (1973)
7. N. Cabibbo, Phys. Rev. Lett. **10**, 531 (1963)
8. M.Y. Han, Y. Nambu, Phys. Rev. **139**(4B), B1006 (1965)
9. S.L. Glashow, Nucl. Phys. **22**, 579 (1961)
10. S. Weinberg, Phys. Rev. Lett. **19**, 1264 (1967); A. Salam, in N. Svartholm: Elementary Particle Theory, *Proceedings of the Nobel Symposium 1968*. Lerum Sweden (1968), p. 367
11. R. Dolen, D. Horn, C. Schmid, Phys. Rev. **166**, 1768 (1968)
12. G. Veneziano, Nuovo Cim. A **57**, 190–197 (1968)
13. B.L. Ioffe, E.P. Shabalin, Yadern. Fiz. **6**, 828 [Soviet J. Nucl. Phys. **6**, 603 (1968)]
14. F. Low, Comm. Nucl. Part. Phys. **2**, 33 (1968)
15. R.N. Mohapatra, J.S. Rao, R.E. Marshak, Phys. Rev. Lett. **20**, 634 (1968)
16. R.N. Mohapatra, P. Olesen, Phys. Rev. **179**, 1417 (1969)
17. L. Maiani, L. Bonolis, Eur. Phys. J. H **42**(4–5), 611–661 (2017). arXiv:1707.01833 [physics.hist-ph]
18. S.L. Glashow, J. Iliopoulos, L. Maiani, Phys. Rev. D **2**, 1285 (1970)
19. M. Kobayashi, T. Maskawa, Progr. Theor. Phys. **49**, 652 (1973)
20. N. Cabibbo, G. Parisi, M. Testa, Lett. Nuovo Cim. **4S1**, 35 (1970)
21. P.A. Zyla, et al., (Particle Data Group), Prog. Theor. Exp. Phys. **2020**, 083C01 (2020) and 2021 update, *Plots of cross sections and related quantities*, Fig. 49.5
22. W.A. Bardeen, H. Fritzsch, M. Gell-Mann, arXiv:hep-ph/0211388 [hep-ph]
23. H. Fritzsch, M. Gell-Mann, in *Proceedings of the XVI international conference on high*, Chicago 1972, ed. by J.D. Jackson, A. Roberts, eConf C 720906V2 (1972), p. 135. [hep-ph/0208010]; H. Fritzsch, M. Gell-Mann, H. Leutwyler, Phys. Lett. B **47**, 365–368 (1973)
24. C. Dominguez, M. Greco, Lettere al Nuovo Cimento **12**, 439 (1975)
25. See the talks at this Conference by G. Giudice, Y. Wang and S. Stapnes
26. G. Altarelli, N. Cabibbo, L. Maiani, Phys. Lett. B **40**, 415 (1972)
27. R. Jackiw, S. Weinberg, Phys. Rev. D **5**, 2396 (1972); I. Bars, M. Yoshimura, Phys. Rev. D **6**, 374 (1972); K. Fujikawa, B.W. Lee, A.I. Sanda, Phys. Rev. D **6**, 2923 (1972)
28. C. Bouchiat, J. Iliopoulos, P. Meyer, Phys. Lett. B **38**, 519 (1972)
29. J. Iliopoulos, Progress in Gauge Theories, ICHEP 1974
30. G. Altarelli, G. Parisi, Nucl. Phys. B **126**, 298 (1977)
31. Y.L. Dokshitzer, Sov. Phys. J.E.T.P. **46**, 691 (1977)
32. V.N. Gribov, L.N. Lipatov, Sov. J. Nucl. Phys. **15**, 438 (1972)
33. G. Altarelli, L. Maiani, Phys. Lett. B **52**, 351 (1974)
34. G. Altarelli, N. Cabibbo, L. Maiani, R. Petronzio, Nucl. Phys. B **69**, 531 (1974)
35. G. Altarelli, N. Cabibbo, L. Maiani, Phys. Lett. B **57**, 277 (1975); G. Altarelli, N. Cabibbo, G. Corbo, L. Maiani, G. Martinelli, Nucl. Phys. B **208**, 365 (1982)
36. N. Cabibbo, G. Parisi, Phys. Lett. B **59**, 67 (1975)
37. N. Cabibbo, L. Maiani, G. Parisi, R. Petronzio, Nucl. Phys. B **158**, 295 (1979)
38. N. Cabibbo, G. Martinelli, R. Petronzio, Nucl. Phys. B **244**, 381 (1984)
39. P. Bacilieri, et al., IFUP-TH84/40, Dec. 1984; M. Albanese, et al. [APE Collaboration], Comput. Phys. Commun. **45**, 345 (1987)
40. M.B. Gavela, L. Maiani, S. Petrarca, F. Rapuano, G. Martinelli, O. Pene, C.T. Sachrajda, Nucl. Phys. B **306**, 677 (1988); C.R. Allton, C.T. Sachrajda, V. Lubicz, L. Maiani, G. Martinelli, Nucl. Phys. B **349**, 598 (1991)
41. UA1 Collaboration, Phys. Lett. B **122**, 103; UA2 Collaboration, Phys. Lett. B **122**, 476
42. UA1 Collaboration, Phys. Lett. B **126**, 398 (1983); UA2 Collaboration. Phys. Lett. B **129**, 130 (1983)
43. G. Aad, et al., [ATLAS Collaboration], Phys. Lett. B **716**, 1 (2012); S. Chatrchyan, et al. [CMS Collaboration], Phys. Lett. B **716**, 30 (2012)

Chapter 8
Detectors and Experiments at the Laboratory for Electro-Strong Physics: A Personal View

Ugo Amaldi

Abstract In the first two Sections I recount few episodes that highlight the unique personality of Bruno Touschek, and I propose to call 'Bruno's domain' the electron–positron energy range dominated by the one-photon channel. Section 8.3, devoted to a presentation of the four LEP detectors, is followed by three Sections describing electroweak precision tests, in particular those involving the production of b-quarks, Higgs searches and the most accurate measurements of the strong coupling. In Sects. 8.7 and 8.8, the unification of the forces is discussed from a personal point of view and the legacy of LEP to the LHC experiments is highlighted.

8.1 Remembering Bruno Touschek

In 1958, at the Physics Department of La Sapienza, I was one of the five or six postgraduate students following a course on Theoretical Physics in the framework of an advanced two-year school called '*Scuola di Perfezionamento in Fisica Nucleare*'. On the first day Bruno Touschek entered the aula smoking his usual cigarette and, in few sentences and short formulas written on the blackboard, defined fields— as systems with an infinite number of degrees of freedom—and their Lagrangian density. Then he introduced a symmetry of the Lagrangian and, in few further steps, derived the existence of a conserved quantity ending with a smile: "This is Noether's theorem". Yet, he did not explain neither who Noether was nor when the theorem had been demonstrated [1].

I was shocked by his simple and direct way of explaining difficult subjects and my first impression was confirmed by the rest of the course. Because of his fascinating lectures—of which I still have the notebook—quantum field theory has enthralled

---

In memory of the members of the DELPHI Collaboration who are no longer with us.

U. Amaldi (✉)
TERA Foundation, Novara, Italy
e-mail: ugo.amaldi@cern.ch

© The Author(s) 2023
L. Bonolis et al. (eds.), *Bruno Touschek 100 Years*,
Springer Proceedings in Physics 287,
https://doi.org/10.1007/978-3-031-23042-4_8

me for the rest of my life, and I even used it in some papers published outside my main activity of experimental physicist. Moreover, in the '80s—at the postgraduate school of Milano University—I taught courses on 'Particle physics and the Standard Model' using the first edition of a very clear book written by Ian Aitchison and Anthony Hey.

Five years after my first introduction to field theory, I was working—with a small group of physicists belonging to the Physics Laboratory of the National Health Institute (ISS) in Rome—at the Frascati electron synchrotron on a new line of research in nuclear physics: the study of 'quasi-free' electron-proton scattering on nuclei. During a coffee break of the 'Congressino dell'INFN', Bruno approached me. In my diary I wrote "Touschek mi chiede se voglio prendere l'incarico di preparare un'esperienza in una delle sezioni dritte di Adone. Rispondo che ci penserò. Tornando a Roma parlo a Giorgio Matthiae di questo." He wanted me to perform an experiment at the electron–positron collider Adone, at the time under construction.

In 1968, also because of Bruno's request, I had changed research field, an important step in my professional life. We were preparing—together with Giorgio Matthiae and some junior collaborators—an Adone experiment to study the production and decay of phi-mesons. Since, to compute radiative corrections to this novel process, I had used a method developed by the French theorist Paul Kessler, I went to Bruno's office at la Sapienza to show him the results. I asked him "Do you know Kessler?" His instantaneous reply was: "Io conosco solo le sorelle Kessler" i.e. "I only know the Kessler sisters", two tall and beautiful German twins, dancers and singers often seen on the main channel of the Italian TV.

In 1977 Bruno was at CERN, where I was then working—having moved in 1973 from Rome to Geneva—and where he was participating in the development of the SPS proton-antiproton collider proposed by Carlo Rubbia. Once, having met him by chance in front of the CERN library, I asked about his first long visit to CERN. I was surprised to hear that he had got convinced that the future of particle physics would be in the proton-antiproton collisions advocated by Carlo and not in his dear electron–positron annihilations.

Few months later he was so sick that he had to go to the La Tour Hospital in Meyrin. As others of his friends, I went to visit him a few times. On July 22, 1978, I wrote on Corriere della Sera an article titled "Who was the man of number 137?" recounting that, in my last visit, I did not find him in the usual room. When, having spotted the room, I excused myself for the delay he said "…because the real problem is the number of this room". After a pause he added "This is the problem around which I have hovered throughout my life without success". Another pause and then "Sai, Ugo, Pauli fu messo in una stanza d'ospedale numero 137 prima di morire", "You know, Ugo, Pauli was brought in a hospital room number 137 before dying". At the end of Sect. 8.7 (p. 106) I discuss the present understanding of the number 137 in the framework of Grand Unified Theories of the strong, electromagnetic and weak interactions.

8.2 Homage to Bruno

In the well-known figure of the normalized hadronic cross-section, R, as a function of the energy $E_{cm} = \sqrt{s}$, which describes the full electron–positron landscape (Fig. 8.1), I have called 'Bruno's domain' the energy range that goes up to $E_{cm} = 40$ GeV and is dominated by the exchange of a virtual photon, with all its radiative corrections. At larger energies, the creation of neutral and charged intermediate boson plays the major role so that this energy range can be called 'electroweak domain'. The figure shows that, at the end of Bruno's domain, R equals the fraction 33/9, expected if the five types of produced quarks come in three colours.

In Fig. 8.2 the total cross-sections of the main LEP processes are plotted as functions of the centre-of-mass energy. The red curve shows that, around $E_{cm} = 210$ GeV, the Standard Model Higgs production cross-section, with a hypothetical mass $M_H = 115$ GeV, is about 100 times smaller than the W^+W^- cross-section.

In 20% (70%) of the cases, a Z-boson decays into invisible neutrinos and antineutrinos pairs (into quark-antiquark pairs); lepton pairs contribute the remaining 10%.

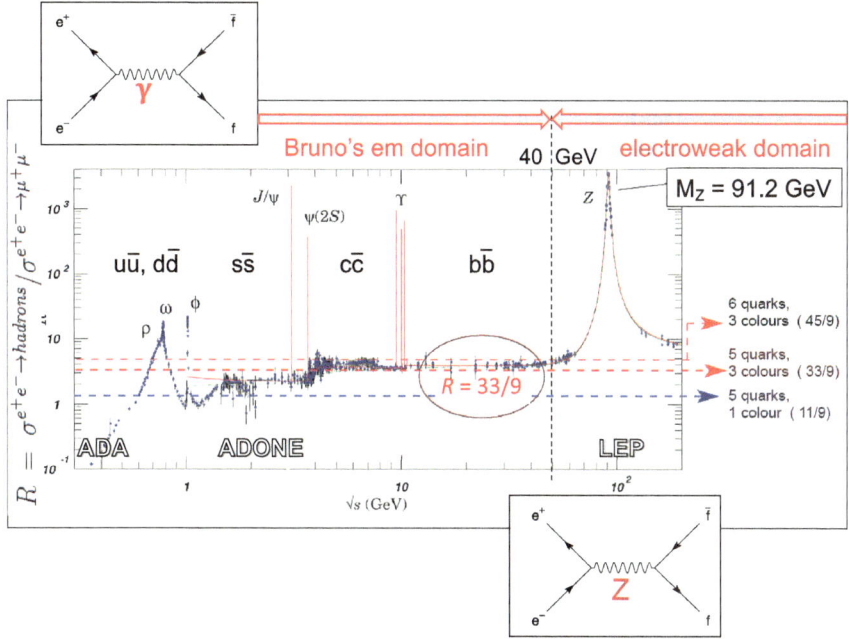

Fig. 8.1 Bruno's domain is dominated by the peaks of the resonant production of quark-antiquark pairs, and the electroweak domain by the Z^0 peak. (The compilation of e^+e^- data is taken from [2])

Fig. 8.2 Values of the
cross-sections measured by
L3 [3] and corresponding
behaviours predicted by the
Standard Model (courtesy of
CERN)

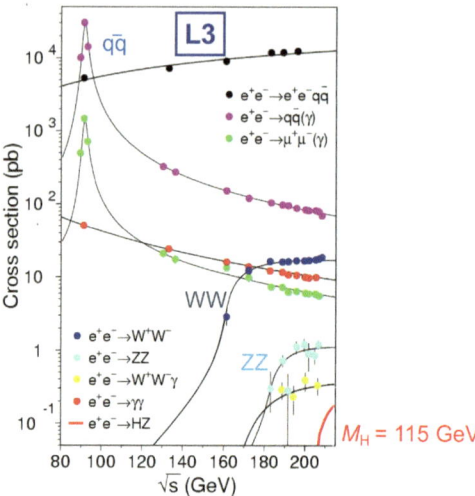

8.3   The Four LEP Detectors

In 1982, the LEP Experiment Committee and the Director General Herwig Schopper
[4] approved two general-purpose detectors (ALEPH and OPAL) and two specialized
detectors (DELPHI and L3). The first three were about 12 m tall while L3, being
20 m high, was definitely larger. The first spokespersons were Jack Steinberger, Aldo
Michelini, Ugo Amaldi, Sam Ting, and technical coordinators were Pierre Lazeyras,
Alasdair Smith, Hans Jürgen Hilke and Alain Hervé.

The superconducting coil of ALEPH produced a 1.5 T field and contained a 2-layer
double-sided micro-vertex silicon detector and a large Time Projection Chamber
(rose in Fig. 8.3—diameter = 3.6 m) which measured—with the typical long longitudinal drifts—the energy deposition of charged particles, recording both the position
of 21 track segments and the corresponding energy losses $\Delta E/\Delta x$ for particle identification. The electromagnetic calorimeter (green in Fig. 8.3)—located inside the
superconducting coil—was based on lead sheets and wire-chambers.

OPAL adapted a conservative design very similar to the one of JADE [5], a very
successful detector built for the DESY PETRA electron–positron collider in the
years 1978–1986. The room-temperature coil produced a 0.4 T magnetic field. Inside
the coil, a 2-layer single-sided microstrip silicon detectors and a Jet Chamber (red
in Fig. 8.3) measured the charged particles. The electromagnetic calorimeter was
made of 9440 lead glass Cherenkov counters. As in the other detectors, a hadron
calorimeter, and many muon chambers (green in Fig. 8.3) covered the full solid
angle.

DELPHI was specialised in hadron tagging. It had a lower field than ALEPH
(1.2 T) but larger diameter so that inside it a Time Projection Chamber (diameter =
2.2 m) was surrounded by two Ring Imaging Cherenkov (RICH) counters (yellow
in Fig. 8.3). This novel detector recorded the rings of photons produced through

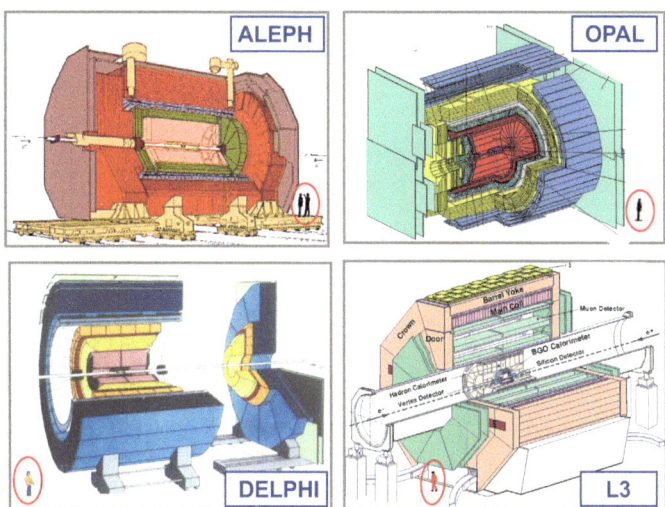

Fig. 8.3 The four LEP detectors (courtesy of CERN)

Cherenkov effect in a liquid and a gas radiator distinguishing kaon from pions of relatively large kinetic energies. The micro-vertex was made of 3 layers of double-sided silicon detectors and silicon pixel detectors covered the forward angles.

L3 was specialised in the accurate measurement of photon/electron and muon energies. The room-temperature solenoid of L3 had a very large diameter: 15 m. The main detector for measuring the curvature of charged particles was a small-radius very precise Time Expansion Chamber (radius = 50 cm) that followed a two-layer double-sided micro-vertex silicon detector. Externally there was the electromagnetic calorimeter made of about 12 000 crystals of bismuth germanium oxide. Outside the hadron calorimeter, three layers of very large drift chambers (green in Fig. 8.3) provided accurate measurements of muon momenta.

In the years 1982–1989 LEP was built, under the direction of Emilio Picasso, 100 m below the plain between the Geneva airport and the Jura Mountain (Fig. 8.4).

The first events were registered in August 1989 and, for eleven years, the four detectors collected data between 89 to 209 GeV. But the LEP Z-decays were not observed for the first time at LEP: four months before at SLAC about 100 events had been registered by the MARKII detector [7]—later substituted by the SLAC Large Detector (SLD)—mounted on the very innovative collider proposed by Burton Richter only ten years before. In the Stanford Linear Collider (SLC) beams of electrons and positrons of about 50 GeV were accelerated by the 2-mile long SLAC linac and brought to the collision point by two large 180° magnetic arcs [8]. Eventually, SLD logged 500 000 Z-events while at LEP I (1989–1995) each of the four CERN detectors registered about 4 10^6 Z decays. In the higher energy run, called LEP II (1996–2000), the centre-of-mass energy increased step-by-step from 180 to 209 GeV and each detector collected about 10 000 events.

Fig. 8.4 Layout of LEP and locations of the four detectors [6]

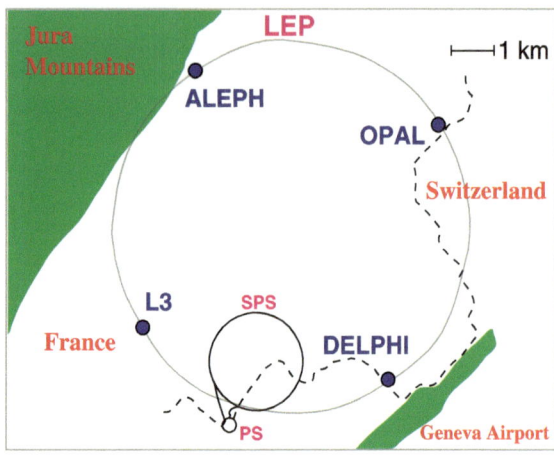

8.4 The Electroweak Sector of the Standard Model

In 1992 Physics Letters published a paper that was signed by 'The LEP collaborations: ALEPH, DELPHI, L3 and OPAL' [9]. The more than thousand authors were referred to in a simple footnote: 'Lists of authors can be found in refs. [1–4]'. The paper was due to a team of experts of the four Collaborations chaired by Jack Steinberger, who originated it and "insisted that the combination was a job for the experimentalists from the four collaborations rather than for the theorists. This led to the establishment of the Electroweak Working Group collaborative effort across the experiments" [10]. This was the first of many LEP Working Groups of the ADLO 'second-order' collaboration; some of them are quoted in the next Sections.

The summary table of the 1992 paper had 11 entries. Twenty years later, the LEP Electroweak Working Group produced Fig. 8.5 with 18 entries [3].

Behind this table there is a second unprecedented feature of the activities developed around LEP: the work of more than hundred theorists who—along the years—computed higher-order processes that contribute to the physical quantities measured by the experimentalists. Examples are given in Fig. 8.6.

This coordinated process started with two CERN Yellow Reports bearing the title 'Physics at LEP' that were distributed in 1986 [11], three years before the first collisions. They were edited by John Ellis and Roberto Peccei who, in their Introduction to the first volume, wrote: "Thanks largely to the initiative of its then Chairman, Günter Wolf, the LEP Experiments Committee asked us, the two theorists on the Committee, to organize this new survey. We identified five principal areas of LEP physics, namely: precision studies at the Z peak; toponium; searches for new particles, QCD, gamma-gamma and heavy quark physics; and high-energy running beyond the $W^+ W^-$ threshold. Working Groups (WG) were set up for each one of these areas." The first contribution on 'Precision tests of the electroweak theory at the Z' was written by Guido Altarelli, chair of the corresponding WG.

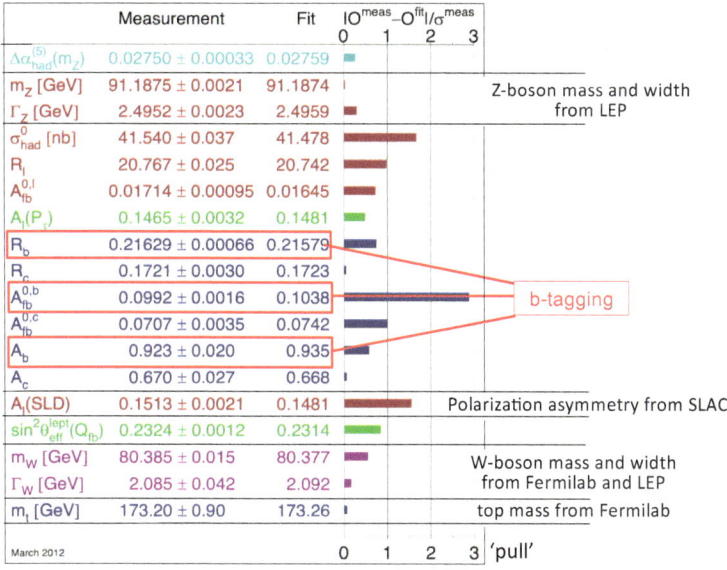

	Measurement	Fit	$|O^{meas}-O^{fit}|/\sigma^{meas}$ 0 1 2 3	
$\Delta\alpha_{had}^{(5)}(m_z)$	0.02750 ± 0.00033	0.02759		
m_Z [GeV]	91.1875 ± 0.0021	91.1874		Z-boson mass and width
Γ_Z [GeV]	2.4952 ± 0.0023	2.4959		from LEP
σ_{had}^0 [nb]	41.540 ± 0.037	41.478		
R_l	20.767 ± 0.025	20.742		
$A_{fb}^{0,l}$	0.01714 ± 0.00095	0.01645		
$A_l(P_\tau)$	0.1465 ± 0.0032	0.1481		
R_b	0.21629 ± 0.00066	0.21579		
R_c	0.1721 ± 0.0030	0.1723		
$A_{fb}^{0,b}$	0.0992 ± 0.0016	0.1038		b-tagging
$A_{fb}^{0,c}$	0.0707 ± 0.0035	0.0742		
A_b	0.923 ± 0.020	0.935		
A_c	0.670 ± 0.027	0.668		
A_l(SLD)	0.1513 ± 0.0021	0.1481		Polarization asymmetry from SLAC
$\sin^2\theta_{eff}^{lept}(Q_{fb})$	0.2324 ± 0.0012	0.2314		
m_W [GeV]	80.385 ± 0.015	80.377		W-boson mass and width
Γ_W [GeV]	2.085 ± 0.042	2.092		from Fermilab and LEP
m_t [GeV]	173.20 ± 0.90	173.26		top mass from Fermilab

March 2012   0   1   2   3 'pull'

Fig. 8.5 Table produced in March 2012 by the LEP Electroweak Working Group [3]

Fig. 8.6 Virtual Higgs and top quarks affect **a** the Z-mass and **b** the Z-decay in b- quarks

All along the LEP lifetime, Guido Altarelli and John Ellis have been the theorists who not only worked themselves on these problems but also urged colleagues to compute new processes and helped the experimentalists to best interpret their data.

Radiative corrections—as the ones depicted in Fig. 8.6—depend only logarithmically on the Higgs mass but are much more sensitive to the top mass m_t; this gave rise to an interesting episode. In March 1994 at the Moriond Meeting, the latest fit to the LEP most precise measurements of the time was presented together with the best value of the top mass: $m_t = (172 \pm 13 \pm 18)$ GeV. Few months later the CDF Collaboration announced the detection at Fermilab of 12 top-decays with a measured mass $m_t = (174 \pm 10 \pm 13)$ GeV that was, within the large errors, superposable with the LEP best fit.

Going back to Fig. 8.5, four entries are contributed by non-LEP experiments: the left–right polarization asymmetry A_l (uniquely measured by the SLAC Large Detector), the mass and width of the W-boson (measured by CDF and D0 at Fermilab and at LEP), and the mass of the top-quark (discovered and measured at Fermilab).

The fourth column of the table gives the best-fit values of the 18 quantities when radiative corrections are properly considered. The histogram to the right of the figure shows by how many standard deviations each result differs from its best fit value (the so-called 'pull'). A glance is enough to state that the fit is very good.

For reasons of space, the meaning of the various quantities and their measurements cannot be treated here. I limit myself to two remarks before discussing in depth a particular subject: b-tagging.

Firstly, among the LEP data the most precise measurements concern the Z mass ($\pm 0.0023\%$), the Z width ($\pm 0.09\%$), the hadronic cross-section σ^0_{had} at the Z peak ($\pm 0.09\%$) and the fraction R_b of b-quark events on all hadronic events ($\pm 0.3\%$). These accuracies surpass any prediction made before data taking.

Secondly, the cross-section σ^0_{had} is so precisely measured because of the enormous amount of work done, theoretically and experimentally, to measure very accurately the luminosity, i.e. to compute the cross-section of very forward electron–positron ('Bhabha') scattering and to construct sophisticated and mechanically accurate electron/positron detectors, which—placed downstream of the collision point—measured very precisely the electron and positron scattering angles, as discussed in [12, 13] for the OPAL and DELPHI detectors.

Considering now 'b-tagging', this novel technique has been very important in the LEP experimental program because it was used not only to measure—as indicated in Fig. 8.5—three of the eighteen parameters (the fraction of b-quark-pairs R_b, the forward–backward asymmetries A_{fb}^0, and the polarization asymmetry parameter A_b) but also to search for the Higgs boson and to measure the running of the mass of the b-quark, subjects that are discussed in the next two Sections.

The four micro-vertex silicon detectors are shown in Fig. 8.7.

Their main feature the four LEP micro-vertex detectors was the 20–30 m μ accuracy in the measurement of the coordinates $R\Phi$ and z, while the z-resolution of the SLD micro-vertex was only 13 μm.

Figure 8.8a explains how the transverse mismatch δ of a track, due to the decay of a hadron containing a b-quark, is measured and Fig. 8.8b compares the experimental and the Monte Carlo event distributions showing how a cut in $S = \delta/\sigma_\delta$. can increase the purity of the sample while reducing the efficiency. The variations of the efficiencies with the purity of the sample are quantitatively shown in Fig. 8.9a.

The first layer of the LEP detectors was at about 65 mm from the centre of the vacuum pipe, while this quantity was 29 mm for SLD (lower part of Fig. 8.7). This, together with the smaller primary vertex resolution, is the reasons for the SLD larger b-tagging efficiency (Fig. 8.9a).

As far as LEP II is concerned, Fig. 8.10 reproduces the measurements by the L3 Collaboration [20] that shows the W^+W^- production cross-section, which is the sum

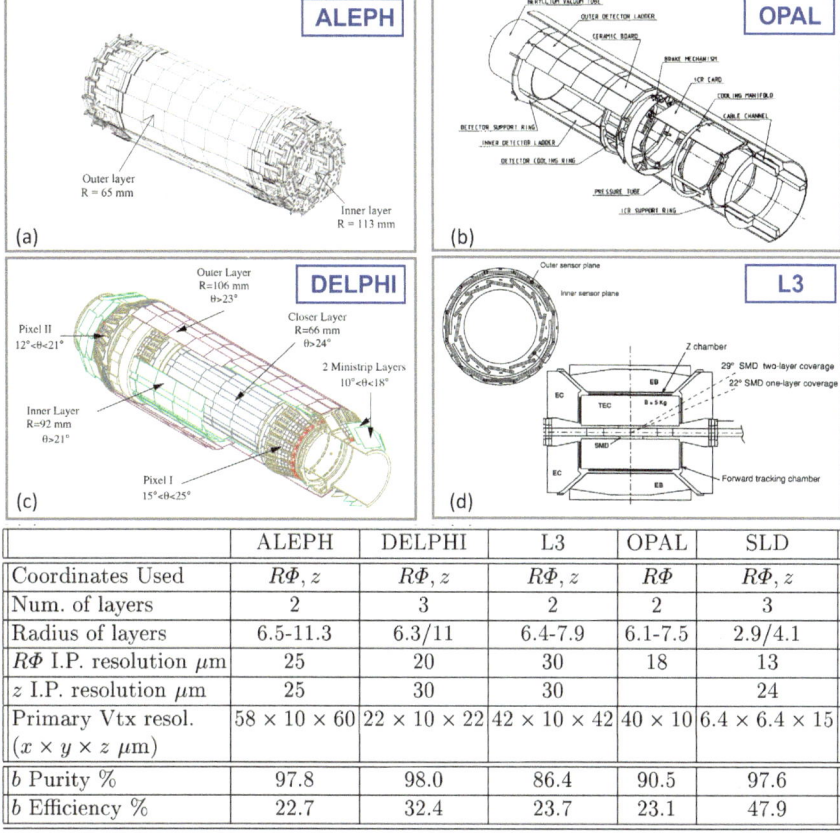

Fig. 8.7 a–d The LEP micro-vertex silicon detectors were located around a thin beryllium beam pipe with a $R_{pipe} = 55$ mm radius [14–17] (courtesy of CERN). **e** Characteristics of the LEP and SLD micro-vertex detectors. The table has been compiled by Chiara Mariotti [18]

of the three contributions depicted in the upper part of Fig. 8.10a. This cross-section would not flatten with energy without the ZWW triple gauge coupling, which in the electroweak theory is due to the non-Abelian nature of the SU(2) group [21]. The figure shows also that the Spin Matrix Elements, plotted in Fig. 8.10b as functions of the W polar angle, perfectly agree with the Standard Model predictions.

8.5 In Search of the Higgs Boson

From the March 2012 best fit of Fig. 8.5, the LEP Electroweak Working Group obtained for the Higgs mass the result [3] $M_H = (94 + 29 - 24)$ GeV, so that M_H was predicted to be *smaller than 152 GeV* with a 95% confidence level. This is an

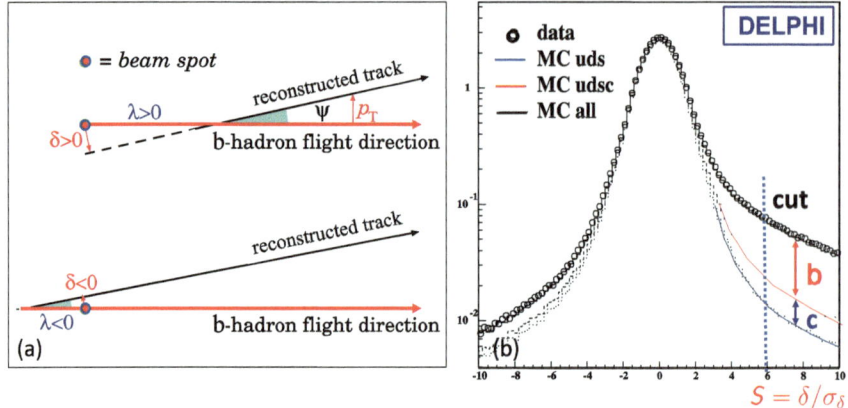

Fig. 8.8 **a** Definition of the mismatch δ. **b** Comparison between data and a Monte Carlo calculation. The x-axis represents the mismatch δ divided by its standard deviation σ_δ [19]

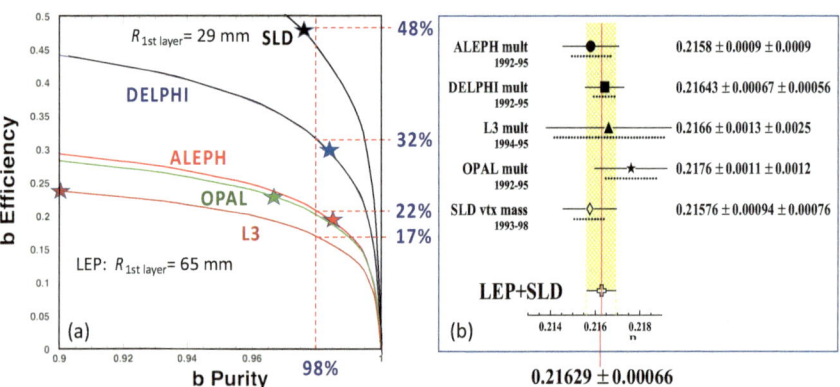

Fig. 8.9 **a** The b-tagging efficiencies decrease with the required purity of the sample. (Courtesy of Chiara Mariotti, CERN) **b** Summary of the data that give the average R_b value of Fig. 8.5 [19]. The $\pm$ 0.0008% LEP error is 20 times smaller than the one estimated before the beginning of data taking

indirect limit but, of course, even before LEP was built the *direct* detection of Higgs bosons decays was on top of the foreseen searches. Figure 8.11 shows the main decay channels; the detection of these events profits from a large b-tagging efficiency.

In 1986, at the Aachen ECFA Workshop on LEP 200, Sau Lan Wu reported the conclusions of the Higgs Working Group [22]: "At centre-of-mass energy of 200 GeV significant signals are certainly observable up to $M_H = 80$ GeV from the missing energy channel and up to $M_H = 70$ GeV from the 4-jet channel". Fourteen years later LEP experiments were engaged in searching for a 115 GeV Higgs.

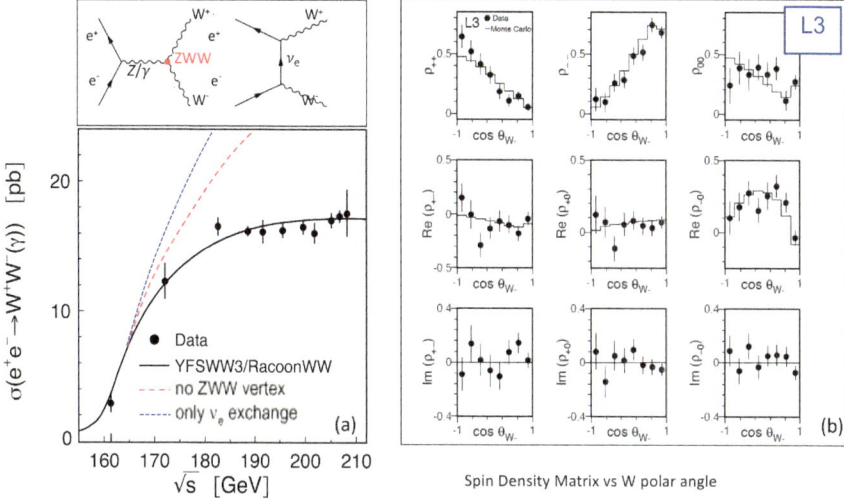

Fig. 8.10 **a** WW total cross-section. **b** Spin Matrix Angle versus the W polar angle

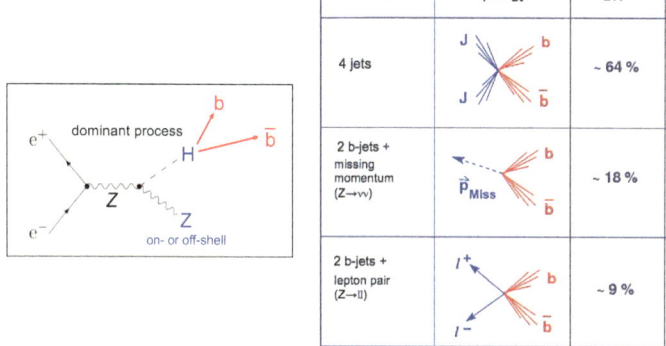

Fig. 8.11 The best channel to detect Higgs bosons is the two-jet decay of both H and Z. To this end, double b-tagging is extremely useful, but the efficiency of Fig. 8.9a enters quadratically

The energy that an electron/positron loses in synchrotron radiation increases as the *fourth power* of the beam energy so that, given the LEP diameter, at 100 GeV the energy loss per turn was about 3 GeV. The losses were replenished by the RF cavity system that, in the years 1996–1999, was continuously upgraded, as shown in Fig. 8.12a. This was made possible by the leadership of Emilio Picasso [23], who had been Project Leader of the superconducting (SC) cavity group, and the invention by C. Benvenuti of SC cavities built by coating, with the 'sputtering' technique, the inner surfaces of copper cavities with a thin niobium layer [24].

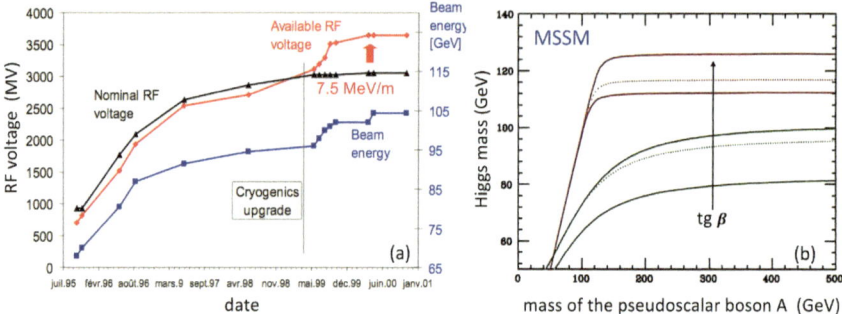

Fig. 8.12 **a** In 1995–99 the LEP toral RF voltage was increased by adding SC cavities and increasing their gradient from 6 to 7.5 MV/m [25] (courtesy of CERN). **b** In MSSM the mass of the lightest neutral Higgs depends on two parameters: the mass of the axial boson A and tgβ [26]

At CERN, the upgrade program was the subject of many animated discussions because it was shown, in 1994–95, that the minimal supersymmetric extension of the Standard Model (MSSM) made a prediction that could be tested at LEP II. In Supersymmetric (SUSY) theories there is one 'superparticle' (often called 'sparticle') for each Standard Model (SM) particle, a fermion for a boson and a boson for a fermion [27]. SUSY is 'broken' because the sparticles are *heavier than about 100* GeV. Radiative corrections cause divergences of the Higgs mass but disappear in SUSY because of the cancellation between the virtual effects of particles and sparticles. However, not to spoil this cancellation, the superparticles must have masses *below about 1000* GeV, so that one speaks of 'low energy' SUSY.

MSSM predicts the existence of a 'light' Higgs boson 'h' and a heavier Higgs boson 'H', of an axial boson 'A' and of two charged Higgs H$^+$ and H. The mass of the lightest Higgs depends on the mass of the axial boson A and a parameter tgβ; as shown in Fig. 8.12b. The limit follows from delicate calculations because, at the lowest order, M_H is lighter than M_Z but large radiative corrections, in which the top mass plays a key role, push it above M_Z [26, 28–30]. For the top mass known at the time the computed M_H did not exceed 125–130 GeV, so that a 220–225 GeV collision energy would have been sufficient for detecting the processes of Fig. 8.11.

For this reason, in those years Daniel Treille—who at the time was DELPHI spokesperson—and others did everything possible to convince the CERN Directorate to invest about 70 million Swiss Francs in the construction of extra SC cavities and reach at least 220 GeV [31, 32]. But the new LHC accelerator—which was to be assembled inside the LEP tunnel—was at an advanced stage of planning and required significant resources, both financial and in personnel; therefore, the decision was finally made to invest in enough superconducting cavities to reach only 200 GeV in the centre of mass, as described by Kurt Hübner in [33].

In the year 2000 the experiments began to collect data, at the maximum energy, knowing that by autumn LEP had to stop. ALEPH observed one, then two, and three events, which could be attributed to the decay of a Higgs boson (Fig. 8.13). L3 also

	EXP	Channel	M (GeV)	w
1	ALEPH	4-jet	114.3	1.73
2	ALEPH	4-jet	112.9	1.21
3	ALEPH	4-jet	110.0	0.64
4	L3	E-miss	115.0	0.53
5	OPAL	4-jet	110.7	0.53
6	DELPHI	4-jet	114.3	0.49
7	ALEPH	Lept	118.1	0.47
8	ALEPH	Tau	115.4	0.41
9	OPAL	4-jet	112.6	0.40
10	ALEPH	4-jet	114.5	0.40

Fig. 8.13 An ALEPH candidate and the first 10 candidates ordered by statistical weight [34] (courtesy of CERN)

observed a candidate in the missing momentum channel (Fig. 8.11), and OPAL and DELPHI joined with 2 and 1 events, compatible with their backgrounds.

CERN Director General Luciano Maiani was required to make a difficult decision. If he were to delay by a year the end of LEP, the thousands of people working on the LHC project would lose enthusiasm and CERN would have had to pay penalty charges of about 100 million Swiss Francs to companies ready to dismantle LEP.

More than 10 years later Luciano Maiani wrote [35]: "It was necessary to kill LEP, the king of CERN, to build a larger giant, the LHC. I did it. There was much stress, which I feel as I write, it was really a transition drenched with great emotion. As well as a stubborn exercise of rationality. [...] I could write [to those who wanted to run for another year] with some justifications: 'The chance of finding ourselves by autumn of next year still with only a 3–3.5 sigma effect is not at all negligible. [...] At this point, we would have spent all our financial reserves, time and credibility on a very, very risky bet. I have never cared for poker.'".

After a one-month prolongation, LEP was switched off on 2 November 2000. Few weeks later, the ALEPH team published a paper that concluded [36]: "The observation is consistent with the production of a Higgs boson with a mass near 114 GeV. More data, or results from other experiments, will be needed to determine whether the observations reported in this letter are the result of a statistical fluctuation or the first sign of direct production of the Higgs boson." In the following years The LEP Working Group for Higgs Boson Searches critically analysed the events of the four Collaborations and combined the data concluding that (i) the signal for a Higgs with 114 GeV mass had a significance of 1.7 standard deviations and (ii) the Higgs mass had to be larger than 114.4 GeV (95% CL) [37].

The lower limit 114.4 GeV (95% CL) must be considered together with the 152 GeV (95% CL) upper limit quoted at the beginning of this Section and obtained in March 2012 with the best fit of Fig. 8.5. The (about 95% CL) interval 114.4–152 GeV—that, with a cavalier approximation, can be written as $M_H = (133 \pm 10)$ GeV—brackets the 125–127 GeV value announced at CERN, four months later, by Fabiola Gianotti and Joe Incandela on behalf of the ATLAS and CMS Collaborations. Ten years later, the best value is $M_H = (125.1 \pm 0.2)$ GeV, which is at the limit of the MSSM parameter space (Fig. 8.12b) and would have been detected with a very long LEP run, if the electron–positron centre of mass energy had reached 220 GeV [31].

The four LEP Collaborations have excluded the existence of many other hypothetical particles, but there is space here to mention only a very topical subject. Dark Matter (DM) candidates of mass smaller than $M_Z/2$ can be excluded if the Z couples to them even with a probability 6–7 orders of magnitude smaller than the coupling to neutrinos [38]. Moreover, LEP data on single-photon events with large missing energy constrain the coupling of DM in the 10 s GeV mass range to electrons, providing limits complementary and competitive to those from direct searches for DM-nucleon scattering and indirect astrophysical searches [39].

8.6 Quantum Chromodynamics

Quantum Chromodynamics (QCD)—the SU(3) colour group theory of quarks and gluons [40]—was well-established before LEP, as written by Guido Altarelli in 1989 [41]: "At present, it is fair to say that the experimental support of QCD is quite solid and quantitative. The forthcoming experiments at pp colliders, at LEP, SLC, and HERA will certainly be very important with their great potential for extending the experimental investigation of the validity of QCD."

The advances brought by LEP I and LEP II to the measurement of the 'running' strong coupling α_s, which becomes feebler with the energy scale Q, are clearly seen by comparing Fig. 8.14b with Fig. 8.14a: in fifteen years the error was reduced by a factor four. As discussed in the rest of this Section, with better calculations and further data analyses, eventually the error shrank by a factor six-seven.

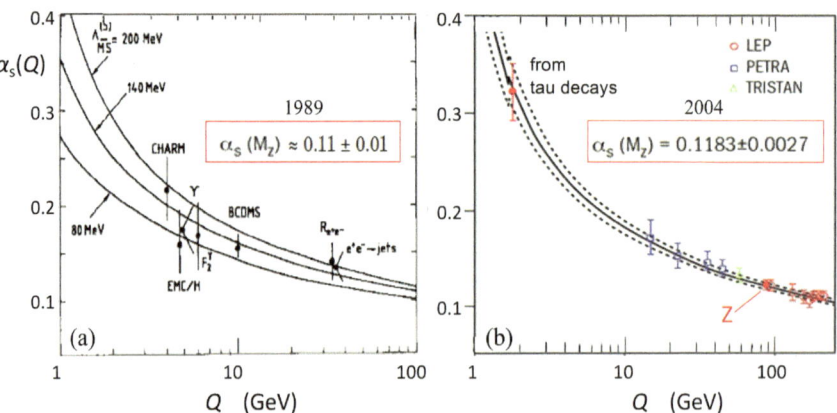

Fig. 8.14 Energy dependence of the strong coupling α_s **a** before the start-up of LEP (1989 review paper by Altarelli [41]). **b** after the stop of LEP (2004 review paper by Bethke [42])

When the energy Q increases the strong coupling $\alpha_s(Q)$ 'runs' towards smaller values so that in hadrons the quarks hit by high-energy mediators are 'asymptotically free'. The running is due to the colour charge of the gluons and can be explained by considering that, in the quantum description of an isolated (electric or colour) 'charge', energy can be borrowed for short times to make evanescent 'virtual' quanta of the force field and 'virtual' particle-antiparticle pairs. These virtual particles disappear rapidly, but others come up so that around a charge there is a dynamical medium in equilibrium, with heavier particles closer to the charge. In the (non-Abelian) U(1) gauge theory, the central negative *electric* charge polarizes this medium in the sense that the positive charged particles (far away mainly virtual positrons) are attracted, and the negative ones (negative electrons) are repelled, while *uncharged* virtual photons are unaffected. Moving from the centre—i.e., probing the source charge with photons of decreasing energy Q—the overall *electric* charge *decreases* because of the *screening* effect, due to a thicker layer of virtual medium.

Differently, in the (Abelian) SU(3) gauge theory, around a *colour* charge, the medium contains quark-antiquark pairs and gluons, which carry *a* colour charge and produce a strong *anti-screening* effect, so that the overall colour charge *becomes stronger* when probed with gluons of smaller and smaller energies Q.

The local slopes of the lines of Fig. 8.14 can be computed with the equations of the renormalization group (considering also small second order corrections [43]) by making hypotheses on the masses of all fermions and bosons that appear and disappear in the virtual medium. More precisely, a mediator of energy Q probes the medium down to distances $\hbar/Q$ so that, at each energy, only virtual particles that have mass smaller than Q influence the slope of the line representing $\alpha_s(Q)$.

In 1989, with reference to Fig. 8.14, Guido Altarelli wrote [41]: "The prediction for α_s to be measured at LEP is very precise: $\alpha_s(M_Z) = 0.110 \pm 0.001$. Establishing that this prediction is experimentally true would be a very quantitative and accurate test of QCD, conceptually equivalent but more reasonable than trying to see the running in a given experiment." This is the approach followed in this Section in discussing the LEP very accurate values of $\alpha_s(Q)$ obtained by (i) measuring quantities that describe the event shape, (ii) determining the fractions of 3-jets and 4-jets events and (iii) performing fits to electroweak data, as the one of Fig. 8.5.

The complex processes involved in hadron production are depicted in Fig. 8.15. The *first step* is the clean creation—through the exchange of a virtual gamma and a Z-boson—of a quark-antiquark pair, which is followed by, as *second step*, the irradiation of gluons and the creation of other pairs. An enormous theoretical effort has gone in the QCD calculation of this second step. The status before LEP was described in a CERN Yellow Report edited by Altarelli et al. [44]. Then the calculations improved from next-to-leading order in perturbation theory (NLO, $O(\alpha_s^2)$) to next-next-to-leading order (NNLO, $O(\alpha_s^3)$), to resummation in next-to-leading-logarithmic approximation (NLLA), arriving—for some processes—to N^3LO.

In the chain of processes, represented in Fig. 8.15a, the energy scale Q decreases and the strong coupling increases getting close to 1, so that to describe the *third step* ('Hadronization') perturbative computations are not possible and Monte Carlo

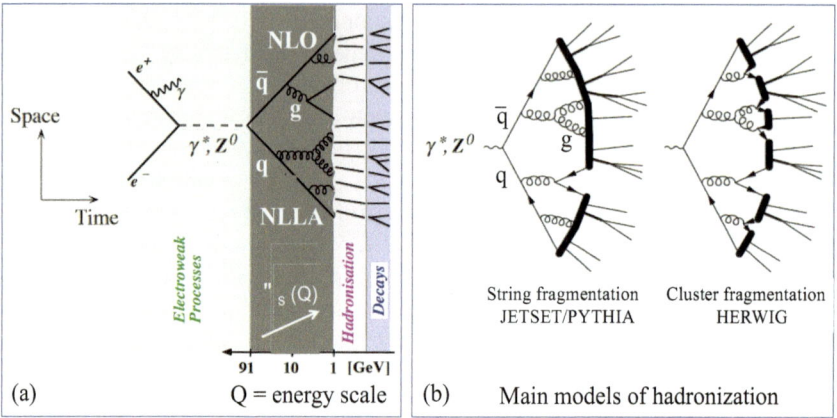

(a) Q = energy scale   (b)   Main models of hadronization

Fig. 8.15 **a** Hadron production is computed by nesting four subprocesses: creation of a quark pair. QCD high order calculations, hadronization and decays. (Figures adapted from *Phenomenology of Particle Physics I*, V. Chiochia, G. Dissertori, Th. Gehrmann, ETH, Zurich.)

models must be used. The main ones, graphically described in Fig. 8.15b, are based on two different approaches: 'String fragmentation' and 'Cluster fragmentation' [45].

Subsequently, the hadrons decay; this *fourth step* ('Decay') is easily computed by using the available experimental data on the various branching ratios.

Considering the *event shape*, the DELPHI measurements of 18 different parameters are summarized in Fig. 8.16 [32, 46].

In 2006 the LEP QCD Working Group computed the averages of the strong coupling from event shapes measured at LEP I and LEP II [47]. The theoretical uncertainty dominates because of the of missing higher order contributions:

Fig. 8.16 In this analysis, 18 event shape parameters have been considered. As an example, 'thrust' is obtained by finding a versor that maximizes the sum of the projected momenta

$$T = \max_{\hat{n}} \left(\frac{\sum_i |\vec{p_i} \cdot \hat{n}|}{\sum_i |\vec{p_i}|} \right)$$

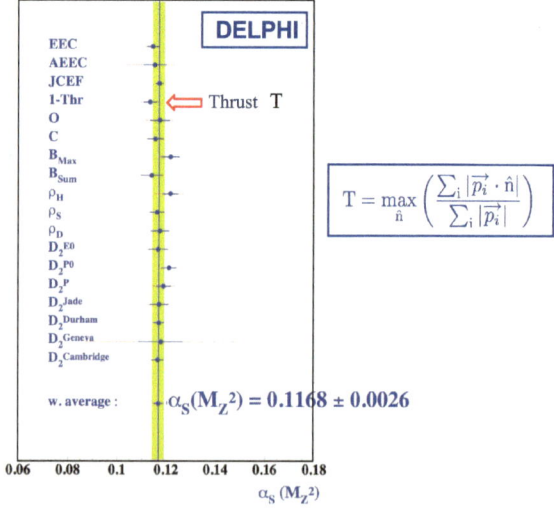

$$\alpha_s(M_Z) = 0.1202 \pm 0.0005(\text{exp}) \pm 0.0042(\text{theo}). \cdots (\text{LEP QCD WG}) \quad (8.1)$$

Jet-rates are more suited than event shape parameters for precise determinations of the strong coupling constant because they have smaller theoretical errors. Figure 8.17 shows the OPAL results on the fraction of 2, 3 4 ... jets. The closeness of the dashed and red curves shows that the hadronization corrections are small.

In the most used 'Durham clustering algorithm', to define y_{cut} one considers, for any two particles, the test variable y_{ij} that is, essentially, the square of the relative transverse momentum. If y_{ij} is smaller than y_{cut}, particles i and j are combined in a single object by summing the two four-momenta. The combination procedure is repeated until no particles can be further combined; the remaining objects are defined as 'jets'. The algorithm is such that it can be applied both to measured tracks in an event and to the partons of a perturbative calculation.

Considering now 4-jet events, I first quote the result obtained by OPAL from a detailed study of both 3-jets and 4-jets events [48]:

$$\alpha_s(M_Z) = 0.1177 \pm 0.013(\text{stat}) \pm 0.0036(\text{sys}). (\text{OPAL}) \quad (8.2)$$

Secondly, in Fig. 8.18 the results of an ALEPH analysis of 4-jets events are plotted versus the logarithm of the resolution parameter y_{cut} [49]. As shown in the figure, an intermediate y_{cut} range was used to fit the experimental data and obtain

$$\alpha_s(M_Z) = 0.1170 \pm 0.0001(\text{stat}) \pm 0.0013(\text{sys}). (\text{ALEPH}) \quad (8.3)$$

The result has a $\pm 1.1\%$ overall error.

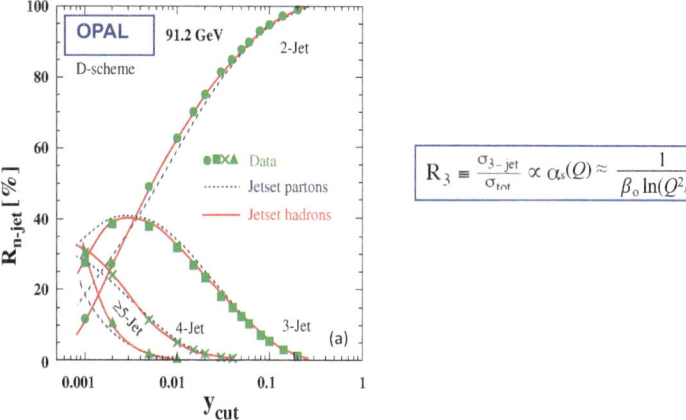

Fig. 8.17 Measured and computed fractions of n-jets plotted versus the resolution parameter y_{cut} [42]. In first order the fraction R_3 is proportional to the strong coupling, which in QCD is inversely proportional to the logarithm of the energy Q divided by the strong scale Λ

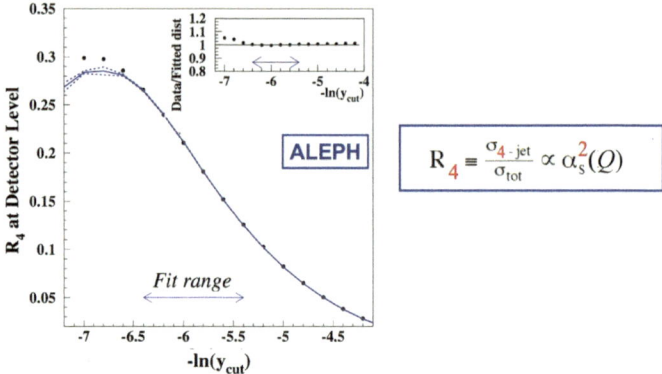

Fig. 8.18 The fraction R_4 is a second order process, proportional to $\alpha_s{}^2$, and its measurement gives smaller errors on $\alpha_s(M_Z)$ than the ones obtained from a measurement of R_3

In the *decays of tau-leptons*, the hadronic branching fractions and the spectral functions are sensitive to the strong coupling. The final ALEPH analysis by Michel Davier and collaborators, published many years after the end of the last LEP run [50], gave $\alpha_s(m_{\mathrm{tau}}) = 0.332 \pm 0.005(\mathrm{exp}) \pm 0.011(\mathrm{theo})$. By evolving this coupling to the Z-mass, the absolute errors on α_s reduces drastically

$$\alpha_s(M_Z) = 0.1199 \pm 0.0006(\mathrm{exp}) \pm 0.0012(\mathrm{theo}) \pm 0.0005(\mathrm{evol}). \ (\mathrm{ALEPH}). \tag{8.4}$$

A fourth method to obtain $\alpha_s(M_Z)$ uses the *electroweak precision fits* discussed in the previous Section. A recent analysis is described in [51].

In 2019 Siggi Bethke summarized the results of the experimental and theoretical work done on all LEP data in a paper written in memory of Guido Altarelli [52]

$$from\ event\ shapes\ and\ jets : \alpha_s(M_Z) = 0.1196 \pm 0.0036.\ (\mathrm{in\ NNLO})$$
$$from\ tau\ decays : \alpha_s(M_Z) = 0.1192 \pm 0.0018\ \left(\mathrm{inN^3LO}\right)$$
$$from\ electroweak\ precision\ fits : \alpha_s(M_Z) = 0.1196 \pm 0.0030.\ \left(\mathrm{in\ N^3LO}\right)$$
$$\tag{8.5}$$

Considering the errors as uncorrelated, these measurements can be combined giving a *single number*, the results of hundreds of experimental and theoretical papers and about 13 million LEP hadronic events recorded in the years 1989–2000:

$$\alpha_s(M_Z) = 0.1194 \pm 0.0014.\ (\mathrm{LEP\ result\ from\ Ref.}[52]) \tag{8.6}$$

Figure 8.19 shows the Review of Particle Physics (RPP) most precise data on $\alpha_s(Q)$ from all the reactions measured at all accelerators.

Fig. 8.19 RPP summary
(2021) of the available
measurements of $\alpha_s(Q)$ [53]

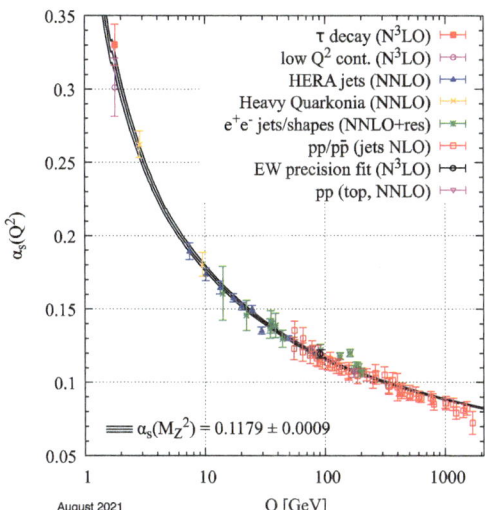

The detailed analysis of the LEP data, presented in the 2021 RPP [53], gives practically the same result as Eq. (8.6) but with a slightly larger error

$$\alpha_s(M_Z) = 0.1186 \pm 0.0016. \text{ (LEP result from Rev. Part. Phys.2021)} \qquad (8.7)$$

The conclusion is that the final LEP error on $\alpha_s(M_Z)$ is *six-seven times* smaller than the error on in 1989, before LEP start-up, which was ± 0.01, as shown in Fig. 8.14a. It is interesting to remark that, from Fig. 8.19, the 2021 world average is

$$\alpha_s(M_Z) = 0.1179 \pm 0.0009. \text{ (world average } - \text{ Rev. Part. Phys.2021).} \qquad (8.8)$$

which has an error about *70% smaller* than the LEP result of Eqs. 8.6 and 8.7.

It has to be added that, in the last years, the lattice calculations of $\alpha_s(M))$ have improved so much that the world average quoted in [54]

$$\alpha_s(M_Z) = 0.11803 + 0.00047 - 0.00068. \text{ (world average of lattice calculations)} \qquad (8.9)$$

has an error that is about half the one of the measured world average of Eq. 8.8.

The determination of the uncertainties is very delicate, as discussed in a recent paper [55]. At any rate, it is easily predictable that, in a few years lattice calculations—which use as input the quark bare masses—will produce a value of $\alpha_s(M_Z)$ with a much smaller error. At that point, the authors of RPP might decide to use the output of lattice calculations as recommended value forgetting all experimentally measured data. After such a decision, the strong sector of the Standard Model will be on a different footing than the electro-weak sector because the parameters of the

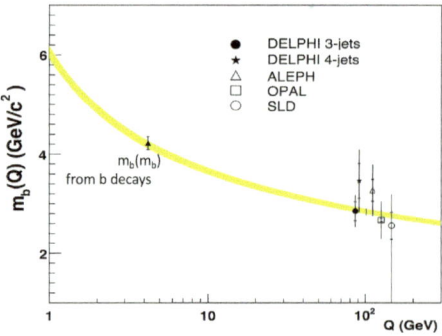

Fig. 8.20 Figure adapted from [56] with the data on the b-quark mass of [56–60].)

U(1)xSU(2) group—including the couplings α_1 and α_2 of the 'pure' electromagnetic interaction U(1) and the 'pure' weak interaction SU(2)—will be obtained from *measurements* of the electric charge, the Fermi constant, and the Z and Higgs masses, while the coupling α_s of the SU(3) group will be given by lattice *calculations*, in which the quarks masses will have to be introduced by hand.

To conclude this Section, I observe that the running of the strong coupling was well established before the LEP start-up. Instead, no information existed on another phenomenon predicted by QCD: the "running" of the quark masses and, in particular, of the b-quark asses. The after-LEP situation is shown in Fig. 8.20.

The values of the b-quark mass (at the scale M_Z) have been computed by measuring the fractions of 3-jets and 4-jets events that contain b-quarks. The results obtained at LEP and at SLC, using the b-tagging methods described in the previous Section, are plotted in the Fig. 8.20 [56–60]. Also in this case, but with less accuracy, the QCD prediction, represented by the yellow band, is experimentally confirmed.

8.7   Unification of the Forces and the First Microsecond

Well before LEP, theorists and experimentalists were performing global fits to the available experimental data on the properties of the intermediate bosons, parity violation in nuclei and neutrino-quark, neutrino-electron, electron-quark, muon-quark, and electron–positron collisions [61–70]. The two most active groups were led by John Ellis [63, 65, 67–69] and Paul Langacker [61, 62, 66, 70]. I had the occasion to contribute to these developments because—while working at CERN on neutrino physics with the CHARM experiment—I gave a talk on neutral currents at the Neutrino79 Bergen Conference [71]. There I discussed precision fits with Paul Langacker who, two years later, asked me to join his research group.

In 1987 the group published a review paper featuring Fig. 8.21a [62], in which $\alpha_1(Q)$ and $\alpha_2(Q)$ are the 'pure' electromagnetic coupling and the 'pure' weak coupling of the U(1) and SU(2) gauge groups; they are analogous to the SU(3) strong coupling $\alpha_s(Q)$. As discussed at the beginning of Sect. 8.6, the couplings

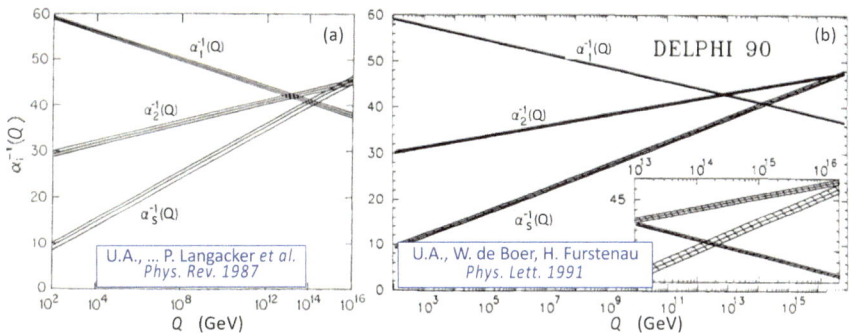

Fig. 8.21 **a** Standard Model extrapolations of the values of the couplings ($\alpha_1, \alpha_2, \alpha_s$) measured in the 80's below 100 GeV [62]. **b** Same graph but drawn with data collected in the first year of LEP: the SM couplings do not cross as in a Grand Unified Theory (GUT) [72]

depend on the polarization of the medium of virtual particles that surrounds the central charge: α_s^{-1} *increases* proportionally, in first order, to the logarithm of Q (as shown in Fig. 8.17) so that in Fig. 8.21 the line is practically straight, while α_1^{-1} *decreases*, almost logarithmically, with Q. As said at the beginning of Sect. 8.6, at each energy Q the local slopes are determined by the virtual particles that have mass smaller than Q.

Figure 8.21a shows that, in 1987 the error bands were large and the only statement that could be made was that, at the level of 2–2.5 standard deviations, the forces did not unify. Four years later LEP data changed the situation (Fig. 8.21b): in the Standard Model unification was not obtained at the level of 7 standard deviations.

I was involved in the production of this figure because, in Fall 1990, I was invited to give a talk at the Texas-ESO-CERN Conference on Astrophysics that had to be held in December in Brighton. Since I wanted to bring some new perspective to the already much publicized LEP data, I visited John Ellis who remarked that the improved quality of the data had to have an influence on the unification of the forces. He knew the problem because he had been working on the paper of Ref. [68] in which, by considering the electroweak parameter $\sin^2(\theta_w)$, it was concluded that the MSSM reproduces the LEP measured value better than the Standard Model.

The day after I showed the graph of the 1987 paper to Wim de Boer, leader of the Karlsruhe group in DELPHI, and to his PhD student Hermann Fürstenau, who had already codes at hand. In the following weeks he modified them following my proposal to (i) introduce in the calculation of the slopes of the three lines the superparticles of MSSM as if they had a single '*effective*' *mass* M_{SUSY} and (ii) to compute M_{SUSY} and its error by imposing the crossing in a *unification point* $Q = M_{GUT}$. The plot, shown in Fig. 8.22a, vividly showed that the LEP data were consistent with the simplest *low-energy* Grand Unified SUSY Theory. Presented in a preliminary form at the Brighton Conference, the plot appears in its final form in the proceedings under the titled '*LEP, the Laboratory for Electrostrong Physics, one year later*' [73].

In 1991, Ugo Amaldi, Wim der Boer and Hermann Fürstenau[88] showed that the cross-over just failed to happen unless a special 'supersymmetry', that had long been suspected to exist in Nature, existed. Its effect was to double the number of elementary-particle types in existence, and so slightly alter the way the force strengths change as energies increased. The result was a more-or-less exact cross-over at high-energy. This cross-over picture brought about huge interest in supersymmetric theories that continues unabated to the present day.

The simple suggestive picture of the three-fold intersection of the strengths of the electromagnetic, weak, and strong forces of Nature, which was first drawn by Jogesh Pati in 1978,[89] has played an inspirational role in the search for a unified description of Nature.[90] The convergence of the running force strengths suggests that unification does exist and led to the exploration of the early history of the Universe, an understanding of the preponderance of matter over antimatter within it today, and the search for the right way to include gravity in the unification scheme. It is a simple symbol of the Universe's deep unity in the face of superficial diversity, which is what we mean by beauty.

Opposite: a calculation of the change in the strengths of the electromagnetic, weak and strong forces' strengths versus increasing energy in the standard theory of grand unification in 1991 (*top*) shows that there is a significant 'miss' of the target of a single cross-over at one energy. By contrast, if the property of supersymmetry exists in Nature, then it changes the number of elementary particles that must exist and produces a convincing single cross-over at high energy (*bottom*). These pictures, produced by Amaldi, de Boer and Fürstenau, created an explosion of interest in theories of supersymmetry that continues to this day.

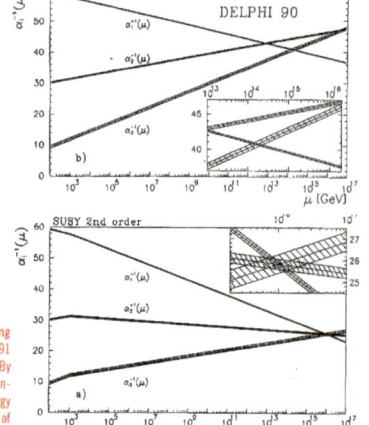

Fig. 8.22 Pages copied from John Barrow's book 'Cosmic Imagery' published in 2008 [74]

Fig. 8.23 Fitted value of M_{SUSY} versus $\alpha_s(M_Z)$ [76]. The band is due to the statistical errors

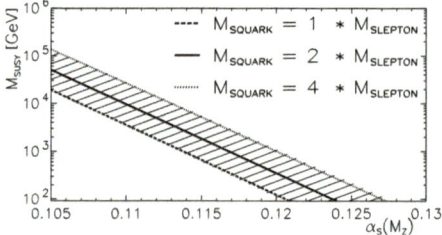

At the beginning of 1991 we published the two figures in a CERN preprint and in a Physics Letters paper [72]. The reactions were overwhelming—I think because of (i) the visual power of the three converging lines and (ii) the novelty of the fitted masses M_{GUT} and M_{SUSY} *with their errors*. These reactions were unexpected because among the experts it was known that the recent LEP data were better fitted by the minimal SUSY model than by the Standard Model [68–70]. The particle physics community got excited, and we received a lot of calls and emails. Wim de Boer and I were interviewed by daily newspapers and TVs [75]. Soon after the publication, many theoretical articles appeared in scientific journals improving our analysis, criticizing our simple approach, and better considering, for instance, threshold effects at the unification energy. Years later, in 2008, John Barrow in his 'Cosmic Imagery' summarized our paper with the two pages of Fig. 8.22 and wrote: "The converging of the running force strengths […] is a simple symbol of the Universe deep unity in face of superficial diversity, which is what we mean by beauty."

Going back to 1991, in July at the Geneva EPS Conference Wim de Boer presented a new analysis [76] in which we had improved the previous parametric study of the

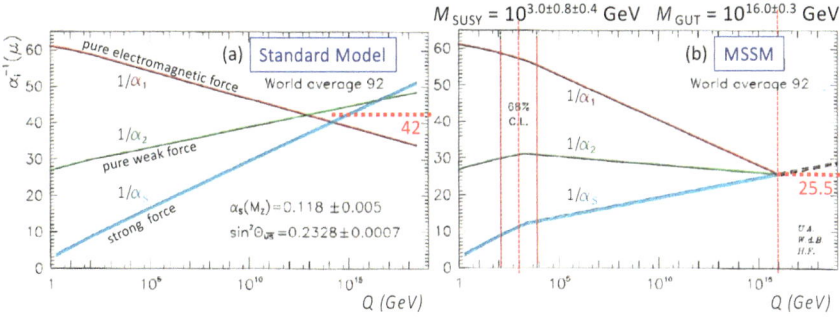

Fig. 8.24 Unification plots computed with the best experimental values of summer 1992

unification parameters [72] by assuming that all the strongly interacting sparticles have mass M_{squark} and all the non-strongly interacting ones have mass $M_{slepton}$.

The results were $M_{GUT} = 10^{15.8\pm0.3\pm0.1}$GeV, $M_{SUSY} = 10^{3.4\pm0.9\pm0.4}$GeV and $\alpha_s(M_{GUT})^{-1} = 26.3 \pm 1.9 \pm 1.0$. The first error was due to the experimental uncertainties of the time—on $\alpha_s(M_Z)$ but also on the electroweak parameter $\sin^2\theta_w = 0.233 \pm 0.008$—and the second errors were the estimate uncertainties due to the SUSY mass spectrum.

One year later the experimental errors on the couplings were further reduced so that also M_{SUSY} and M_{GUT} were slightly better determined, as shown in Fig. 8.24b.

Today, with $\alpha_s(M_Z)$ from Eq. (8.8) (2021) and the latest $\sin^2\theta_w$ error, Fig. 8.23 gives

$$M_{SUSY} = 10^{2.7\pm0.35\pm0.4}\text{GeV},\qquad(8.10)$$

so that, by combining the errors quadratically, $M_{SUSY} = 10^{2.7\pm0.5}$ GeV, which says that, in the framework of this simple model, the spectrum of the supersymmetric particles has an effective mass $M_{SUSY} \approx 500$ GeV—the logarithmic centre of the 95 % CL range 50–5000 GeV. Such a statement is weak but nontrivial: M_{SUSY} could have come out orders of magnitude larger than 1000 GeV, which is the upper limit for the cancellation of the divergences in the Higgs mass due to the opposite virtual effects of particles and their supersymmetric partners. Moreover, M_{GUT} is well below the Planck mass and its numerical value does not violate proton decay bounds.

It is worth noting that plots as the one of Fig. 8.24b indicate that MSSM may be valid, but, of course, many non-supersymmetric *unified* models can be constructed [77]. The plot of Fig. 8.24b can also be used to describe the phenomena that happened at the beginning of the Universe by (i) reading the x-axis from left to right, (ii) identifying the energy scale Q with the temperature of the primordial medium, and (iii) recalling the simple thermodynamical relation $Q_{GeV} \cong T_{\mu s}^{-\frac{1}{2}}$, where $T_{\mu s}$ is the cosmic time measured in microseconds.

The drawing of Fig. 8.25 is a figure that I have been using for many years [78, 79], and appears in my Springer book '*Particle accelerators: from Big Bang physics*

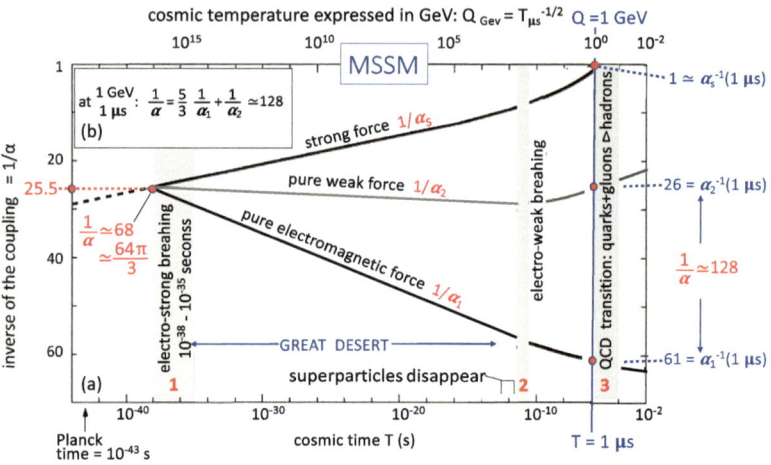

Fig. 8.25 Time evolution of the couplings in the framework of the minimal SUSY model

to hadron therapy' [80].The grey areas represent three transition regions: (1) the phenomena that originated the electro-strong breaking are unknown; (2) the phase transition at $T \approx 10^{-11}$ s was caused by the electro-weak symmetry breaking due to the Higgs field; (3) the disappearance of the quark-gluon plasma and the appearance of hadrons, happened when the increasing strong coupling got close to 1.

As shown in Fig. 8.25, at the divergence time the inverse of the electromagnetic coupling α^{-1}—a linear combination of $\alpha_1{}^{-1}$ and $\alpha_2{}^{-1}$—had the value $\alpha^{-1} \approx 68$, at the cosmic time $T = 1$ μs was $\alpha^{-1} \approx 128$ and, in the present very cold Universe, is 137, twice greater than at the beginning. This evolution is dictated by the masses of all the particles (whatever their nature) that virtually exist around each charge. In this simple GUT model, all the sparticles have masses smaller that 5 TeV and, in the fast running towards their destination, the three couplings traverse a Great Desert.

Going back to my recollections of Sect. 8.1, in a Grand Unified Theory, even without SUSY, the number $\alpha^{-1} \approx 137$—which occupied so much the mind of Bruno Touschek—is not so important because it is, at least in principle, calculable from the initial coupling $\alpha_s(M_{GUT})^{-1}$—the really fundamental quantity in a Grand Unified Theory, which in Fig. 8.25 is 8π[81]—and the masses of the particles (whatever their nature) in the enormous range that goes from zero to M_{GUT}. In future such a calculation may be feasible IF a Great Desert occupies the central part of Fig. 8.25.

In the last years the LHC experiments have excluded a large fraction of the MSSM 5-parameters phase space so that many theorists are convinced that a low-energy minimal SUSY theory is no longer defendable. However, even if the detailed behaviours of the curves of Figs. 8.24b and 8.25 are not supported by the present experimental situation, many experts think that there are still corners of the enormous available phase space for more elaborated version of a low-energy supersymmetric theory. For instance, John Ellis and collaborators have studied a minimal supersymmetric extension of MSSM with 11 parameters (pMSSM11) [82].

Personally, I believe that, even if simplest MSSM is not realized in Nature, some form of low-energy supersymmetry, with a Great Desert, is still a viable Grand Unified theory so that plots, as the ones of Figs. 8.24b and 8.25, will be drawn and used also in future for both scientific purposes and science popularization.

8.8 LEP Highlights and Its Legacy to the LHC Experiments

An enormous amount of coordinated experimental and theoretical work has been invested in the writing of the about 2600 papers published by the four LEP collaboration, of which about 15% have been produced after 2004 [83]. The quality and the amount of the results were such that a big effort has gone also in keeping the data available for future analyses [83]. Moreover, the main protagonists of this endeavour wrote hundreds review papers of the many experimental results; for space reasons, I have discussed only a small personal selection. In many of these reviews—see, for instance, [31, 32, 84]—it has been underlined that the precisions achieved are by far better than what was foreseen before LEP start-up. This is the opinion expressed by Wilbur Venus [32] few months after the end of the last run.

What did LEP achieve? The new physics initially anticipated (W, Z) was there. Due to the clean initial situation, hermetic detectors, etc, it was probed with unprecedented precision, typically 2 orders of magnitude better than before LEP started (e.g. M_Z was measured to ± 2.1 MeV, Γ_Z to ± 2.3 MeV, the number of neutrinos $N_\nu = 3$ to 1 part in 350, R_b to $\pm$ 0.3%, which is 20 times better than initial hopes, M_W to ± 39 MeV [in the final analysis ± 33 MeV]; and m_t was predicted correctly, universality was tested at the ~ 1 per mille level in electroweak interactions and to 1% in QCD, the cancellation of WW production amplitudes required by gauge theory was tested at the 1% level, and purely weak loop corrections at the $\sim 10\%$ level.

LEP also brought deeper knowledge of heavy flavours, deeper understanding of QCD, and showed that GUTs work with SUSY but not without. And the new particle searches were remarkably complete and rigorous, leaving very few corners still unexplored (and squeezing minimal SUSY into a very tight one!). But there were no further surprises. Apparently, nature chose to be at her most boring.

Frank Wilczek at the CERN LEPfest of November 2000 said: 'The historic achievement of LEP has been to establish with an astonishing degree of rigor and beyond all reasonable doubt what will stand for the foreseeable future - perhaps for all time - as the *working Theory of Matter* ..... and to give us some very definite and specific clues for what lies beyond.'

The reasons for the successes of LEP experiments were many and each LEP physicist has his own list. I like the one proposed by Jurgen Drees [84]:

"Why was LEP so successful? Many fortunate facts had to come together:

- A highly dedicated machine group responsible for the excellent performance of LEP,
- low background in the detectors,
- good performance of all detectors from the pilot run in August 1989 till the end of data taking,
- effective division of work between CERN and the outside laboratories,

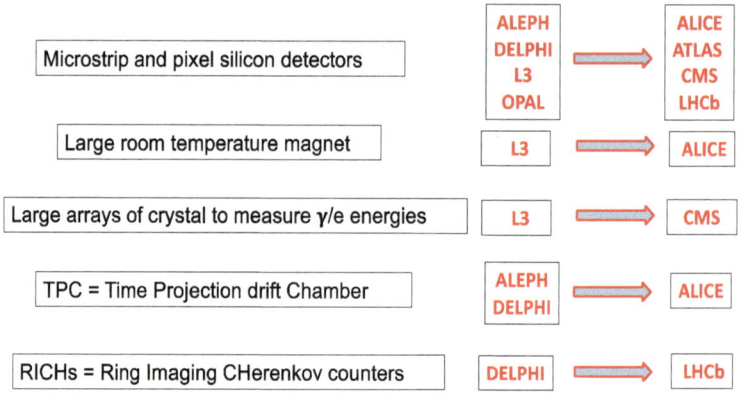

Fig. 8.26 LEP techniques, methods and hardware components used by the LHC experiments

- close cooperation between the 4 collaborations and, also, between LEP and SLD (without avoiding competition),
- close cooperation between experiments and the machine group,
- and, very important, close cooperation with theory groups."

The LEP detectors developed novel techniques and methods that worked better than initially foreseen, in particular the micro-vertex detectors discussed in Sect. 8.4. As shown in Fig. 8.26, these techniques have been left as a *material* legacy to the Large Hadron collider experiments, which used them but had to introduce substantial improvements because the running conditions, the event rate and the backgrounds are harsher than at LEP.

However, the main legacy of LEP to LHC experiments is *immaterial*: the Standard Model, which was checked from all points of view in the finest details and with accuracies unforeseen before the start-up of the largest electron–positron collider ever built, which has its origin in the minuscule ADA ring, built sixty years ago by Bruno Touschek and collaborators in less than one year. The LEP Standard Model legacy was accompanied by sophisticated software codes, describing hadronization processes and hadron decays, which are essential for computing at LHC the signatures of novel phenomena and their backgrounds.

Acknowledgements I thank Siggi Bethke, Alessandro de Angelis and Daniel Treille for the critical readings of the drafts of this paper, constructive criticisms, corrections, and many suggestions of useful improvements. Together with all the members of DELPHI, I am grateful to Jean-Eudes Augustin, Daniel Treille, Wilbur Venus, Tiziano Camporesi and Jan Timmermans for their guidance as spokespersons of our Collaboration.

References

1. The episodes of this Section are described with more details in U. Amaldi, Remembering Bruno Touschek, in 'Bruno Touschek Memorial Lectures', ed. by M. Greco, G. Pancheri, Frascati Physics Series, vol. 33, p. 89 (2004)
2. Particle Data Book 2007, Plots of cross-sections. https://pdg.lbl.gov/2007/reviews/hadronicrpp. pdf. Accessed 5 Jan. 2023
3. The ALEPH, DELPHI, L3 and OPAL Collaborations, The LEP Electroweak Working Group, S. Schael et al., The Electroweak Measurements in Electron-Positron Collisions at W-Boson-Pair Energies at LEP, Phys. Rep. 532, 119 (2013)
4. H. Schopper, LEP—The lord of the collider rings at CERN 1980–2000 (Springer, 2016)
5. B. Naroska, Physics with the JADE detector at PETRA. Phys. Rep. 148 (1987)
6. The ALEPH, DELPHI, L3 an OPAL Collaboration, SLD Collaboration, LEP Electroweak Working Group, SLD Electroweak and Heavy Flavour Groups. Phys. Rep. 257 (2006)
7. The SLD Collaboration, S.D. Abrams et al., Measurements of Z-boson resonance parameters in e+e− annihilations. Phys. Rev. Lett. 63, 2173 (1989)
8. P.C. Rowson, D. Su, S. Willocq, Highlights of the SLD physics program at the sLAC linear collider Ann. Rev. Nucl. Part. Sci. 51, 345 (2001)
9. The LEP collaborations, ALEPH, DELPHI, L3 and OPAL, Electroweak parameters of the Z° resonance and the standard model. Phys. Lett. B 276, 247 (1992)
10. M. Pepe Altarelli, The number of neutrino species, Special Colloquium for the Thirty Anniversary for the Start of Operations of LEE, https://indico.cern.ch/event/858488/. Accessed 5 Jan. 2023
11. J. Ellis, R. Peccei (eds.), Physics at LEP, vols. 1 and 2, CERN Yellow Reports 86–02 and 86–03 (1986)
12. See for instance: The OPAL Collaboration., G. Abbiendi et al., Precision luminosity for Z^0 line-shape measurements with a silicon-tungsten calorimeter. Eur. Phys. J. C 14, 373 (2000). Marcello Mannelli was Project Leader
13. See for instance: S.J. Alvsvaag et al., The DELPHI small angle calorimeter. IEEE Trans. Nucl. Sci. 42(4), 478 (1995). Tiziano Camporesi was Project Leader
14. D. Creanza et al., The new ALEPH silicon vertex detector. Nucl Instrum. Meth. Phys. Res. A409, 157 (1998)
15. P.P. Allport et al., The OPAL silicon micro-vertex detector. Nucl Instrum. Meth. Phys. Res. A324, 34 (1993)
16. V. Chabaud et al., The DELPHI silicon strip micro-vertex detector with double sided readout. Nucl. Instrum. Meth. in Phys. Res. A368, 314 (1996)
17. M. Acciarri et al., The L3 silicon micro-vertex detector. Nucl. Instrum. Meth. Phys. Res. A351, 300 (1994)
18. C. Mariotti, The measurement of R_{uds}, R_c and R_b at LEP and SLD. Proc. 1997 Eur. Conf. HEP 667 (1997)
19. Ref. 6, p. 122
20. Figure taken from: E. Delmeire, L3 Collaboration, Measurement of e+ e− to W+ W− cross-section, Thèse, Université de Geneve (2004)
21. J. Iliopoulos, Introduction to the standard model of the electro-weak interactions, CERN Summer School of Part. Phys. (2012). Angers, France. hal-00827554
22. S.L. Wu et al., Search for neutral Higgs at LEP 200, CERN EP-87–40, 312
23. E. Picasso, A few memories from the days at LEP. Eur. Phys. J. H36, 551 (2012)
24. C. Benvenuti, Superconducting coatings for accelerating rf cavities: past, present, future. Part. Accel. 40, 43 (1992). Chris Benvenuti invented also the non-evaporable getter (NEG) pumps used at LEP: C. Benvenuti et al., Vacuum 53 (1999) 219
25. R. Assmann, M. Lamont, S. Myers, A brief history of the LEP collider. Nucl. Phys. B Proc. Suppl. 109, 17 (2002)

26. D. Treille, LEP in the year 2000. *Europhysics News*, vol. 58, Mar.–Apr. 1999. Figure from M. Carena, J. Espinosa, M. Quiros, C. Wagner, Expressions for radiatively corrected Higgs masses and couplings in the MSSM, vol. 209 (1995)

27. For MSSM, now called pMSSM for 'phenomenological' MSSS, see for instance: The MSSM: Group summary report (1999). arXiv:hep-ph/9901246. Accessed 5 Jan. 2023

28. J. Ellis, G. Ridolfi, F. Zwirner, On radiative corrections to supersymmetric Higgs boson masses and their implications for LEP searches. Phys. Lett. B257, 83 (1991) and Phys. Lett. B262, 477 (1991)

29. Y. Okada, M. Yamaguchi, T. Yanagida, Upper bound of the lightest higgs boson mass in the minimal supersymmetric standard model. Prog. Theor. Phys. **85**, 1 (1991)

30. H.E. Haber, R. Hempfling, Can the mass of the lightest Higgs boson of the minimal supersymmetric model be larger than M_Z? Phys. Rev. Lett. **66**, 1815 (1991)

31. D. Treille, LEP/SLC: what did we expect? What did we achieve? A quick historical review. Nucl Phys. B Proc. Suppl. **109**, 1 (2002)

32. W. Venus, A LEP Summary, in *Proceedings of the International Europhysics Conference on High-Energy Physics*, Budapest, PoS 007 (2001). https://pos.sissa.it/007/284/. Accessed 5 Jan. 2023

33. K. Hübner, Designing and building LEP. Phys. Rep. (2004)

34. M.M. Kado, The searches for Higgs bosons at LEP. Ann. Rev. Nucl. Sci **52**, 65 (2002)

35. L. Maiani, R. Bassoli, *A caccia del bosone di Higgs* (Mondadori, 2013)

36. The ALEPH Collaboration, P. Barate et al., Observation of an excess in the search for the standard model higgs boson at ALEPH. Phys. Lett. B **495**, 1 (2000)

37. The ALEPH, DELPHI, L3, and OPAL Collaborations, The LEP Working Group for Higgs Boson Searches, G. Abbiendi et al, Search for the standard model higgs boson at LEP. Phys. Lett. B **565**, 61 (2003)

38. G. Arcadi et al., The waning of the WIMP? A review of models, searches, and constraints. Eur. Phys. J. C **78**, 203 (2018)

39. P.J. Fox, R. Harnik, J. Kopp, Y. Tsai, LEP shines light on dark matter. Phys. Rev. D **84**, 014028 (2011)

40. See for instance: F. Wilczek, QCD made simple. Phys. Today **22** (2000)

41. G. Altarelli, Experimental tests of perturbative QCD. Ann. Rev. Nucl. Part. Sci. **89**, 357 (1989)

42. S. Bethke, QCD studies at LEP. Phys. Rep. **404**, 203 (2004)

43. See for instance: M.B. Einhorn, D.R.T. Jones, The weak mixing angle and unification mass in supersymmetric SU(5). Nucl. Phys. B **96**(3), 475 (1982)

44. G. Altarelli, R. Kleiss, C. Verzegnassi (eds.), Physics at LEP-I, CERN Yellow Report, CERN 89–08 (1989)

45. See, for instance, Cnet Collaboration, A. Buckley et al., General-purpose event generators for LHC physics. Phys. Rep. **504**, 145 (2011)

46. The DELPHI Collaboration, P. Abreu et al., Consistent measurements of alpha-s from precise oriented event shape distributions. Eur. Phys. J. C **14**, 557 (2000) and erratum Eur. Phys. J. C **19**, 761 (2001)

47. R.W.L. Jones, Final α_s combinations from the LEP QCD Working Group. Nucl. Phys. B Proc. Suppl. **152**(1), 15 (2006)

48. G. Abbiendi et al., Determination of alpha(s) using jet rates at LEP with the OPAL detector. Eur. Phys. J. C **45**, 547–568 (2006). (hep-ex/0507047)

49. The ALEPH Collaboration, A. Heister et al., Measurements of the strong coupling constant and the QCD colour factors using four-jet observables from hadronic Z decays. Eur. Phys. J. C **27**, 1 (2003)

50. M. Davier, A. Hocker, B. Malaescu, C.-Z. Yuan, Z. Zhang, Update of the ALEPH non-strange spectral functions from hadronic τ decays. Eur. Phys. J. C **74**(3), 2803 (2014)

51. The Gfitter Group., J. Haller, A. Hoecker et al., Update of the global electroweak fit and constraints on two-Higgs-doublet models. Eur. Phys. J. C **78**, 675 (2018)

52. S. Bethke, Precision physics at LEP, in From my vast repertoire—the legacy of Guido Altarelli, ed. by S. Forte, A. Levy, G. Ridolfi (World Scientific, 2019)

53. P.A. Zyla et al. Particle Data Group, Prog. Theor. Exp. Phys. 2020, 083C01 (2020) and 2021 update
54. J. Komijania, P. Petreczky, J.H. Weber, Strong coupling constant and quark masses from lattice QCD. Prog. Part. Nucl. Phys. **113**, 103788 (2020)
55. L. Del Debbio and A. Ramos, Lattice determinations of the strong coupling. Phys. Rep. **920**, 1–71 (2021). https://arxiv.org/abs/2101.04762. Accessed 5 Jan. 2023
56. The DELPHI Collaboration, Abdallah et al., Study of b-quark mass effects in multi-jet topologies with the DELPHI detector at LEP. Eur. Phys. J. C **55**, 525 (2008)
57. The DELPHI Collaboration, P. Abreu et al., m_b at Mz. Phys. Lett. B **418**, 430–442 (1998)
58. Brandenburg et al., SLD Collaboration, Measurement of the running b-quark mass using e^+e^- to quark-antiquark-gluon events. Phys. Lett. B **468**, 168 (1999)
59. The ALEPH Collaboration, R. Barate et al., A measurement of the b-quark mass from hadronic Z decays. Eur. Phys. J. C **18**, 1 (2000)
60. The OPAL Collaboration, G. Abbiendi et al., Determination of the b quark mass at the Z mass scale. Eur. Phys. J. C **21**, 411–422 (2001)
61. E. Kim, P. Langacker et al., A theoretical and experimental review of the weak neutral current. Rev. Mod. Phys. **53**, 211 (1981)
62. U. Amaldi, P. Langacker et al., A comprehensive analysis of data pertaining to the weak neutral current and the intermediate vector boson masses. Phys. Rev. D **36**, 1385 (1987)
63. G. Costa, J.R. Ellis et al., Neutral currents within and beyond the Standard Model. Nucl. Phys. B **297**, 244 (1988)
64. E.G. Fogli, Electroweak radiative corrections and parameters of the standard model: a comparative analysis. Zeit. für Phys. C Part. Fields **43**, 229 (1989)
65. J. Ellis, G.L. Fogli, The implications of recent electroweak data for m_t and M_H. Phys. Lett. B **232**, 139 (1989)
66. P. Langacker, Implications of recent M_Z, W and neutral-current measurements for the top-quark mass. Phys Rev. Lett. **63**, 1920 (1989)
67. J. Ellis, G.L. Fogli, New bounds on m_t and M_H from precision electroweak data. Phys. Lett. B **249**, 543 (1990)
68. J.R. Ellis JR, S. Kelley, D.V. Nanopoulos, Precision LEP data, supersymmetric GUTs and string unification. Phys. Lett. B **249**, 441 (1990)
69. J.R. Ellis JR, S. Kelley, D.V. Nanopoulos, Probing the desert using gauge coupling unification. Phys. Lett. B **260**, 131 (1991)
70. P. Langacker P, M-X Luo, Implications of precision electroweak experiments for m_t, ρ_0, $\sin^2\theta_W$ and grand unification. Phys. Rev. D **44**, 817 (1991)
71. U. Amaldi, Advances in neutral currents, in Proceedings of the Bergen Conference in Processes Neutrino-79, Bergen, ed. by A. Haatuft, C. Jarlskog (Bergen University), p. 376
72. U. Amaldi, W. de Boer, H. Fürstenau, Comparison of grand unified theories with electroweak and strong coupling constants measured at LEP. Phys. Lett. B **260**, 47 (1991)
73. U. Amaldi, LEP, the Laboratory for Electrostrong Physics, one year later, Texas-ESO-CERN Conferences Astrophys., Brighton, Dec 1990. Ann. N.Y. Acad. Sci. **647**, 244 (1991)
74. J. Barrow, Cosmic Imagery: Key Images in the History of Science (W. W. Norton & Company, 2008)
75. Feature articles were published. See for instance: N. Hall, New Scientist, Apr. 1763, 13 (1991); J. S. Stirling, Phys. World May 4/5) **19** (1991); P. Hamilton, Science **253**, 272 (1991); G.G. Ross, Nature **52**, 21 (1991). S. Dimopulos, A.A. Raby and F. Wilczek, Phys. Today **25**, 25 (1991).
76. U. Amaldi, W. de Boer, H. Fürstenau, Consistency checks of GUTs with LEP data, Proc. Lept.. Ph. Symp. Eur. Conf. HEP, Geneva, ed. by S. Hegarthy, K. Potter and E. Quercigh (World Scientific, 1991), Conf. Proc. C 910725V1, p. 690 (1991)
77. See, for instance, the non-supersymmetric 'split models' with multiplets of quarks and leptons that are split between the TeV scale and the unification energy M_{GUT}: U. Amaldi, W. de Boer, P.H. Frampton, H. Fürstenau and J.T. Liu, Consistency checks of grand unified theories. Phys. Lett. B **281**, 374 (1992)

78. U. Amaldi, Le forze fondamentali e il primo secondo di vita dell'Universo, in *'Scienza e Vita nel momento Attuale' IV*. (Mem. Ist. Lomb. Sci. Lett, Milan, 1994), p.52
79. U. Amaldi, The importance of particle accelerators. Europhys. News **31**, 5 (2000)
80. U. Amaldi, Particle Accelerators: from Big Bang Physics to Hadron Therapy (Springer, 2015), p. 259.
81. It is interesting to remark that in MSSM the fit of Figure 25, which gives Eq. (10), implies an inverse of the fundamental unified coupling at M_{GUT} $\alpha_{GUT}^{-1} = 25.5$ that can be written (by chance?) as $\alpha_{GUT}^{-1} = 8\pi = 25.1327$, within one third of one standard deviation. With this value the electromagnetic coupling at M_{GUT} would be $\alpha^{-1}(M_{GUT}) = 8^2 \pi/3 = 67.02064$
82. J. Ellis, Searching for supersymmetry and its avatars. Phil. Trans. R. Soc. **A377**, 20190069 (2019)
83. Z. Akopov et al., ICFA Study Group on Data Preservation and Long-Term Analysis in HEP— DPHEP, Towards a global effort for sustainable data preservation in High Energy Physics, DPHEP-2012–001. https://arxiv.org/pdf/1205.4667.pdf. Accessed 5 Jan. 2023
84. J. Drees, Review of final LEP results or a tribute to LEP. Int. J. Mod. Phys. A **17**(23), 3259 (2002)

Chapter 9
From the Hadronic String to Quantum Gravity ... and Back

Gabriele Veneziano

Abstract I will outline the conceptual developments that led, through half a century, to the present formulation of string theory. The phenomenological observation of Dolen-Horn-Schmit "duality" in 1967, the formulation, a year later, of the Dual Resonance Model, and its eventual interpretation as a theory of quantum relativistic strings, marked the birth of the hadronic string. Immediately after, however, this elegant *S*-matrix theory of the strong interactions lost its phenomenological battle against a more, but not quite, conventional quantum field theory risking, around 1974, total oblivion. Nonetheless, ten years later, upon a huge rescaling of its intrinsic length scale, string theory made an impressive comeback as a candidate unified and finite quantum theory of all interactions, including gravity. This dream of a "Theory of Everything" has not come true yet, but new developments have uncovered an amazing new "duality" between gauge and gravitational interactions making it conceivable that the real hadronic string (the one implied by the confinement of quarks and gluons in QCD) will be eventually understood by addressing an easier gravitational problem.

9.1 Strong Interactions in the Mid Sixties

Having graduated from the University of Florence in 1965, I had the enormous luck of entering the field just at the beginning of that period which, a posteriori, can be rightly called the "golden decade" of elementary particle physics. At that time the status of the theory of strong (nuclear) interactions was not in very good shape. Data were abundant, but we could only confront them with a handful of models each one capturing one or another aspect of the complicated hadronic (hadron is a generic name for any particle feeling the strong force) world. Many hadrons had been identified, most of them metastable (resonances), and with large mass and angular momentum (spin): the "hadronic zoo" seemed to be increasing in size every day.

G. Veneziano (✉)
Theory Department, CERN, CH-1211, Geneva 23, Switzerland

Collège de France, 11 place M. Berthelot, 75005 Paris, France
e-mail: gabriele.veneziano@cern.ch

© The Author(s) 2023   113
L. Bonolis et al. (eds.), *Bruno Touschek 100 Years*,
Springer Proceedings in Physics 287,
https://doi.org/10.1007/978-3-031-23042-4_9

Today, with hindsight, we can easily assert that, in the late sixties, we took the wrong way by rejecting, a priori, a description of these phenomena based on quantum field theory (QFT) the framework that had already been so successful for the electromagnetic interactions via quantum electrodynamics (QED). There were (at least) two very good excuses for having chosen the wrong way:

- Unlike in QED, the theory of just photons and electrons, there were too many particles to deal with, actually, as I just said, an ever increasing number;
- QFTs of particles with high angular momentum were known to be very difficult, if not impossible, to deal with in a QFT framework.

Instead, a so-called S-matrix approach looked much more promising.

The constraint of relativistic causality forces the S-matrix elements to be analytic functions of the kinematical variables they depend upon, like the energy of the collision. Also, the symmetries of the strong interactions can easily be implemented at the level of the S-matrix. These symmetries could also be used to put some order in the hadronic zoo by grouping particles with the *same* spin into multiplets (with respect to symmetries such as $SU(2)$ of isospin or its $SU(3)$ extension to include strange particles).

Also, the recently developed Regge theory [1] was also able to assemble together particles of *different* angular momentum. One amazing empirical observation at the time was that the masses M and angular momenta J of particles lying on the same "Regge trajectory" approximately satisfied a simple relation:

$$J = \alpha(M^2) = \alpha_0 + \alpha' M^2 , \tag{9.1}$$

with α_0 a parameter depending of the particular Regge-family under consideration and α' a universal constant ($\alpha' \sim 0.9 \text{ GeV}^{-2}$ in natural units where $c = \hbar = 1$).

Regge theory had a second important facet, pointed out later by Gribov, Chew, Mandelstam and others [2]: it could be used to describe the behaviour of the S-matrix at high energy. These two uses of Regge theory are illustrated in Fig. 9.1, where we see the linear and parallel Regge trajectories (with one exception, the so-called vacuum or Pomeranchuk trajectory) and the fact that the trajectory interpolates among different particles at positive J, M^2 while it determines high-energy scattering at negative M^2.

Chew [3] had invoked these two appealing feature of Regge's theory to formulate what I will call (for reasons that will become clear later) an "expensive bootstrap". Chew's idea was to add to the already mentioned constraints (unitarity, analyticity, symmetry) the assumption of "Nuclear" Democracy" according to which:

- *All* hadrons, whether stable or unstable, lie on Regge trajectories (at $M^2 \geq 0$) and are on the same footing;
- The high-energy behaviour of the S-matrix is *entirely* given in terms of the same Regge trajectories (at $M^2 \leq 0$).

In Chew's bootstrap Unitarity (i.e. conservation of probability) played a crucial role. It represented a non linear and thus very non-trivial, constraint. Would that give

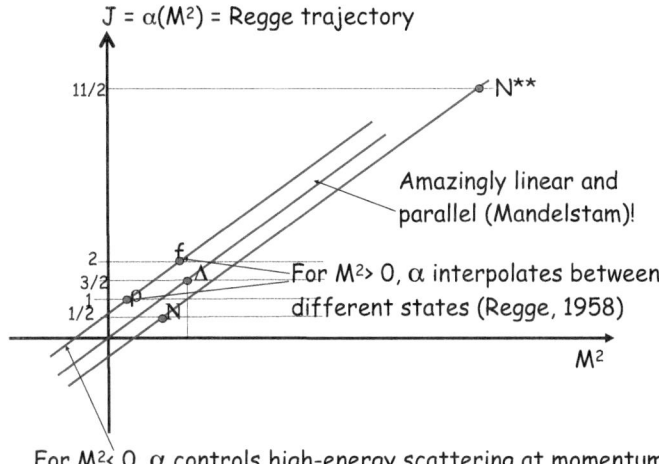

Fig. 9.1   Regge trajectories at positive and negative values of M^2

a unique solution to the bootstrap? The S-matrix knew about both uses of Regge theory:

$$S = S_{s-channel} + S_{t-channel} \,, \tag{9.2}$$

Considering, for instance, $\pi^+\pi^-$ scattering we would expect to find a contribution from both the formation of s-channel resonances (ρ^0 and the like) and from the exchange of the (ρ^0 and the like) Regge trajectory in the t-channel. This would mimic, for the strong interactions, the situation for e^+e^- scattering in QED, with the exchange of either a photon or a Z^0 in both channels.

However, an interesting surprise came out in 1967 through a fundamental observation made by Dolen, Horn and Schmit [4] who, after looking carefully at some pion-nucleon scattering data, concluded that contributions from resonance formation and those from particle exchange should *not* be added but were actually each one a complete representation of the process. This property became known as DHS duality.

In the summer of 1967, at a summer school in Erice, I was strongly influenced by a talk given by Murray Gell Mann reporting about DHS duality and stressing that such a framework could lead to what he defined as a "cheap bootstrap" as opposed to Chew's expensive one. In order to get interesting constraints on the Regge trajectories themselves it was enough to require that the two dual descriptions of a process would produce the same answer. This was a non trivial constraint, yet a linear one, thus providing a "cheaper" bootstrap.

DHS duality prompted Harari and Rosner [5] to introduce "Duality Diagrams" (see Fig. 9.2) where hadrons are represented by a set of quark lines (two for the mesons, three for the baryons) and the scattering process is described in terms of the flow of these quark lines through the diagram. By looking at the diagram in different

Duality Diagrams

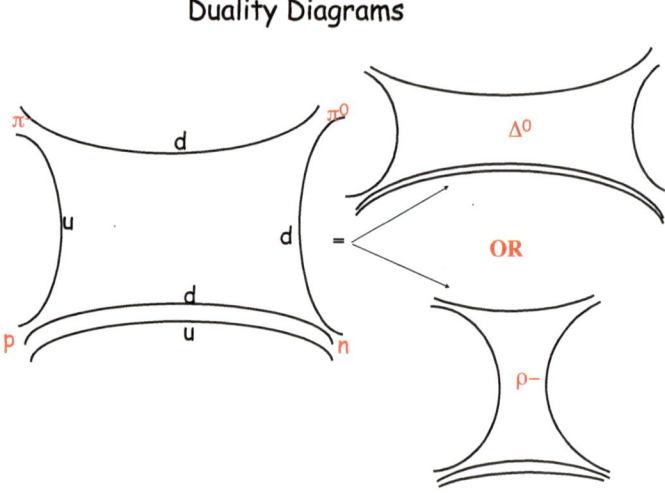

Fig. 9.2 Duality diagrams illustrating DHS duality

directions (channels), the process is seen to proceed in different -but equivalent in the sense of DHS duality- ways. Note that in those days quarks were just a mnemonic to keep track of quantum numbers and internal symmetries: they were not considered as having any real substance.

9.2 Dual Resonance Models

The crucial question was: Can we associate a precise mathematical expression to duality diagrams like we do with Feynman diagrams in quantum field theory?

A tentative answer to that question was found in 1968 [6] for a very simple and convenient (theoretically speaking!) process: $\pi\pi \to \pi\omega$, represented pictorially by three duality diagrams (two of which are shown in Fig. 9.3).

The educated guess for this process was in terms of the well-known Euler Beta-function:

$$A = \beta \left[B\left(1 - \alpha(u), 1 - \alpha(t)\right) + B\left(1 - \alpha(s), 1 - \alpha(u)\right) + B\left(1 - \alpha(s), 1 - \alpha(t)\right) \right]$$

$$B(x, y) \equiv \frac{\Gamma(x)\Gamma(y)}{\Gamma(x + y)} \ ; \ \alpha(t) = \alpha_0 + \alpha' t \ ; \ s + t + u = 3m_\pi^2 + m_\omega^2 \qquad (9.3)$$

where $\Gamma(x)$ denotes Euler's Gamma-function and the three terms in (9.3) are in one-to-one correspondence with the three duality diagrams of Fig. 9.3. Note the exact linearity of the Regge trajectory and the consequent appearance there of a dimensionful constant, the Regge slope α'.

Fig. 9.3 Duality diagrams
for $\pi\pi \to \pi\omega$

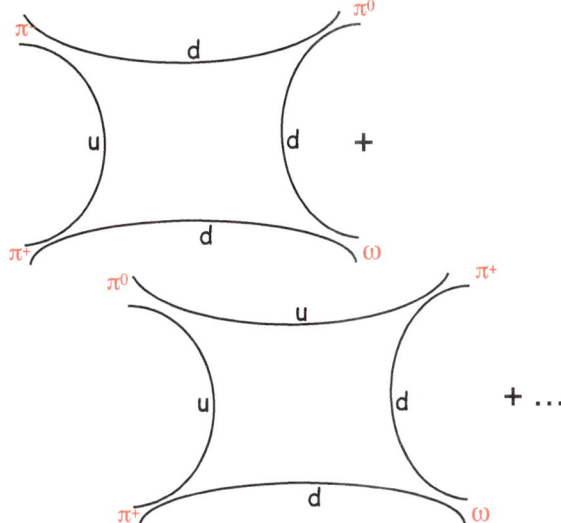

Although measuring the process $\pi\pi \to \pi\omega$ is challenging the same amplitude (9.3) can be used, by analytic continuation, to describe the decay $\omega \to 3\pi$. The result turned out to be very satisfactory (in particular the presence of zeroes [7] due to the Γ-functions in the denominators). The amplitude was also successfully extended to describe $\pi\pi \to \pi\pi$ scattering in the so-called Lovelace-Shapiro model [8].

Finally, (9.3) was generalized to production processes i.e. to amplitudes with more than four external legs. This last generalization became known as the Dual Resonance Model (DRM), the progenitor of string theory as we know it today.

9.3   The Dual Resonance Model and Relativistic Quantum Strings: From Hints to Proof

Since the early days of DRM research there were definite hints of some sort of underlying vibrating string (as particularly emphasized by H. Nielsen, L. Susskind and Y. Nambu). We can list some of them:

- The linear Regge trajectories imply a constant ratio between angular momentum and squared mass. That fits very well with an object that has a mass M proportional to its size L (then $J \sim M \cdot L \sim M^2$) where the constant of proportionality (α') has dimensions of length per unit mass, i.e. of the inverse of a string tension. In this reasoning we took the characteristic speed to be of $\mathcal{O}(c)$, hence the string is supposed to have relativistic motion.

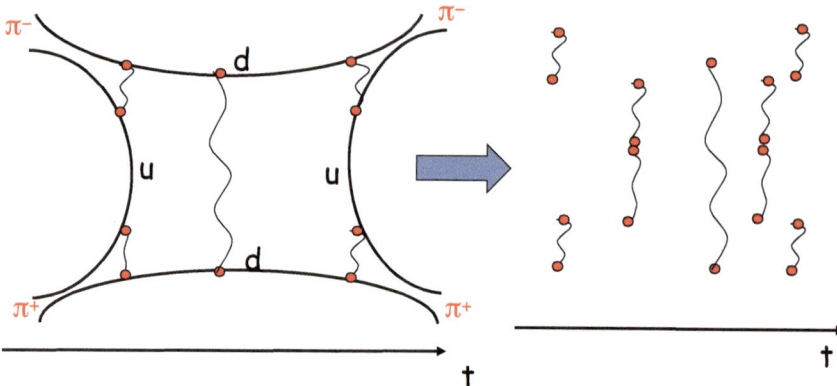

Fig. 9.4 Strings joining and splitting?

- The duality diagrams (say for meson-meson scattering) can be visualized (Fig. 9.4) in terms of strings connecting quark-antiquark pairs first joining to form a single string (by quark antiquark annihilation) and then splitting again (by pair creation).
- The spectrum of the DRM could be described [9] in terms of an infinite set of (quantized) harmonic oscillators having integer frequencies in terms of a fundamental one. This latter property is typical of a classical (say violin) string, but it was also obvious that the putative string had to be quantum-mechanical.
- There was a two-dimensional (conformal) field theory underlying the DRM with its Virasoro operators [10] and algebra [11]. And this would be the natural description of the dynamics of one-dimensional objects (in analogy with the world-line description of one-dimensional objects).
- …

The remaining hints were not missed, but the connection with strings remained qualitative for sometime. Eventually, it was established on solid grounds through a precise formulation of the classical relativistic string by Nambu and Goto [12] in 1970–1971. But its first correct (light-cone) quantization by Goddard, Goldstone, Rebbi and Thorn [13] had to wait till 1972. I refer to P. Di Vecchia's contribution for more details on this part.

9.4   Beautiful, Elegant, But Not the Right Theory!

Paradoxically, now that the DRM had been raised to the level of a respectable theory, it became apparent that it was not the right one for the (strong) interactions it had been conceived for! There were actually both good and bad news for the newly born string!

The good news (mainly theoretical)

- The Neveu-Scherk-Ramond extensions for adding fermions,
- The Gliozzi-Scherk-Olive (GSO) projection, leading to supersymmetry discovery (in the west).
- The combination of all these developments gave fully consistent superstring theories, with neither negative norm states (ghosts) nor imaginary mass states (tachyons).

The bad news (basically phenomenological):

- Unwanted massless states giving problems at large distance (strong interactions are short range forces)
- Softness giving problems at short distance (see below)
- Need for six extra dimensions of space for a total of ten space-time dimensions.

On the other hand the following experimental facts:

- The constant high-energy limit of $R = \sigma(e^+e^- \to$ hadrons$)/\sigma(e^+e^- \to \mu^+\mu^-)$,
- Bjorken scaling in deep inelastic lepton-hadron collisions,
- The relative abundance of large p_t events at CERN's pp collisions at the Intersecting Storage Ring (ISR),

were providing strong evidence for the existence of point-like structures inside the hadrons, structures completely absent in the Nambu-Goto string.

9.5   QCD Takes over

Around 1973–1974 QCD clearly took the upper hand on the hadronic strings. The points in its favor were many:

- Its proven ultraviolet (asymptotic) freedom explaining the abundance of hard collisions;
- Its conjectured, and later proven (see Guido Martinelli's talk), infrared slavery (confinement) leading to string-like excitations via chromo-electric flux tubes. The string tension is a well-defined quantity in QCD, via the behavior of large Wilson loops;
- Its reinterpretation of duality diagrams (and their higher order topologies) in terms of large-N expansions [14]. In large-N_c-QCD (at fixed 't Hooft coupling $\lambda = g^2 N_c$) duality diagrams take up a precise meaning: they are the sum of planar Feynman diagrams bounded by quark propagators and filled with gluons (as shown in Fig. 9.5). In this approximation resonances have zero width, the scattering amplitude is meromorphic, exhibits (most likely) DHS duality, and generates a scale ($\Lambda^{-2} \sim \alpha'$) via a renormalization-group phenomenon known as dimensional transmutation.

120   G. Veneziano

Fig. 9.5 Reinterpretation of
a duality diagram in the
't-Hooft limit [14]

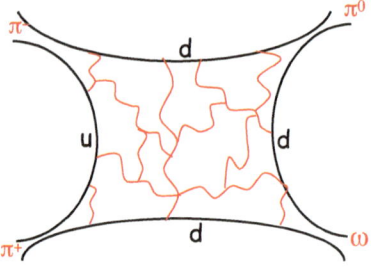

With the exception of the first one, these properties of what we now believe to be the correct theory of strong interactions explain why, starting from a bottom up approach, we landed on a string theory of hadrons albeit not on the right one! Strings are there in QCD, and possibly represent the best description of its large-distance confining dynamics, but its precise formulation (even in the large-N_c limit) is still missing.

9.6 Turning a Defeat into a Victory?

Around 1974 most people working in string theory turned their attention to the newly constructed Standard Model (of which QCD is a basic component). An important proposal by Scherk and Schwarz [15] went almost unnoticed for a full decade. In retrospect, perhaps too daring for the time, it was as follows.

Upon a rescaling of the string tension by some twenty orders of magnitude string theory could be perhaps reinterpreted as a candidate theory of *all truly elementary* particles: not of hadrons, but of their constituents (quarks and gluons), as well as of leptons, gauge bosons, and all the way including the graviton.

Under this reinterpretation the shortcomings of the hadronic string became advantages:

- Massless particles of spin $J = 1, 2$ are very needed for gauge interactions and gravity.
- Softness cures the long-standing problem of QFT's UV divergences, making quantum string gravity well defined (at least perturbatively).
- Extra dimensions, if compact, can be used to generate new gauge interactions through (a stringy version of) the Kaluza-Klein idea.

The combination of these properties could possibly provide a *finite quantum* theory of all interactions, including gravity.

It took however till 1984 before a breakthrough paper by Green and Schwarz [16] made it possible for people to take seriously such a dream. Their paper showed how to eliminate (almost miraculously) the only remaining inconsistency, a quantum gauge/gravitational anomaly (a well known constraint in more conventional quantum field theories such as the standard model) upon severely restricting the underlying gauge symmetry.

Overnight many theorists went back to (if old enough) or jumped in (if young enough) the new adventure and many new results quickly followed. For lack of space I will mention just three of them.

9.6.1 Stringy Symmetries

The stringy version of Kaluza-Klein theory leads to new kinds of symmetries, known as T-dualities: large and small compactification radii (with respect to $\sqrt{\alpha'}$) are equivalent for closed strings (upon the swapping of momentum and winding modes for closed strings) implying a minimal compactification radius $R_c \sim \sqrt{\alpha'}$.

At that minimal (self-dual) radius, compactification gives *non-abelian* gauge interactions with R_c playing the role of the Higgs field. A cosmological variant of T-duality is also at the basis of new (big bounce) cosmologies [17].

9.6.2 The D-brane revolution

T-duality looked only possible for closed strings since open strings can carry momentum but, apparently, no winding. That looked suspicious to J. Polchinski, who found a way out of the puzzle in 1994 [18]. It went as follows.

T-duality is deeply rooted in the canonical transformation [19] $P \leftrightarrow X'$ (the latter being related to winding). For open strings such a transformation relates open strings with Neumann boundary conditions (i.e. with free ends as one had assumed to be the case till then) to open strings with Dirichlet boundary conditions (i.e. with fixed ends). The latter were called D-strings. Note that different boundary conditions can be specified in different spatial coordinates.

While Neumann open strings (N-strings) carry momentum but no winding, D-strings carry winding but no momentum. T-duality then simply connects N- to D-strings. Instead, as we have already discussed, it relates closed strings to themselves (as they move/wind in apparently different but equivalent compact spaces).

D-branes is the name given to sub-manifolds of the full (typically 9-dimensional) space on which the ends of D-strings are, by definition, stuck. Their dimensionality, p, is thus related to the number of Neumann directions along which those end can freely move. One thus talks about D_p-branes.

The brane revolution led to many important results e.g.

- The first example, by Strominger and Vafa [20] of black-holes whose Bekenstein-Hawking entropy can be given a statistical mechanics interpretation by counting their micro-states.

- Apparently unrelated string theories are actually connected to each other through a web of dualities so that, eventually, they all appear to descend from different limits of a common ancestor, a mysterious M-theory in eleven dimensions [21], with the finite size of the 11th dimension playing the role of the string coupling (the string analog of the fine structure constant).
- The most recent (and amazing) use of D-branes came however in 1997.

9.6.3   Gauge-Gravity Duality

A stack of N coincident D_3-branes has an associated $U(N)$ gauge theory living on their four-dimensional (in general $(p + 1)$-dimensional) space-time. One can then take the large-N limit, keeping $\lambda = g^2 N$ fixed (cf. the already mentioned 't-Hooft limit in QCD).

In the ambient ten-dimensional space-time the branes (whose energy density is known) generate a geometry which approaches asymptotically five-dimensional Anti-de Sitter space time (AdS_5) times a five-dimensional sphere (S_5) with AdS and sphere radii both fixed (in string units) in terms of λ.

In 1997 Maldacena [22] conjectured an equivalence (made precise soon after by E. Witten) between a maximally supersymmetric gauge theory in four-dimensions (the boundary of AdS_5) and a ten-dimensional supergravity theory in $AdS_5 \otimes S_5$. The large-λ limit of the gauge theory gets related to the large-AdS radius limit of the gravity theory. Difficult non-perturbative phenomena on the gauge-theory side get thus mapped into an "easy" small-curvature regime on the gravity side.

Example: a lower bound on the ratio of shear viscosity and entropy density ($\frac{\eta}{s} > \frac{1}{4\pi}$) was predicted and is apparently nearly saturated by the quark-gluon plasma produced at Brookhaven and LHC. There is by now overwhelming evidence for the validity of Maldacena's conjecture.

9.7   Back to Square One?

Maldacena's conjecture has been generalized to other gauge-gravity pairs. Attempts have been made, with some success, to extend the correspondence to less supersymmetric theories and even to (large-N_c) QCD.

We seem to be back to the problem we mentioned earlier: Can we find out, at least in 't Hooft's limit, how to describe the true string lurking behind the hadronic world? Perhaps a simple gravity problem can shed light on a hard gauge theory problem …

That would close a 50-years-old circle!

Acknowledgements I wish to thank Luisa Bonolis, Luciano Maiani and Lia Pancheri for having organized such an interesting event, and for the invitation. I wish to dedicate this talk to the memory of two collaborators and friends who gave crucial contributions to the topics I have discussed: Sergio Fubini and Miguel Virasoro.

References

1. T. Regge, Nuovo Cim. **14**, 951 (1959)
2. V.N. Gribov, Nucl. Phys. **22**, 249 (1961); M. Froissart, Phys. Rev. **123**, 1053 (1961); G.F. Chew, S.C. Frautschi, Phys. Rev. Lett. **8**, 41 (1962); G.F. Chew, S.C. Frautschi, S. Mandelstam, Phys. Rev. **126**, 1202 (1961)
3. G.F. Chew, Nuclear democracy and bootstrap dynamics. University of California Radiation Laboratory preprint, UCRL-11163,12 (1963)
4. R. Dolen, D. Horn, C. Schmid, Phys. Rev. Lett. **19**, 402 (1967); Phys. Rev. **166**, 1768 (1968)
5. H. Harari, Phys. Rev. Lett. **22**, 562 (1969); J.L. Rosner, Phys. Rev. Lett. **22**, 689 (1969)
6. G. Veneziano, Nuovo Cim. A **57**, 190 (1968)
7. R. Odorico, Nucl. Phys. B **37**, 509 (1972)
8. C. Lovelace, Phys. Lett. **28B**, 264 (1968); J.A. Shapiro, Phys. Rev. **179**, 1345 (1969)
9. S. Fubini, D. Gordon, G. Veneziano, Phys. Lett. B **29**, 679 (1969)
10. M.A. Virasoro, Phys. Rev. D **1**, 2933 (1970)
11. S. Fubini, G. Veneziano, Ann. Phys. **63**, 12 (1971); J.H. Weis, private communication to the authors
12. Y. Nambu, Lectures at the Copenhagen Symposium (1970) (unpublished); T. Goto, Progr. Theor. Phys. **46**,1560 (1971)
13. P. Goddard, J. Goldstone, C. Rebbi, C.B. Thorn, Nucl. Phys. B **56**, 109 (1973)
14. G. 't Hooft, Nucl. Phys. B **72**, 461 (1974); G. Veneziano, Nucl. Phys. B **117**, 519 (1976)
15. J. Scherk, J.H. Schwarz, Nucl. Phys. B **81**, 118 (1974)
16. M.B. Green, J.H. Schwarz, Phys. Lett. **149B**, 117 (1984)
17. M. Gasperini, G. Veneziano, Astropart. Phys. **1**, 317 (1993)
18. J. Polchinski, Phys. Rev. Lett. **75**, 4724 (1995)
19. G. Veneziano, Europhys. Lett. **2**, 199 (1986)
20. A. Strominger, C. Vafa, Phys. Lett. B **379**, 99 (1996)
21. E. Witten, Nucl. Phys. B **443**, 85 (1995)
22. J.M. Maldacena, Adv. Theor. Math. Phys. **2**, 231 (1998)

Chapter 10
QCD and Supercomputers

Guido Martinelli

Abstract The title of this talk should rather have been Lattice QCD and Supercomputers. I will introduce Lattice QCD as the fundamental tool to predict (postdict) the hadron spectrum and most of the matrix elements relevant for hadronic physics in the non-perturbative regime. Lattice calculations are used to study the dynamics of QCD at large temperature or chemical potential, the anomalous magnetic moment of the muon, $g-2$, the nucleon structure functions, the meson scattering amplitudes at low and intermediate energies and, last but not least, the weak matrix elements relevant in flavour physics and CP violation. In this presentation only some example particularly illustrative of the present sophistication and accuracy of lattice QCD calculations will be discussed in some detail.

10.1 Introduction

In the last 40 years numerical simulations of Lattice QCD (LQCD) allowed an unpreceded progress in understanding the non-perturbative dynamics of strong interactions. Precise calculations of the hadron spectrum and accurate predictions of hadronic matrix elements are now a reality and more and more quantities relevant to the phenomenology of the Standard Model (SM) and for searches of signals of new physics beyond the SM (BSM) will soon become available. This progress has been possible thanks to the development of very sophisticated theoretical tools coupled to an extraordinary increase of the computer power and memory. In this talk a brief description of the methods of LQCD and of the main achievements obtained in recent years will be presented. The plan of this review is the following: after a general introduction, the derivation of the hadron spectrum and of the simplest matrix elements will be presented, followed by the description of the calculation of more complicated amplitudes such as those entering semi-leptonic decays or neutral

G. Martinelli (✉)
Dipartimento di Fisica, Università di Roma La Sapienza, P.le A. Moro 2, I-00185, Roma, Italy
e-mail: guido.martinelli@roma1.infn.it

INFN, Sezione di Roma, P.le A. Moro 2, 00185 Roma, Italy

© The Author(s) 2023
L. Bonolis et al. (eds.), *Bruno Touschek 100 Years*,
Springer Proceedings in Physics 287,
https://doi.org/10.1007/978-3-031-23042-4_10

meson mixing and non leptonic decays. Given the precision of the present calculations, radiative corrections and isospin breaking effects become relevant and they
will also be discussed. A presentation of some *anomalies* in B decays which are
difficult to be explained within the SM will then be given. The review is closed by
an outlook on future developments.

10.2 Perturbative and Non-perturbative QCD

The QCD Lagrangian has indeed a very simple form

$$\mathcal{L} = -\frac{1}{2} Tr \left[G_{\mu\nu} G^{\mu\nu} \right] + \sum_f \bar{q}_f (\slashed{D} - m_f) q_f + \theta \, Tr \left[G_{\mu\nu} \tilde{G}^{\mu\nu} \right], \qquad (10.1)$$

where $G_{\mu\nu} = G^A_{\mu\nu} t^A$ is the gluon tensor, $\tilde{G}^{\mu\nu} = \epsilon^{\mu\nu\rho\sigma} G^A_{\rho\sigma} t^A$, q_f are the quark
fields and the last term represents the strong CP violating term which still waits for
a satisfactory explanation. This term is related to a very interesting phenomenology
but it will not be discussed in the following. Although the form of the Lagrangian
is very simple, it gives rise to a very rich and complicated dynamics. In particular,
because of asymptotic freedom [1, 2], the effective coupling decreases and quarks
and gluons behave as almost free particles at large energies, see Fig. 10.1 [3]. In the
high energy regime physical quantities can be computed in perturbation theory and
the main limitations come from the order at which the amplitudes are computed and
the accurary of the Montecarlos describing the hadronization processes for quark
and gluons.

At low energy in order to study the dynamics of QCD, like the hadron spectrum or
the weak matrix elements, a non-perturbative approach is necessary. Among all the
possible methods the one which resulted the most reliable, with systematic effects
that can be *systematically* reduced in time is QCD on the lattice, which consists in
constructing the theory on a space-time that is a finite cubic lattice of points, which
reduces to QCD when the mesh of the lattice is infinitely fine, that is the lattice
spacing $a \to 0$, and simultaneously the physical volume goes to infinity, namely
when the volume is much larger than the range of the interactions [4]. On a finite
lattice, to describe ordinary matter, QCD requires more than 104 numbers for each
lattice point, and this complexity explains why in LQCD it was necessary to reach
an enormous computer power in order to be able to make realistic calculations, with
small lattice spacings and physical volumes large enough. In the '80s we started with
computers with a power of one GigaFlops (1 GigaFlops $= 10^9$ operations/second) to
arrive to the actual power of 0.1–1.0 Exaflops (1 ExaFlops $= 10^{18}$ operations/second)
today ! A large part of this progress was due to the use of GPUs which were invented
for video games [5, 6].

On the lattice all the physical information can be extracted from the Green functions of the theory, schematically

Asymptotic freedom and infrared slavery

PDG Summary 2019

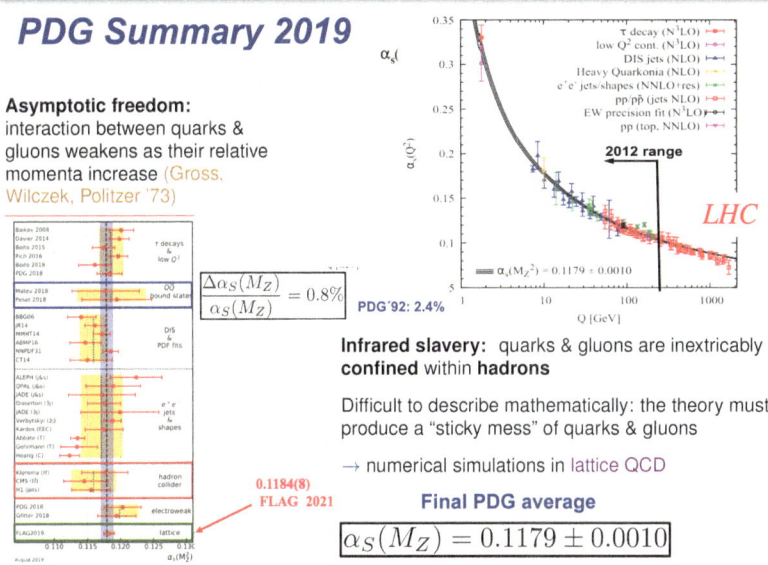

Asymptotic freedom: interaction between quarks & gluons weakens as their relative momenta increase (Gross, Wilczek, Politzer '73)

$$\frac{\Delta \alpha_S(M_Z)}{\alpha_S(M_Z)} = 0.8\%$$

PDG'92: 2.4%

Infrared slavery: quarks & gluons are inextricably **confined** within **hadrons**

Difficult to describe mathematically: the theory must produce a "sticky mess" of quarks & gluons

→ numerical simulations in lattice QCD

Final PDG average

$$\boxed{\alpha_S(M_Z) = 0.1179 \pm 0.0010}$$

Fig. 10.1 Values of the strong coupling constant $\alpha_s(M_Z^2) = g_s^2(M_Z^2)/(4\pi)$ at the scale of the mass of the Z^0, left hand side, and $\alpha_s(Q^2)$ as a function of the scale, right hand side, from a wide set of experiments ranging from τ decays to jets at collider energies. The figures have been taken from [3]

$$\langle 0|\phi(x_1)\phi(x_2)\dots\phi(x_N)|0\rangle = \frac{1}{Z}\int [d\phi]\,\phi(x_1)\phi(x_2)\dots\phi(x_N)\exp\left[i\,S(\phi)\right]\,, \tag{10.2}$$

where $Z = \int [d\phi]\exp\left[i\,S(\phi)\right]$ is a suitable normalisation factor and some regularisation must be introduced to make the expression in (10.2) finite. On a lattice with a finite number of lattice points (L^4) and a finite lattice spacing a, the functional integral in (10.2) becomes an integral on L^4 variables which can be performed using Important Sampling techniques, which require, however, the use of a mesh of points in an Euclidean space-time. Many of the present limitations in computing amplitudes with many particles in the final state derive from the unavoidable analytic continuation of the theory from the Minkowskian space-time to the Euclidean four-dimensional space.

Let us consider now the calculation the simplest possible Green-function, namely the two point function

$$G(t,\vec{q}) = \frac{1}{Z}\int d^3x\,\exp\left[i\vec{q}\cdot\vec{x}\right]\langle 0|\phi^\dagger(t,\vec{x})\phi(0,\vec{0})|0\rangle = \sum_n \langle 0|\phi^\dagger|n\rangle\langle n|\phi|0\rangle\frac{e^{-E_n t}}{2E_n}\,, \tag{10.3}$$

where now the funtional integrals are all performed after a Wick rotation in the Euclidean space-time. At large time distances only the state with the smallest energy, $|E_{min}\rangle$, will contribute to $G(t, \vec{q})$ and we may thus extract the energy of this state and the matrix element of the operator ϕ, $\langle E_{min}|\phi|0\rangle$

$$\lim_{t\to\infty} G(t, \vec{q}) = \langle 0|\phi^\dagger|E_{min}\rangle\langle E_{min}|\phi|0\rangle \frac{e^{-E_{min}t}}{2E_{min}}. \qquad (10.4)$$

If we consider the case $\vec{q} = 0$ and the four component of the axial current, $\phi = A_0 = \bar{u}\gamma_0\gamma_5$, as interpolating operator, for example, we can extract the mass of the pion, m_π, and its decay constant f_π, $\langle\pi|A_\mu|0\rangle = if_\pi p_\mu$. Indeed all the quantities are obtained in dimensionless units, namely in units of the lattice spacing, $M_\pi = m_\pi a$ and we have to fix the mass of a set of hadrons to determine the value of the lattice spacing in physical units (GeV^{-1} or fermi) and the physical masses of the quarks. For a recent discussion see [7] and references therein. By changing the lattice coupling, and by readjusting the lattice bare quark masses, we can make the dimensionless correlations lengths, $\xi_H = 1/M_H$, corresponding to the inverse dimensionless hadron masses, larger and larger thus converging to the continuum limit. In this limit the physical volume must remain large i.e. the number of lattice point must increase, thus requiring larger and larger computer resources. Only quite recently it became possible to work at light quark masses very close to the physical point with lattice volumes large enough to avoid finite volume effects, Fig. 10.2. The agreement of the most recent lattice calculations with the experimental hadron spectrum is impressive. The results of the

Physics Reach (Mainly Heavy Flavor Physics)
many slides from Lattice Conferences

- charm physics directly accessible for some time now
- fraction of available ensembles used for HQ physics still limited

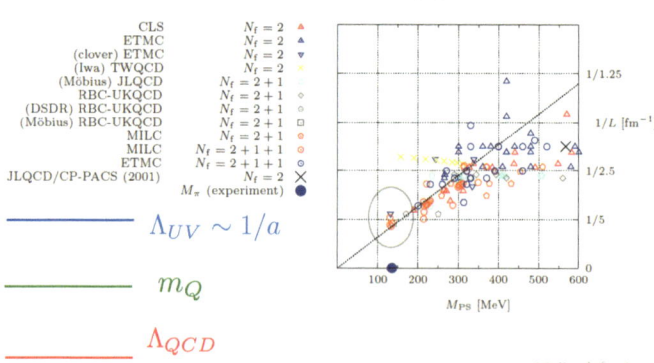

Fig. 10.2 Values of the lightest pseudo-scalar mass, corresponding to the *pion* in the limit of physical quark masses, in LQCD simulations versus the physical volume, starting from the year 2001. The figure is an updated version of the figure in [8] by G. Herdoiza

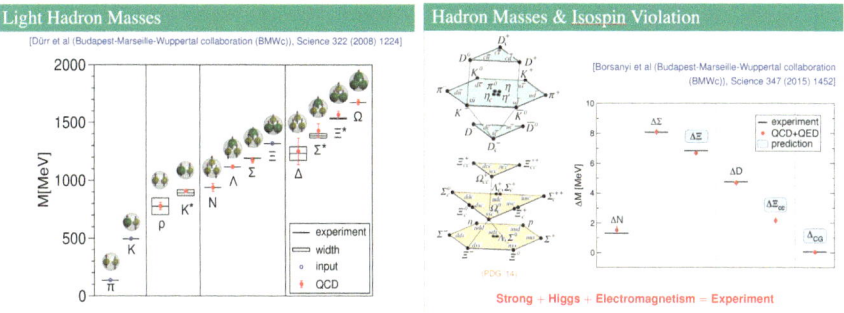

Fig. 10.3 Comparison of lattice calculations and experimental values for the masses of several hadrons with different spin and flavour quantum numbers. Courtesy of L. Lellouch, after [9, 10]

pioneering work of the BMW collaboration [9, 10] which first reached a sufficient accuracy by including isospin and electromagnetic corrections are given in Fig. 10.3.

In order to compute more complicate amplitudes, for example the matrix elements of the vector and axial vector currents entering weak hadronic decays, one generalizes the method used to compute the pseudo-scalar decay constants. Thus for example one can define suitable sources/sinks to create/annihilate pseudo-scalar mesons in analogy with the axial current mentioned above

$$B^\dagger(t_1, \vec{p}_B) = \sum_{\vec{x}} B^\dagger(t_1, \vec{x}) e^{-i \vec{p}_B \cdot \vec{x}}, \qquad \pi(t_2, \vec{p}_\pi) = \sum_{\vec{x}} \pi(t_2, \vec{x}) e^{+i \vec{p}_\pi \cdot \vec{x}}, \quad (10.5)$$

and study the 3-point function

$$\langle 0 | \pi(t_2, \vec{p}_\pi) J_\mu^{\text{weak}}(0) B^\dagger(t_1, \vec{p}_B) | 0 \rangle \rightarrow \left[\frac{\langle 0 | \pi | \pi(\vec{p}_\pi) \rangle \langle B(\vec{p}_B) | B^\dagger | 0 \rangle}{2 E_B 2 E_\pi} \right] \langle \pi(\vec{p}_\pi) | J_\mu^{\text{weak}} | B(\vec{p}_B) \rangle,$$
$$(10.6)$$

in the limit $t_1 \rightarrow -\infty$ and $t_2 \rightarrow +\infty$. The source/sink matrix elements and energies can be extracted from the two-point functions, thus we easily obtain the weak current matrix element $\langle \pi(\vec{p}_\pi) | J_\mu^{\text{weak}} | B(\vec{p}_B) \rangle$. A similar procedure can be used for the matrix elements of the four fermion operators of the weak Hamiltonian and also to study more complicated final states as for examples two-pion states below the inelastic threshold. More complicated 4-particles final states or two pions above the inelastic threshold cannot be studied yet because the theory for the analytical continuation from the Minkowski to the Euclidean space in a finite volume has not been developed yet for these cases. A quick summary of main weak amplitudes which are computed in LQCD and used for flavour phenomenology are shown in Figs. 10.4.

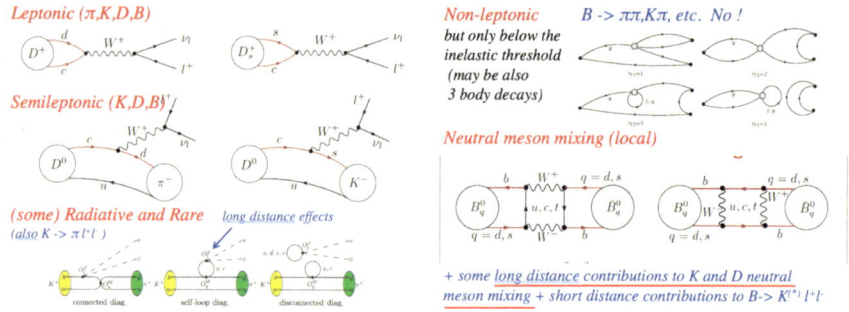

Fig. 10.4 A synthetic overview of the amplitudes which are most frequently computed for weak interaction phenomenology is displayed in these figures. Left panel: Leptonic, Semi-leptonic and Radiative decays (also for baryons and electromagnetic transitions not shown in the figure). Right Panel: Non-leptonic decays of Kaons, Neutral meson mixing of B_q mesons (also neutral D meson and Kaon mixing not shown in the figure). LQCD computed also some long distance contributions to K and D neutral meson mixing and short distance contributions to $B \to K^{(*)} \ell^+ \ell^-$ decays, not shown in the figure

10.3   Lattice QCD and Flavour Physics

It would be very interesting to describe all the possible quantities that have been computed in LQCD so far, from QCD at finite temperature to structure functions, from two nucleon states to $g - 2$. For lack of time I will restrict in the following to a few selected examples taken from weak interactions.

Our starting point is the CKM matrix [11, 12] in the SM (first term on the left-hand-side):

$$\mathbf{V}_{\text{CKM}} = \begin{pmatrix} V_{ud} & V_{us} & V_{ub} \\ V_{cd} & V_{cs} & V_{cb} \\ V_{td} & V_{ts} & V_{tb} \end{pmatrix}, \quad \mathbf{V}^{\mathbf{W}}_{\text{CKM}} = \begin{pmatrix} 1 - 1/2\lambda^2 & \lambda & A\lambda^3(\rho - i\eta) \\ -\lambda & 1 - 1/2\lambda^2 & A\lambda^2 \\ A\lambda^3(1 - \rho - i\eta) & -A\lambda^2 & 1 \end{pmatrix} + O(\lambda^4), \lambda = \sin\theta_c ,$$

(10.7)

where θ_c is the Cabibbo angle. The absolute values of the matrix elements, $|V_{ij}|$, are mostly determined by studying leptonic and semi-leptonic decays whereas only one independent phase, related to η, controls CP violating effects, for example in $K^0 \to \pi\pi$ or $B \to D^{(*)} K^{(*)}$ decays. From the observation that the CKM matrix $\mathbf{V}_{\text{CKM}}$ is very close to the identity, Wolfenstein [13] suggested to expand it in powers of the sine of the Cabibbo angle. That defines the parameters ρ and η ($\bar{\rho} = \rho(1 - 1/2\lambda^2)$ and $\bar{\eta} = \eta(1 - 1/2\lambda^2)$) which will be used in the following (second term on the right-hand-side of (10.7)). Other important quantities are the unitarity triangles that can be defined using the unitarity of the CKM matrix. From the phenomenological point of view, the most renown of these triangles, because its sites correspond to well measurable quantities, is the one defined from the product of the first and third columns of the CKM matrix

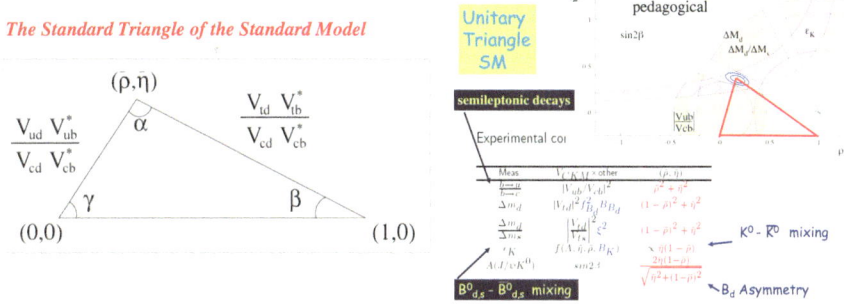

Fig. 10.5 Left panel: the *Standard* triangle of the Standard Model; Right panel: pedagogical representation of the Unitarity Triangle, normalised to $V_{cd}V_{cb}^*$, in the $\bar{\rho} - \bar{\eta}$ plane. Each measurement corresponds to a curve and in the SM all the curves should meet in a point corresponding to one of the vertices of the unitarity triangle. The curves become bands, due to the experimental and theoretical uncertainties, in the $\bar{\rho} - \bar{\eta}$ plane. The overlap of the different bands is the allowed SM region in this plane

$$V_{ud}V_{ub}^* + V_{cd}V_{cb}^* + V_{td}V_{tb}^* \,. \tag{10.8}$$

The position of the vertex of the triangle in the $\bar{\rho} - \bar{\eta}$ plane can be determined by combining the measurements of several processes, e.g. semi-leptonic decays of heavy mesons, B^0-$\bar{B}^0$ and K^0-$\bar{K}^0$ mixing, the asymmetry in $B_d \rightarrow J/\psi K^0$ decays and many others [14], see Fig. 10.5.

In order to compare experimental measurements and theoretical predictions we need the hadronic matrix elements of the weak currents or of the local operators of the Fermi-like weak Hamiltonian, schematically

$$Q^{\mathrm{EXP}} = V_{CKM}\langle F|\hat{O}|I\rangle \,. \tag{10.9}$$

From the measurement Q^{EXP} and the matrix element $\langle F|\hat{O}|I\rangle$ computed in LQCD simulations we can determine a given combination of CKM matrix elements denoted here as V_{CKM}. The high quality (small uncertainties) of the lattice calculations of the hadronic matrix elements $\langle F|\hat{O}|I\rangle$ is illustrated by the examples given in Table 10.1.

Beyond the SM, (10.9) generalizes into

$$Q^{\mathrm{EXP}} = \sum_i C^i_{SM}(M_W, m_t, \alpha_s, V_{CKM})\langle F|\hat{O}_i|I\rangle + \sum_{i'} C^{i'}_{BSM}(\tilde{m}_\beta, \alpha_s)\langle F|\hat{O}_{i'}|I\rangle \,. \tag{10.10}$$

New physics effects can modify the value of the Wilson coefficients $C^i_{SM}(M_W, m_t, \alpha_s, V_{CKM})$ of the operators already present in the SM or generate the contributions of new operators $O_{i'}$ which do not appear in the SM. In this game the calculation of the hadronic matrix elements from lattice QCD is essential and no other non-perturbative approach (QCD-sum rules, chiral Lagrangians etc.) can compete with LQCD.

Table 10.1 Full lattice inputs. The values of the different quantities have been obtained by mediating the $N_f = 2 + 1$ and $N_f = 2 + 1 + 1$ FLAG numbers [15]

Input	Lattice	Input	Lattice
$\hat{B}_K$	0.756(16)	$f^{K\pi}(0)$	0.9698(17)
f_K/f_π	1.1932(19)	f_{B_s}	230.1(1.2) MeV
f_{B_s}/f_B	1.208(5)	$\hat{B}_{B_s}$	1.284(59)
$\hat{B}_{B_s}/\hat{B}_B$	1.015(21)	$m_{ud}^{\overline{MS}}(2\,\text{GeV})$	3.394(29) MeV
$m_s^{\overline{MS}}(2\,\text{GeV})$	93.11(52) MeV	$m_c^{\overline{MS}}(3\,\text{GeV})$	991(5) MeV
$m_c^{\overline{MS}}(m_c^{\overline{MS}})$	1290(7) MeV	$m_b^{\overline{MS}}(m_b^{\overline{MS}})$	4196(14) MeV

Fig. 10.6 $\bar{\rho} - \bar{\eta}$ plane with the SM global fit results in various configurations. The black contours display the 68% and 95% probability regions selected by the given global fit

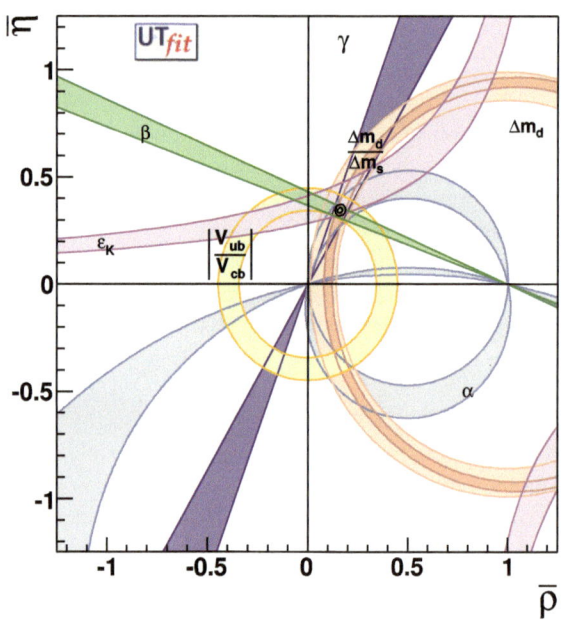

In the SM, in the absence of theoretical and experimental uncertainties, all the curves, corresponding to different physical processes, should meet in a single point of the $\bar{\rho} - \bar{\eta}$ plane. With the uncertainties, instead, the curves become bands which should overlap in the same region, Fig. 10.5. The results of the latest UTfit analysis [14, 16] are given in Fig. 10.6. We observe the impressive agreement between a very large set of experimental measurents and the SM expectations. Possible new physics effects, if present, must be rather tiny and models of physics BSM must cope with these results.

Although the overall picture shows a very good agreement of the experimental measurements with the SM predictions, there remain a few quantities, called anomalies or tensions, for which important differences have been observed between the theoretical expectations and the data. Some of them have been related to a possible failure of the Lepton Flavour Universality (LFU) with respect to weak interactions. The most difficult to explain are the ratios of the branching fractions of the rare-decays $R_{K^{(*)}} = BR(B \to K^{(*)}\mu^+\mu^-)/BR(B \to K^{(*)}e^+e^-)$ which differ from the expected value of about one by at least 2.6 σ [17–19]. For these processes the lattice is not yet in the position to make reliable predictions and I will not discuss them in the following. Other tensions are observed in the difference in the value of $|V_{cb}|$ as determined from exclusive [20–27] or inclusive [28–30] B meson semi-leptonic decays and in the ratios $R_{D^{(*)}} = BR(B \to D^{(*)}\tau\nu_\tau)/BR(B \to D^{(*)}\ell\nu_\ell)$, where ℓ is one of the light leptons (μ or e) [34–43]. Finally we may consider the unitarity test

$$|V_{ud}|^2 + |V_{us}|^2 + |V_{ub}|^2 = 1 . \tag{10.11}$$

$|V_{ud}|^2$ accounts for 95% of this sum and for this reason a precise determination of this CKM matrix element from β decays is of fundamental importance. Moreover, since the contribution from $|V_{ub}|^2$ is very small, it is also very important, besides $|V_{ud}|^2$, an accurate determination of $|V_{us}|^2$ using LQCD. It turns out that there are strong hints that the currently accepted data for $|V_{ud}|$ and $|V_{us}|$ fall short of unitarity by 2σ or even more, although a definite conclusion is still out of reach. One of the missing ingredients is a better control of the radiative electromagnetic corrections in β decays. A accurate calculation of these corrections in LQCD is for this reason of the utmost importance. For both $B \to D^{(*)}$ semi-leptonic decays and the radiative corrections in weak decays the lattice will certainly play the role of protagonist in the near future and, for this reason, I will discuss these two cases in the following.

10.4   The Inclusive-Exclusive V_{cb} Saga

Semi-leptonic B decays are very challenging processes from a phenomenological point of view because of two issues. The first one is the so-called $|V_{cb}|$ puzzle, namely the observation of a tension between the exclusive [20–27] and the inclusive [28–30] determination of $|V_{cb}|$ at the level of 3.3 standard deviations. The second one is the discrepancy between the Standard Model predictions and experiments in the determinations of the $\tau/(\mu, e)$ ratios of the branching fractions, the so called $R(D^{(*)})$ anomalies, which represent an important test of Lepton Flavour Universality (LFU). Some important novelties have, however, recently changed the previous situation: on the one hand the inclusive predictions were recently reconsidered and the uncertainties of the calculation performed in the Heavy Quark Effective Theory were reevaluated [31, 32]. On the other hand, new lattice calculations of the relevant form

Table 10.2 Values of $|V_{cb}|$ from inclusive or exclusive determinations. [a] This value of $|V_{cb}| \times 10^3$ has been derived using the value of the form factor at zero recoil given in (267) of [15]. DM in the rows 2-4-5-11 denotes the values obtained by using the Dispersive Matrix approach mentioned in the text

	Process	Reference	$|V_{cb}| \cdot 10^3$
1	$b \to c$ inclusive	[31]	42.16 (50)
2	$B \to D$	[46] DM	41.0 (1.2)
3	$B \to D \; N_f = 2 + 1$	[15]	40.0 (1.0)
4	$B_s \to D_s \; N_f = 2 + 1$	[47] DM	41.7 (1.9)
5	$B \to D^*$	[48] DM	41.3 (1.7)
6	$B \to D^*$	[49]	39.6 (1.1)
7	$B \to D^*$	[50]	39.6 (1.1)
8	$B \to D^* \; N_f = 2 + 1$	[15]	38.9 (0.9)
9	$B \to D^*$ $N_f = 2 + 1 + 1$	[15]	39.9 (1.4)[a]
10	$B \to D^*$ and $B \to D$ $N_f = 2 + 1$	[15]	39.4 (0.7)
11	$B_s \to D_s^* \; N_f = 2 + 1$	[47] DM	40.7 (2.4)

factors in the small recoil region [33], new approaches to their determination in the full kinematical range [44–48] and new measurements of the exclusive differential decay rates appeared. We think that it is possible to argue that for $|V_{cb}|$, although some difference remains, the tension is finally resolved, see the recent average from [47] given in (10.15) below. In the case of the value of $|V_{ub}|$ a difference between the inclusive and exclusive determinations at the 1.7 σ still persists, although with large relative errors. A set of values from different estimates of $|V_{cb}|$ from inclusive and exclusive decays are given in Table 10.2. Note that all the form factors relevant to determine $|V_{cb}|$ from exclusive decays have been computed in lattice QCD.

10.4.1 The Classical Determination of $|V_{cb}|$

For $B \to D^*$ semi-leptonic decays, one averages the form factor $F(1)$ obtained from $N_f = 2 + 1$, $F(1) = 0.906(13)$, and $N_f = 2 + 1 + 1$, $F(1) = 0.895(10)(24)$ [15], obtaining $F(1) = 0.904(11)$; then, using the formula derived from the rate, $F(1) \eta_{EW} |V_{cb}| = 35.44(64)10^{-3}$, and $\eta_{EW} = 1.00662$ one gets $|V_{cb}| = 38.95(86)$ 10^{-3}; for $B \to D$, following [15], the result is $|V_{cb}| = 40.0(1.0)10^{-3}$.

Averaging the above values of $|V_{cb}|$ from $B \to D^*$ and $B \to D$ one obtains

$$|V_{cb}| \cdot 10^3 \; \text{(excl.)} = 39.44(65) \, . \tag{10.12}$$

This procedure uses all the available information from $B \to D^*$ but neglects the correlation of the lattice determination of form factors for $B \to D^*$ and $B \to D$ decays obtained using the same gauge field configurations. Taking into account the correlation of the lattice determination of form factors for these decays, the final result is

$$|V_{cb}| \cdot 10^3 \ (\text{excl.}) = 39.44(63) \,, \tag{10.13}$$

which differs by 3.3 σ from the inclusive value in Table 10.2. We may combine the inclusive value of $|V_{cb}|$ in Table 10.2 with the result in (10.13) obtaining

$$|V_{cb}| \cdot 10^3 \ = 41.1(1.3) \ (\text{incl.} + \text{excl.}) \,. \tag{10.14}$$

10.4.2   The Dispersive Matrix Determination

An alternative determination of the exclusive value of $|V_{cb}|$ can be obtained by using the values obtained using the Dispersive Matrix (DM) approach of [44] and given in Table 10.2, [46–48]. Besides the use of the DM approach, in the analysis of $B \to D^{(*)}$ decays it was essential a critical reappraisal of the experimental differential distributions and correlations among data and of the difference between the slope of $d\Gamma/dq^2$, where q^2 is the momentum transfer, between the lattice calculations and the experimental data. By combining these results, which include $B_s \to D_s^{(*)}$ decays, the exclusive value was

$$|V_{cb}| \cdot 10^3 \ (\text{DM excl.}) = 41.2(8) \,, \tag{10.15}$$

namely a value much closer and compatible at the 1 σ level with the inclusive one, with an uncertainty comparable to the uncertainty quoted in (10.13) . By combining the inclusive value of Table 10.2 with the DM result in (10.15) one obtains the (more precise) result

$$|V_{cb}| \cdot 10^3 \ = 41.9(4)(\text{incl.} + \text{excl.} - \text{DM}) \,. \tag{10.16}$$

10.4.2.1 $|V_{ub}|$

The matrix element V_{ub} is determined from the measurements of the branching ratios of leptonic $B \to \tau \nu_\tau$ decays, using the lattice determination of the B meson decay constant f_B, and from *exclusive* and *inclusive* semi-leptonic $b \to u$ decays. Its precision is limited by the uncertainty of the theoretical calculations of the B meson decay constant and of the relevant form factors, for leptonic and exclusive semi-leptonic decays, and of the matrix elements of the operators appearing in the HQET expansion of the inclusive rate. For $B \to \tau \nu_\tau$, which is very interesting because it is particularly sensitive to physics beyond the SM, a further source of large uncertainty comes for the large error in the experimental measurement of the rate. Although the determinations from inclusive semi-leptonic decays are systematically higher than the exclusive ones, the two values are compatible, once the spread of the inclusive determinations using different theoretical models is considered.

For the exclusive semi-leptonic decays we take the lattice number of Table 57 of [15], quoted in b). Finally, for inclusive semi-leptonic decays we use the value b) in (10.17) from the same reference. We give the average of a) and b) in c).

$$V_{ub}^{B \to \pi} = 3.74(17) \cdot 10^{-3} \quad a)$$
$$V_{ub}^{\text{incl.}} = 4.32(29) \cdot 10^{-3} \quad b)$$
$$V_{ub}^{a+b} = 3.89(25) \cdot 10^{-3} \quad c) \qquad (10.17)$$

A percent precision is expected to be reached by LQCD using Exaflops CPUs for f_B and for the form factors entering the exclusive determination of $|V_{ub}|$. A higher precision will require the non-perturbative calculation of the radiative corrections to the decay rates [7]. The progress of lattice calculations allow us to use in the analysis also the constraint coming from the ratio $|V_{ub}|/|V_{cb}|$ determined either from $\Lambda_b \to (p, \Lambda_c)\mu^- \bar{\nu}_\mu$ or $B_s \to (K^-, D_s^-)\mu^+ \nu_\mu$ decays. We use only the latter decays since the lattice form factors relevant in Λ_b decays do not satisfy the quality criteria of FLAG [15]. Following [15] we quote

$$\frac{|V_{ub}|}{|V_{cb}|} = 0.0844(56) . \qquad (10.18)$$

We note here that the DM method for $|V_{ub}|$ [51] gives, within the errors, substantially the same result which have been reported in this subsection, namely $|V_{ub}^{\text{excl}}| = 3.69(34)$.

10.4.3 $|V_{ub}|$, $|V_{cb}|$ and UTfit

The above numbers can be compared with the results of the Global UTfit analysis or with the values obtained by predicting the values of V_{cb} and V_{ub} by making the UTfit analysis without including at all semi-leptonic decays, denoted as UTfit Prediction [16]:

Global SM Fit	UTfit Prediction
$|V_{cb}| \times 10^3 = 42.4(0.4)$	$|V_{cb}| \times 10^3 = 42.6(0.5)$
$|V_{ub}| \times 10^3 = 3.72(0.09)$	$|V_{ub}| \times 10^3 = 3.70(0.10) .$

$$(10.19)$$

The situation is illustrated in Fig. 10.7: on the left panel the input values of the classical determinations of $|V_{cb}|$ are shown together with $|V_{ub}|$, and the value of $|V_{ub}|/|V_{cb}|$ from $B_s \to K$ semi-leptonic decays. The ratio $|V_{ub}|/|V_{cb}|$ from Λ_b

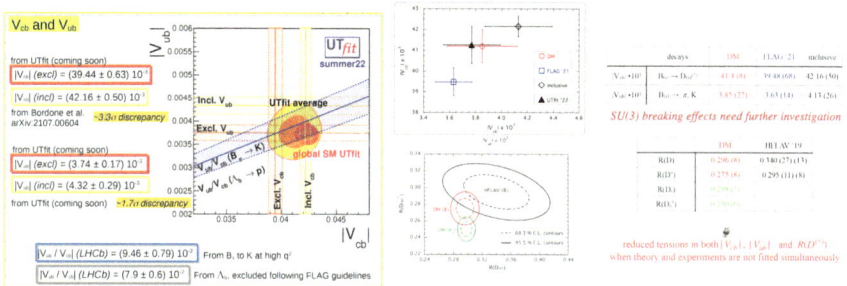

Fig. 10.7 Comparison of experimental results and lattice predictions for $|V_{cb}|$-$|V_{ub}|$ (left panel) and $R(D)$-$R(D^*)$ (right-lower panel). In the case of $|V_{cb}|$-$|V_{ub}|$ the results of the global UTfit analysis is also displayed (right-upper panel). The figures have been taken from [16]

decays has not been used because the lattice results do not satisfy the FLAG quality requests [15]. In this panel also the results of the global UTfit analysis are given, showing that the inclusive value of $|V_{cb}|$ and the exclusive value of $|V_{ub}|$ are preferred. On the right panel we compare the results in the $|V_{cb}|$-$|V_{ub}|$ plane with the values of the standard determination of these CKM matrix elements, the values obtained with the DM approach [44, 46, 48] and UTfit. The agreement between the DM results and the global UTfit analysis is remarkable. On the right panel we also compare the predictions for $R(D)$ and $R(D^*)$ using the calssical approach and the DM results. In the latter case the tension between experimental results and the theoretical predictions is strongly reduced.

10.5   Radiative Corrections to Weak Decays

The precision in determining hadron masses and weak amplitudes is such that it is no more possible to ignore isospin breaking effects or radiative corrections, simply denoted in the following as isospin corrections. From Table 10.1 we see that the precision in the determination of f_K/f_π is 0.16% and on the semi-leptonic form factor for $K \to \pi$ decays is about 0.18%, in both cases smaller than the size of the isospin corrections which are expected of the order of 1.0%. For this reason the Rome-Southampton group developed a strategy to include isospin corrections in the amplitudes/rates relevant to weak decays [7, 52–57]. The recent progress in this field gave also the possibility of studying decays of light or heavy hadrons accompanied by the emission of a real photon or a virtual photon of mass q^2 that then materialises as a lepton pairs in the final state [55–57]. In this section we discuss the present status of the calculation of radiative corrections to weak decays in LQCD.

Let us start from the inclusive decay rate of a pseudo-scalar photon. The formula which includes isopin breaking and radiative corrections can be written as

$$\Gamma(P^+ \rightarrow \ell^+ \nu_\ell + (\gamma)) = \frac{G_F^2}{8\pi} |V_{q_1 q_2}|^2 m_\ell^2 m_P \left(1 - \frac{m_\ell^2}{m_P^2}\right)^2 f_P^2 S_{ew} \left(1 + \delta R_{IB}^P + \delta R_{QED}^P\right),$$

$$(10.20)$$

where f_P is the leptonic decay constant in isoQCD ($m_u = m_d, e_f = 0$); R_{IB}^P are the strong isospin breaking corrections $\propto (m_u - m_d)/\Lambda_{QCD} \sim O(1\%)$; R_{QED}^P are the QED corrections $\propto \alpha_{em} \sim O(1\%)$. Note that, at order α_{em}, the separation between R_{IB}^P and R_{QED}^P is artificial and depends on the convention. In the calculation of Γ the problem is the appearence, in the intermediate steps of the calculation, of infrared divergences in the zero-photon, Γ_0, or one-photon, Γ_1, emission rates at $O(\alpha_{em})$. The combination of the two, however, is infrared finite and a strategy to regularise the infrared divergences in the intermediate steps has been developed by the Rome-Southampton Collaboration. The result has been used to compute the correction to the ratio $\Gamma(K^- \rightarrow \mu^- \bar{\nu}_\mu + (\gamma))/\Gamma(\pi^- \rightarrow \mu^- \bar{\nu}_\mu + (\gamma))$ and extract the most precise value of $|V_{us}|/|V_{ud}|$, namely

$$\frac{|V_{us}|}{|V_{ud}|} = 0.23134(24)_{exp}(30)_{th} = 0.23134(38), \qquad (10.21)$$

which, using $|V_{ud}|$ from nuclear decays, corresponds to $|V_{us}| = 0.2254(4)$. With the latest values and reevaluation of the uncertainties due to radiative corrections to β-decays [58–62], i.e. to $|V_{ud}| = 0.97373(31)$, the most updated number is now $|V_{us}| = 0.2251(8)$.

We have seen that the decays of charged pseudo-scalar mesons into light leptons, $P \rightarrow \ell \nu_\ell \gamma$ represent an important contribution to flavour physics since they give access to the CKM matrix elements [11, 12]. At tree level, i.e. without a photon in the final state, these decays are helicity suppressed in the SM due to the V - A structure of the leptonic weak charged current, while the helicity suppression can be overcome by the radiated photons. Therefore, radiative leptonic decays may provide sensitive probes of possible SM extensions inducing non-standard currents and/or non-universal corrections to the lepton couplings. Radiative leptonic decays also provide a powerful tool with which to investigate the internal structure of the decaying meson. In addition to the leptonic decay constant f_P, there are indeed two other structure-dependent (SD) amplitudes describing the emission of real photons from hadronic states, usually parametrized in terms of the vector and axial-vector form factors, FV and FA respectively. Thus, a first-principle calculation of radiative leptonic decays requires a non-perturbative accuracy, which can be provided by numerical QCD+QED simulations on the lattice.

In [56] a comparison between the theoretical predictions based on the non-perturbative determination of the SD form factors FV and FA and the experimental data available on the leptonic radiative decay $K \rightarrow e\nu_e \gamma$ from the KLOE Collaboration [63], on the decay $K \rightarrow \mu \nu_\mu \gamma$ from E787 [64], ISTRA+ [65] and OKA [66] collaborations and on the decay $\pi^+ \rightarrow e^+ \nu_e \gamma$ from the PIBETA Collaboration [67] was presented. An example of the comparison of the lattice prediction with the KLOE measurements is given in Fig. 10.8. There is good consistency between the theoretical predictions and the experimental results from the KLOE experiment on $K \rightarrow e\nu_e \gamma$

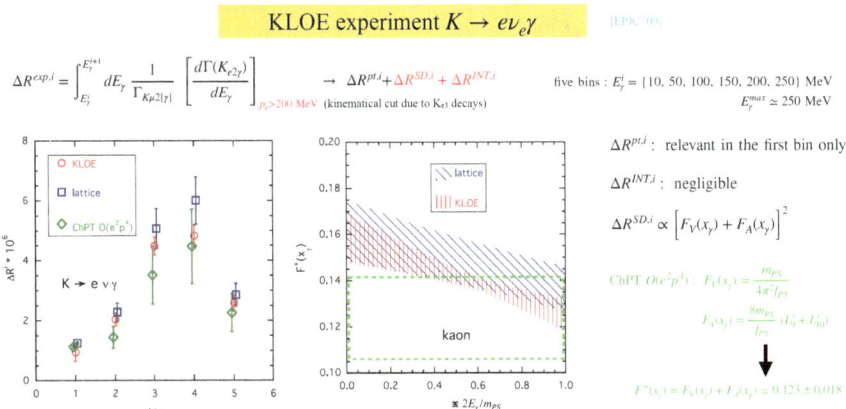

Fig. 10.8 Left panel: comparison of the KLOE experimental data $\Delta R^{exp,i}$ [63] (red circles) with the theoretical predictions $\Delta R^{th,i}$ (blue squares) evaluated with the vector and axial form factors of [56]. The green diamonds correspond to the prediction of ChPT at order $O(e^2 p^4)$. Right panel: Comparison of the form-factor $F^+(x_\gamma)$ extracted by the KLOE collaboration in [63] and the theoretical prediction from lattice QCD [56]. The shaded areas represent uncertainties at the level of 1 standard deviation

decays [63], but a discrepancy at the level of about 2 standard deviations for the data at large x_γ from the E787 experiment on $K \to \mu\nu_\mu\gamma$ decays. $x_\gamma = 2P \cdot /m_P^2$ where P is the four-momentum of the decaying meson with mass m_P and k is the four-momentum of the photon. Indeed the results from the two experiments do not agree. There are differences of up to 3–4 standard deviations at large photon energies in the comparison of the predictions with the E787, ISTRA+ and OKA data on radiative kaon decays as well as for some kinematical regions of the PIBETA experiment on the radiative pion decay. These conclusions call for improvements in the determination of the structure-dependent form factors $F^+(x_\gamma)$ and $F^-(x_\gamma)$ from both experiment and theory (for a definition of the different form factors see [56]).

The study of radiative decays with both real and virtual photons open the road to predict, and compare with experiments, many rare-decay rates, with the possibility of putting interesting bounds on physics BSM, for example in $B \to \mu^+\mu^-\gamma$ decays. It also give us the possibility of computing the radiative corrections to the neutron β decay, the Holy Grail of these kind of calculations.

10.6 Conclusions

Thanks to the impressive development of computer resources and to the progress in the theoretical methods, Lattice QCD is now in the position of providing very accurate and reliable predictions for a large variety of hadronic quantities and of giving the possibility to detect even smallish effects of physics beyond the Standard Model. We have described in more detail two cases, $B \rightarrow D^{(*)}$ semi-leptonic decays and radiative corrections, where the precision of the theoretical predictions has sensibly improved in the recent past and shown the phenomenological implications of this improvements. More results on the anomalous magnetic moment of the muon, inclusive processes, axion physics and weak interaction are foreseen in the near future.

Acknowledgements I wish to thank A. Di Domenico, V. Lubicz, C. Sachrajda, L. Silvestrini, S. Simula and L. Vittorio, for very useful discussions.

References

1. D.J. Gross, F. Wilczek, Phys. Rev. Lett. **30**, 1343–1346 (1973). https://doi.org/10.1103/PhysRevLett.30.1343
2. H.D. Politzer, Phys. Rev. Lett. **30**, 1346–1349 (1973). https://doi.org/10.1103/PhysRevLett.30.1346
3. P.A. Zyla et al., [Particle Data Group], PTEP **2020**(8), 083C01 (2020). https://doi.org/10.1093/ptep/ptaa104
4. K.G. Wilson, Phys. Rev. D **10**, 2445–2459 (1974). https://doi.org/10.1103/PhysRevD.10.2445
5. Exascale computing, wikipedia
6. C.W. Bauer, Z. Davoudi, A.B. Balantekin, T. Bhattacharya, M. Carena, W.A. de Jong, P. Draper, A. El-Khadra, N. Gemelke, M. Hanada, et al. arXiv:2204.03381 [quant-ph]
7. M. Di Carlo, D. Giusti, V. Lubicz, G. Martinelli, C.T. Sachrajda, F. Sanfilippo, S. Simula, N. Tantalo, Phys. Rev. D **100**(3), 034514 (2019). https://doi.org/10.1103/PhysRevD.100.034514, arXiv:1904.08731 [hep-lat]
8. G. Herdoiza, PoS **LATTICE2010**, 010 (2010). arXiv:1103.1523 [hep-lat]
9. S. Durr, Z. Fodor, J. Frison, C. Hoelbling, R. Hoffmann, S.D. Katz, S. Krieg, T. Kurth, L. Lellouch, T. Lippert et al., Science **322**, 1224–1227 (2008). https://doi.org/10.1126/science.1163233, arXiv:0906.3599 [hep-lat]
10. S. Borsanyi, S. Durr, Z. Fodor, C. Hoelbling, S.D. Katz, S. Krieg, L. Lellouch, T. Lippert, A. Portelli, K.K. Szabo et al., Science **347**, 1452–1455 (2015). https://doi.org/10.1126/science.1257050, arXiv:1406.4088 [hep-lat]
11. N. Cabibbo, Phys. Rev. Lett. **10**, 531–533 (1963). https://doi.org/10.1103/PhysRevLett.10.531
12. M. Kobayashi, T. Maskawa, Prog. Theor. Phys. **49**, 652–657 (1973). https://doi.org/10.1143/PTP.49.652
13. L. Wolfenstein, Phys. Rev. Lett. **51**, 1945 (1983). https://doi.org/10.1103/PhysRevLett.51.1945
14. M. Ciuchini, G. D'Agostini, E. Franco, V. Lubicz, G. Martinelli, F. Parodi, P. Roudeau, A. Stocchi, JHEP **07**, 013 (2001). https://doi.org/10.1088/1126-6708/2001/07/013, arXiv:hep-ph/0012308 [hep-ph]
15. Y. Aoki, T. Blum, G. Colangelo, S. Collins, M. Della Morte, P. Dimopoulos, S. Dürr, X. Feng, H. Fukaya and M. Golterman, et al. [arXiv:2111.09849 [hep-lat]]
16. M. Bona, M. Ciuchini, D. Derkach, F. Ferrari, E. Franco, V. Lubicz, G. Martinelli, M. Pierini, L. Silvestrini, C. Tarantino et al. PoS **EPS-HEP2021**, 512 (2022). https://doi.org/10.22323/1.398.0512
17. R. Aaij et al. [LHCb], arXiv:2110.09501 [hep-ex]
18. R. Aaij et al., LHCb. JHEP **08**, 055 (2017). https://doi.org/10.1007/JHEP08(2017)055. ([arXiv:1705.05802 [hep-ex]].)

19. A. Abdesselam et al. [Belle], Phys. Rev. Lett. **126**(16), 161801 (2021). https://doi.org/10.1103/ PhysRevLett.126.161801, arXiv:1904.02440 [hep-ex]
20. B. Aubert et al., BaBar. Phys. Rev. Lett. **100**, 231803 (2008). https://doi.org/10.1103/ PhysRevLett.100.231803. ([arXiv:0712.3493 [hep-ex]].)
21. B. Aubert et al., BaBar. Phys. Rev. D **77**, 032002 (2008). https://doi.org/10.1103/PhysRevD. 77.032002. ([arXiv:0705.4008 [hep-ex]].)
22. B. Aubert et al., BaBar. Phys. Rev. D **79**, 012002 (2009). https://doi.org/10.1103/PhysRevD. 79.012002. ([arXiv:0809.0828 [hep-ex]].)
23. B. Aubert et al., BaBar. Phys. Rev. Lett. **104**, 011802 (2010). https://doi.org/10.1103/ PhysRevLett.104.011802. ([arXiv:0904.4063 [hep-ex]].)
24. W. Dungel et al., Belle. Phys. Rev. D **82**, 112007 (2010). https://doi.org/10.1103/PhysRevD. 82.112007. ([arXiv:1010.5620 [hep-ex]].)
25. R. Glattauer et al. [Belle], Phys. Rev. D **93**(3), 032006 (2016). https://doi.org/10.1103/ PhysRevD.93.032006, arXiv:1510.03657 [hep-ex]
26. A. Abdesselam et al. [Belle], arXiv:1702.01521 [hep-ex]
27. E. Waheed et al. [Belle], Phys. Rev. D **100**(5), 052007 (2019). [erratum: Phys. Rev. D **103**(7), 079901 (2021)]. https://doi.org/10.1103/PhysRevD.100.052007,arXiv:1809.03290 [hep-ex]
28. P. Gambino, C. Schwanda, Phys. Rev. D **89**(1), 014022 (2014). https://doi.org/10.1103/ PhysRevD.89.014022, arXiv:1307.4551 [hep-ph]
29. A. Alberti, P. Gambino, K.J. Healey, S. Nandi, Phys. Rev. Lett. **114**(6), 061802 (2015). https:// doi.org/10.1103/PhysRevLett.114.061802, arXiv:1411.6560 [hep-ph]
30. P. Gambino, K.J. Healey, S. Turczyk, Phys. Lett. B **763**, 60–65 (2016). https://doi.org/10.1016/ j.physletb.2016.10.023, arXiv:1606.06174 [hep-ph]
31. M. Bordone, B. Capdevila, P. Gambino, Phys. Lett. B **822**, 136679 (2021). https://doi.org/10. 1016/j.physletb.2021.136679. ([arXiv:2107.00604 [hep-ph]].)
32. F. Bernlochner, M. Fael, K. Olschewsky, E. Persson, R. van Tonder, K.K. Vos, M. Welsch, arXiv:2205.10274 [hep-ph]
33. A. Bazavov et al. [Fermilab Lattice and MILC], arXiv:2105.14019 [hep-lat]
34. J.P. Lees et al., BaBar. Phys. Rev. Lett. **109**, 101802 (2012). https://doi.org/10.1103/ PhysRevLett.109.101802, arXiv:1205.5442 [hep-ex]
35. J.P. Lees et al. [BaBar], Phys. Rev. D **88**(7), 072012 (2013). https://doi.org/10.1103/PhysRevD. 88.072012, arXiv:1303.0571 [hep-ex]
36. R. Aaij et al. [LHCb], Phys. Rev. Lett. **115**(11), 111803 (2015). [erratum: Phys. Rev. Lett. **115**(15), 159901 (2015)]. https://doi.org/10.1103/PhysRevLett.115.111803, arXiv:1506.08614 [hep-ex]
37. M. Huschle et al. [Belle], Phys. Rev. D **92**(7), 072014 (2015). https://doi.org/10.1103/ PhysRevD.92.072014, arXiv:1507.03233 [hep-ex]
38. Y. Sato et al. [Belle], Phys. Rev. D **94**(7), 072007 (2016). https://doi.org/10.1103/PhysRevD. 94.072007, arXiv:1607.07923 [hep-ex]
39. S. Hirose et al. [Belle], Phys. Rev. Lett. **118**(21), 211801 (2017). https://doi.org/10.1103/ PhysRevLett.118.211801, arXiv:1612.00529 [hep-ex]
40. R. Aaij et al. [LHCb], Phys. Rev. Lett. **120**(17), 171802 (2018). https://doi.org/10.1103/ PhysRevLett.120.171802, arXiv:1708.08856 [hep-ex]
41. S. Hirose et al. [Belle], Phys. Rev. D **97**(1), 012004 (2018). https://doi.org/10.1103/PhysRevD. 97.012004, arXiv:1709.00129 [hep-ex]
42. R. Aaij et al. [LHCb], Phys. Rev. D **97**(7), 072013 (2018). https://doi.org/10.1103/PhysRevD. 97.072013, arXiv:1711.02505 [hep-ex]
43. R. Aaij et al. [LHCb], arXiv:2103.11769 [hep-ex]
44. M. Di Carlo, G. Martinelli, M. Naviglio, F. Sanfilippo, S. Simula, L. Vittorio, Phys. Rev. D **104**(5), 054502 (2021). https://doi.org/10.1103/PhysRevD.104.054502, arXiv:2105.02497 [hep-lat]
45. G. Martinelli, S. Simula, L. Vittorio, Phys. Rev. D **104**(9), 094512 (2021).https://doi.org/10. 1103/PhysRevD.104.094512, arXiv:2105.07851 [hep-lat]

46. G. Martinelli, S. Simula, L. Vittorio, Phys. Rev. D **105**(3), 034503 (2022). https://doi.org/10. 1103/PhysRevD.105.034503
47. G. Martinelli, M. Naviglio, S. Simula, L. Vittorio, arXiv:2204.05925 [hep-ph]
48. G. Martinelli, S. Simula, L. Vittorio, arXiv:2109.15248 [hep-ph]
49. P. Gambino, M. Jung, S. Schacht, Phys. Lett. B **795**, 386–390 (2019).https://doi.org/10.1016/ j.physletb.2019.06.039, arXiv:1905.08209 [hep-ph]
50. S. Jaiswal, S. Nandi, S.K. Patra, JHEP **06**, 165 (2020). https://doi.org/10.1007/ JHEP06(2020)165, arXiv:2002.05726 [hep-ph]
51. G. Martinelli, S. Simula, L. Vittorio, JHEP **08**, 022 (2022). https://doi.org/10.1007/ JHEP08(2022)022, arXiv:2202.10285 [hep-ph]
52. N. Carrasco, V. Lubicz, G. Martinelli, C.T. Sachrajda, N. Tantalo, C. Tarantino, M. Testa, Phys. Rev. D **91**(7), 074506 (2015). https://doi.org/10.1103/PhysRevD.91.074506, arXiv:1502.00257 [hep-lat]
53. V. Lubicz, G. Martinelli, C.T. Sachrajda, F. Sanfilippo, S. Simula, N. Tantalo, Phys. Rev. D **95**(3), 034504 (2017), https://doi.org/10.1103/PhysRevD.95.034504, arXiv:1611.08497 [hep-lat]
54. D. Giusti, V. Lubicz, G. Martinelli, C.T. Sachrajda, F. Sanfilippo, S. Simula, N. Tantalo, C. Tarantino, Phys. Rev. Lett. **120**(7), 072001 (2018). https://doi.org/10.1103/PhysRevLett. 120.072001, arXiv:1711.06537 [hep-lat]
55. A. Desiderio, R. Frezzotti, M. Garofalo, D. Giusti, M. Hansen, V. Lubicz, G. Martinelli, C.T. Sachrajda, F. Sanfilippo, S. Simula et al. Phys. Rev. D **103**(1), 014502 (2021). https:// doi.org/10.1103/PhysRevD.103.014502, arXiv:2006.05358 [hep-lat]
56. R. Frezzotti, M. Garofalo, V. Lubicz, G. Martinelli, C.T. Sachrajda, F. Sanfilippo, S. Simula, N. Tantalo, Phys. Rev. D **103**(5), 053005 (2021). https://doi.org/10.1103/PhysRevD.103. 053005, arXiv:2012.02120 [hep-ph]
57. G. Gagliardi, V. Lubicz, G. Martinelli, F. Mazzetti, C.T. Sachrajda, F. Sanfilippo, S. Simula, N. Tantalo, Phys. Rev. D **105**(11), 114507 (2022). https://doi.org/10.1103/PhysRevD.105. 114507, arXiv:2202.03833 [hep-lat]
58. W.J. Marciano, A. Sirlin, Phys. Rev. Lett. **96**, 032002 (2006). https://doi.org/10.1103/ PhysRevLett.96.032002. ([arXiv:hep-ph/0510099 [hep-ph]].)
59. C.Y. Seng, M. Gorchtein, H.H. Patel, M.J. Ramsey-Musolf, Phys. Rev. Lett. **121**(24), 241804 (2018). https://doi.org/10.1103/PhysRevLett.121.241804, arXiv:1807.10197 [hep-ph]
60. C.Y. Seng, M. Gorchtein, M.J. Ramsey-Musolf, Phys. Rev. D **100**(1), 013001 (2019). https:// doi.org/10.1103/PhysRevD.100.013001, arXiv:1812.03352 [nucl-th]
61. A. Czarnecki, W.J. Marciano, A. Sirlin, Phys. Rev. D **100**(7), 073008 (2019). https://doi.org/ 10.1103/PhysRevD.100.073008, arXiv:1907.06737 [hep-ph]
62. J.C. Hardy, I.S. Towner, Phys. Rev. C **102**(4), 045501 (2020). https://doi.org/10.1103/ PhysRevC.102.045501
63. F. Ambrosino et al. [KLOE], Eur. Phys. J. C **64**, 627–636 (2009) [erratum: Eur. Phys. J. **65**, 703 (2010)]. https://doi.org/10.1140/epjc/s10052-009-1217-6, arXiv:0907.3594 [hep-ex]
64. S. Adler et al. [E787], Phys. Rev. Lett. **85**, 2256–2259 (2000). https://doi.org/10.1103/ PhysRevLett.85.2256, arXiv:0003019 [hep-ex]
65. V.A. Duk et al. [ISTRA+], Phys. Lett. B **695**, 59–66 (2011). https://doi.org/10.1016/j.physletb. 2010.10.043, arXiv:1005.3517 [hep-ex]
66. V.I. Kravtsov et al. [OKA], Eur. Phys. J. C **79**(7), 635 (2019). https://doi.org/10.1140/epjc/ s10052-019-7145-1, arXiv:1904.10078 [hep-ex]
67. M. Bychkov, D. Pocanic, B.A. VanDevender, V.A. Baranov, W.H. Bertl, Y.M. Bystritsky, E. Frlez, V.A. Kalinnikov, N.V. Khomutov, A.S. Korenchenko et al., Phys. Rev. Lett. **103**, 051802 (2009). https://doi.org/10.1103/PhysRevLett.103.051802, arXiv:0804.1815 [hep-ex]

Part II
Talks Given at National Laboratories of INFN, Frascati

Chapter 11
Bruno Touschek and the Physics at Frascati at the Time of AdA and ADONE

Mario Greco

Abstract The physics at Frascati in the years 60–70s is reviewed together with the role played by Bruno Touschek.

This is for me a kind of "Amarcord" because I arrived to Frascati in January 1965 joining the ADONE Group, and stayed there for 25 years. AdA had successfully completed its cycle in Orsay [1] and the last publication [2] was at the end of 1964. On the contrary the ADONE Group was in a great turmoil, there was a lot of excitement and a general feeling of sharing a new adventure. The 5th International Conference on High Energy Accelerators was held in Frascati in 1965 and Fernando Amman, the ADONE Group leader, gave an optimistic status report on the construction [3]. Claudio Pellegrini, who was coordinating the theoretical activities of the Group, asked me to face the calculation of the double bremsstrahlung process as a monitor reaction for the luminosity, since the process of single bremsstrahlung, which had been studied by Altarelli and Buccella [4] a year before, couldn't be used due to the large background coming from the bremsstrahlung on the residual gas. That wasn't an easy task, the calculation was not simple, and the two previous estimates of the cross section, by Bayer and Galitsky [5] and Bander [6] differed by a factor of two. At that time this type of calculation required the appropriate handling of Feynman diagrams, traces and integrals totally by hand, and a constant and patient work.

Bruno Touschek at the time was collaborating with Claudio Pellegrini and Enrico Ferlenghi on the instability of the beams [7] and sometimes he was visiting ADONE. In one of those occasions Claudio introduced me to him, also informing him of my work. That was my first personal meeting with him. Bruno was very kind, as usual, and happy that I was proceeding well in my calculation. Indeed he told me that he considered the study of the double bremsstrahlung process quite important and had proposed to Paolo Di Vecchia the study of the soft photon limit as a thesis for his bachelor's degree. Later we joined forces with Paolo and got the final full result using two different approaches [8]. Since that first meeting with Bruno, we started

M. Greco (✉)
Department of Mathematics and Physics and INFN, University of Roma Tre, Rome, Italy
e-mail: mario.greco@roma3.infn.it

© The Author(s) 2023
L. Bonolis et al. (eds.), *Bruno Touschek 100 Years*,
Springer Proceedings in Physics 287,
https://doi.org/10.1007/978-3-031-23042-4_11

seeing each other regularly in Frascati and also in Rome, where we were living quite closely, and that was the starting of a long collaboration and friendship which lasted many years until 1978, when both of us were visiting CERN and he was hospitalized at the Hôpital de la Tour.

At the end of 1965 my fellowship was over and Bruno proposed me to join a new theory group he was going to found in Frascati. That took place in a short while by adding together the new young forces of Bruno's students and some newly graduates in Rome. Gian De Franceschi, a mathematician who was already in Frascati with Raoul Gatto before the latter moved to Florence, also joined the group. In more detail, Gian De Franceschi, M.G., Etim Etim, Giulia Pancheri, Paolo Di Vecchia, Giancarlo Rossi, Francesco Drago and Pucci Di Stefano composed the new theory group. The radiative corrections for ADONE experiments were the main problem Bruno had in his mind, because of their importance in the new electron-positron collider. Indeed he had already given a number of theses to his students, namely "Proposal for the administration of radiative corrections" to Etim Etim, "The double bremsstrahlung process in the soft photon approximation" to Paolo di Vecchia and "Application of the Block-Nordsieck theorem to radiative corrections in Adone experiments" to Giancarlo Rossi. To summarize the further theoretical efforts in that direction, two main research lines had been followed.

First, the infrared corrections to be applied in an electron-positron collider experiment are obtained with the help of the Bloch-Nordsieck theorem, using a statistical approach to define the probability for the four-momentum to be carried away by the electromagnetic radiation [9]. Alternatively, from a field theoretical point of view, a new finite S-Matrix is defined using a realistic definition of initial and final states, by "dressing" the charged particle's states with a phase containing the electromagnetic operator in the exponential, in order to create an undetermined number of soft photons. The new S-Matrix was explicitly shown to be equivalent to all orders in α to the conventional perturbative result [10]. In other words this approach corresponds to the introduction of the coherent states in QED. It is extraordinary that both approaches led exactly to the same result for the soft radiative effect, namely the observed cross section can be written as

$$d\sigma = \frac{1}{\gamma^\beta \Gamma(1 + \beta)} \left(\frac{\Delta\omega}{E} \right)^\beta d\sigma_E$$

where 2E is the total c.m. energy, $(\Delta\omega/E)$ is the relative energy resolution of the experiment, γ is the Euler constant, $d\sigma_E$ differs from the lowest cross section $d\sigma_0$ by finite terms of $\mathcal{O}(\alpha)$, and β is the famous Bond-factor, so named by Bruno because its numerical value at ADONE was 0.07, and more generally $\beta = \frac{4}{\pi}\alpha[\ln(2E/m) - 1/2]$.

The coherent states approach played a major role later in the description of the radiative effects in case of the production of the J/Ψ and of the Z boson, as we'll discuss later. Also that was first extended to QCD in the late '70s [11] and studied further [12, 13]. Many and important QCD results concerning exponentiation, resummation formulae, K-factors, transverse momentum distributions of DY pairs,

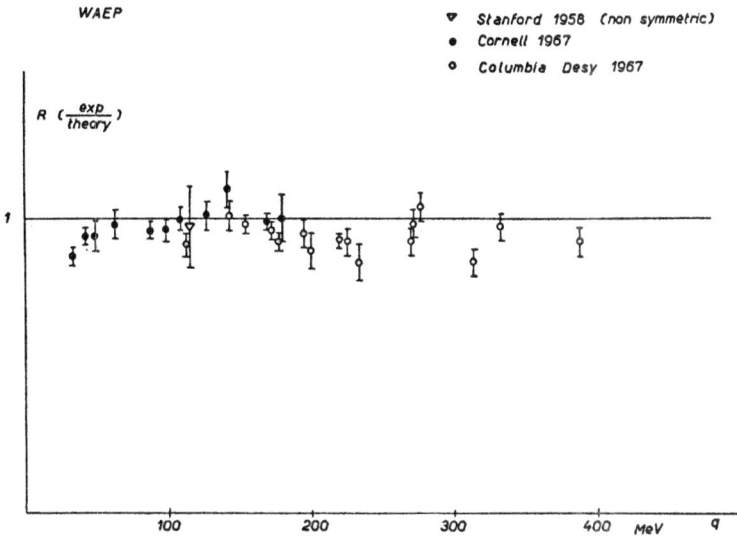

Fig. 11.1 Wide angle electron positron pairs experiments [18], from Ref. [17]

W/Z and H production, have their roots in Bruno Touschek's ideas on the exponentiation and resummation formulae in QED. We give here a reference list [14] of the papers that were written later in the Frascati-Rome area and certainly inspired by his ideas.

Coming back to the late '60, a strong shock was inflicted to the physics community due to an apparent violation of QED reported by a Harvard group, R.B. Blumenthal et al., in the wide-angle production of electron-positron pairs [15]. New experiments immediately took place everywhere, and also Carlo Bernardini at Frascati started an experiment of wide-angle bremsstrahlung. On the theory side Bruno Touschek suggested a simple method of modifying QED by introducing $ee\gamma\gamma$ vertices in addition to the usual minimal $ee\gamma$ interaction, and asked me to study the possible constraints coming from the known effects and experiments. When the draft of our paper was ready, the news arrived of the confirmation of the validity of QED, and the paper got unpublished [16]. The new experiments were summarised by Bernardini [17] and reported in Fig. 11.1.

It's amusing to notice that the first and last authors in WAEP experiments [18] are B. Richter and S.C.C. Ting who will share in a few years the discovery of the J/Ψ. Let's discuss now the theoretical framework and the expectations concerning ADONE and the experimental results. At the time the Vector Meson Dominance (VMD) model of Sakurai [19] was quite successful in describing the e.m. interaction of hadrons as being mediated by the vector mesons ρ, ω and ϕ. That led T.D. Lee, N. Kroll and B. Zumino to try to give a field theoretical approach to VMD [20]. In this framework the total hadronic annihilation cross section was expected to behave at large s as

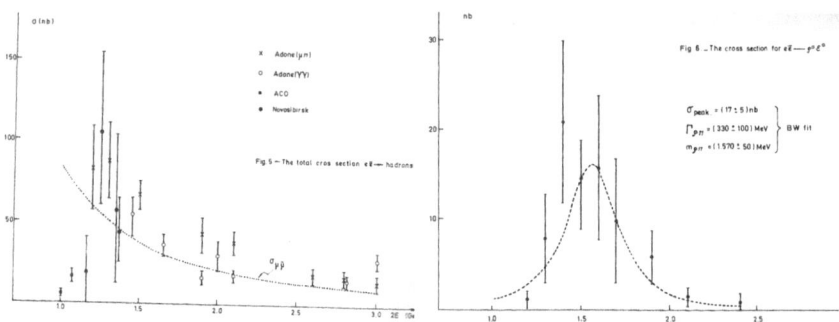

Fig. 11.2 ADONE experimental results, from Ref. [29]

$$\sigma(s) = \left(\frac{1}{s}\right)^2$$

However departures from the simple VMD model were observed in some radiative decays of mesons and the possible existence of new vector mesons was suggested by Bramon and myself [21], as also predicted by dual resonance models and the Veneziano model [22]. On the other hand the results of Deep Inelastic Scattering (DIS) experiments at SLAC, with the idea of Bjorken scaling and the Feynman parton model were naturally leading to

$$\sigma(s) = \frac{1}{s}$$

and indeed Cabibbo et al. [23] suggested that the ratio R of the hadronic to the point-like cross section would asymptotically behave as

$$R = \frac{\sigma_{had}(s)}{\sigma_{\mu\mu}} \rightarrow \sum_i Q_i^2$$

where the sum extends to all spin 1/2 elementary constituents, neglecting scalars.

As it's well known, the results of all experiments, namely the MEA Group [24], the $\gamma\gamma$ Group [25], the $\mu\pi$. Group [26] and the Bologna-CERN-Frascati Collaboration [27] showed a clear evidence of a large multihadron production with $R \approx 2$, pointing to the coloured quark model. On the other hand they also indicated evidence for a new vector meson $\rho'(1.6)$ with a dominant decay in four charged pions, which had been suggested by A. Bramon and myself [28] The experimental data are shown in Fig. 11.2, taken from a review paper of Bernardini and Paoluzi [29].

The ADONE results together with the request of scaling, both in DIS and e^+e^- annihilation, and the Veneziano's duality idea led us to propose a new scheme where the asymptotic scaling is reached through the low energy resonances mediating the asymptotic behaviour [30]. Thus the value of R is also connected to the low energy resonances's couplings, and in the 3 coloured quark model, led us to the prediction $R \approx 2.4$. This scheme—named duality in e^+e^- annihilation—was immediately shared by Sakurai [31]. Later J. Bell and collaborators also studied a potential model where the bound states could be solved analytically and verified this idea of duality [32]. In addition a set of e^+e^- duality sum rules was derived from the canonical trace anomaly of the energy momentum tensor by Etim and myself [33] much earlier than the Russian sum rules of Shifman et al. [34]. The lowest order sum rule gives

$$\int_{s_0}^{\bar{s}} ds \left(Im\Pi(s) - \frac{\alpha R}{3} \right) = 0$$

where $Im\Pi(s) = s\sigma_{had}(s)/4\pi\alpha$ and clearly it relates the asymptotic value of R to the low energy behaviour. One has to stress here that QCD wasn't there yet at that time. Fifty years later, by comparing now the value of R from the Particle Data Group with all the experimental information, as shown in Fig. 11.3, with the theoretical prediction of QCD with $\mathcal{O}(\alpha_s)$, $\mathcal{O}(\alpha_s^2)$ and $\mathcal{O}(\alpha_s^3)$ corrections included—as indicated by the continuous red line—one easily concludes the duality is very well satisfied. The average value of R in particular, in the ADONE region, is about 2.4 as it was also confirmed by the SPEAR data at the c.m. energy just below the J/Ψ. Let's consider in detail the J/Ψ discovery, or what was called the November Revolution, from a Frascati point of perspective. As it's well known, on November 11th 1974, B. Richter and S.C.C. Ting jointly announced in Stanford the discovery of the J/Ψ both at SLAC and at Brookhaven [35, 36]. I had the terrific chance of arriving at SLAC the day after, with an invitation by Sid Drell to give a seminar on our duality works, on the way for a visit of a few weeks to Mexico City. Sid had been on a sabbatical leave the year before at Frascati and Rome, so we knew each other pretty well. There was a great excitement in the theory discussion room and once I was informed of the details of the discovery I realized immediately that the J/Ψ could be seen possibly also at ADONE. I asked Sid to let me call Frascati, and from his confidential office—he was scientific advisor of the President of United States—I gave to Giorgio Bellettini, the director of the Laboratory, the exact position of the J/Ψ.

The night after, the resonance was also observed at Frascati. Giorgio Salvini communicated the results to the Phys. Rev. Letters over the telephone and the paper was published [37] in the same issue of the American results.

As far as the theoretical interpretation of the J/Ψ, hundreds of papers had been published on the argument, as it's well known. In a recent review article on this subject, Alvaro De Rújula has reported [38] the papers published on the first issue of Phys. Rev. Letters after the discovery, as shown in Fig. 11.4.

Of course only two of them, both from Harvard, had the right interpretation: by Applequist and Politzer [35], who related the reason for the very narrow width of the J/Ψ to the asymptotic freedom of QCD just discovered, and by De Rújula and Glashow [40] because of the GIM mechanism and charm suggested earlier [41]. On the other hand Alvaro is also quoting two papers published on Lett. Nuovo Cimento,

Fig. 11.3 The ratio
$R = \sigma_{had}(s)/\sigma_{\mu\mu}(s)$ as a
function of $\sqrt{s}$, from particle
data group

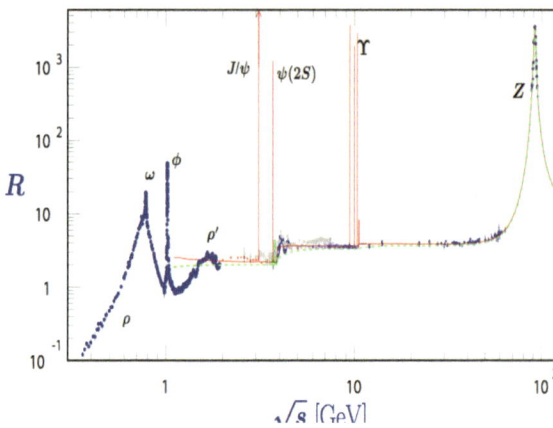

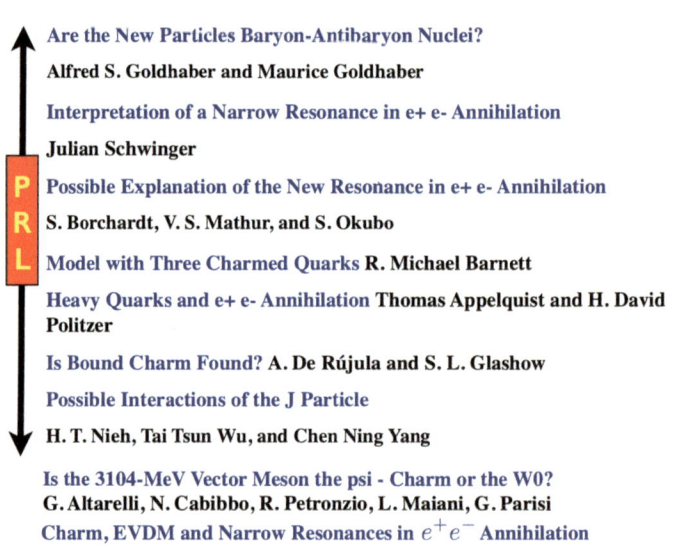

Fig. 11.4 The immediate interpretation of J/Ψ from Ref. [38]. PRL is Phys. Rev. Letts. 34, Jan. 16th, 1975

by Altarelli et al. [42], who had the wrong interpretation in favour of the weak boson, and C. Dominguez and myself [43], written in Mexico City a few days after I had left Stanford, who also had the right interpretation in favour of charm. In more detail, as soon as I found the news on a local newspaper of the subsequent discovery of the Ψ', by using the duality ideas discussed above, we arrived at the conclusion that the new series of resonances was indeed composed by $c - \bar{c}$ pairs. An enjoyable note concerning this paper came forty years later. On December 2013, S.C.C. Ting was invited to Frascati for a Bruno Touschek Memorial Lecture and commenting the J/Ψ saga, he said that our paper was the first one to give the right interpretation. I was very surprised because I had never compared the exact dates of the three papers, and our original preprint had been lost and not easily available. However, searching virtually at the CERN library archives I indeed found that our paper was preceding by one day that one of Applequist and Politzer and by one week the other by De Rújula and Glashow.

The problem of the radiative corrections to the J/Ψ line-shape, in virtue of the very narrow width, showed the crucial role played by the theoretical ideas of the early times on the infrared behaviour of QED, namely the exponentiation results and the approach of the coherent states. The detailed analysis by Pancheri, Srivastava and myself [44], showed that the main infrared correction factor was of the type

$$C_{infra} \approx \left(\frac{\Gamma}{M} \right)^{\beta}$$

where Γ and M are the width and the mass of the resonance respectively, and β the Bond-factor. The detailed result of this analysis, compared with the SLAC and Frascati data, is shown in Fig. 11.5. When we showed our result to Bruno Touschek, he immediately commented that the experimental errors of the Frascati data had been overestimated. At this point I should add that the SLAC analysis of their data had been based on a paper by Yennie [45] that contained a wrong dependence on the width Γ and the parameter σ of the Gaussian energy distribution of the beams, with a resulting difference with respect to our analysis on the leptonic width of the J/Ψ.

Fig. 11.5 Experimental results for J/Ψ and Ψ' production in e^+e^- annihilation. The data are from SPEAR and ADONE (see text). The full lines refer to the theoretical analysis including radiative corrections of Ref. [44]

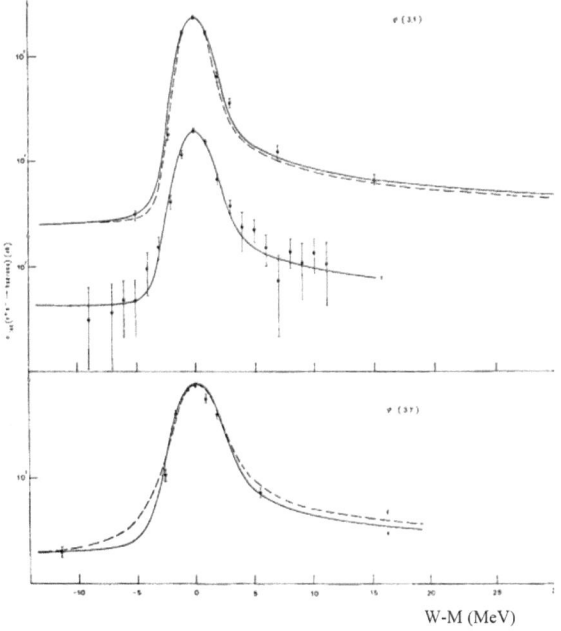

W-M (MeV)

It was only in 1987, in occasion of the first La Thuile meeting, that I convinced Burt Richter to update the SLAC radiative corrections codes with the right formulae, in perspective of the coming data on the Z boson physics at SLC. Indeed the re-analysis of all charm data at SLAC caused a change of many properties of the charm particles in the Particle Data Group in 1988.

The above treatment of the radiative corrections for the J/Ψ production was extended a few years later to study the radiative effects in the case of Z production at LEP/LHC [46]. Our work was the first study to all orders in the infrared corrections, with a complete evaluation of all finite terms of $\mathcal{O}(\alpha)$ and at the base of the later analyses of the experiments. Very recently, within the general discussion on the possibility of constructing a muon collider Higgs factory to study with great care on resonance the properties of the H, the line-shape has been studied [47] in the same way, as in the old times. As a result we have shown that the radiative effects put very stringent bounds on the energy spread of the beams, and make this project very tough.

To conclude, from AdA/ADONE to LEP/LHC and the future linear and/or circular colliders, the seminal idea of Bruno Touschek has contributed with so many discoveries to the assessment and the progress of the Standard Model. This certainly constitutes his main legacy. In addition he has given many ideas in theoretical physics, from QED to other aspects of the Standard Model, and that is also an important legacy to us.

References

1. C. Bernardini, The first electron-positron collider. Phys. Perspect. **6**, 156 (2004)
2. C. Bernardini, G. Corazza, G. Di Giugno, J. Haissinski, P. Marin, R. Querzoli and B. Touschek Measurements of the rate of interaction between stored electrons and positrons. Nuovo Cimento **34**, 1473 (1964)
3. F. Amman et al., Adone - The Frascati 1.5 GeV electron positron storage ring, in *5th International Conference on High Energy Accelerators, HEACC '65, Frascati* (1965)
4. G. Altarelli, F. Buccella, Nuovo Cimento **34**, 1337 (1964)
5. V.N. Bayer, V.M. Galitsky, JETP Lett. **2**, 165 (1965)
6. M. Bander, SLAC-TN-64-93 (1964)
7. E. Ferlenghi, C. Pellegrini, B. Touschek, The transverse resistivity wall instability of extremely relativistic beams of electrons and positrons, Contribution to HEACC '65, 378 (1965)
8. P. Di Vecchia, M. Greco, Nuovo Cimento **50**, 319 (1967)
9. E. Etim, G. Pancheri, B. Touschek, Nuovo Cimento **51B**, 276 (1967)
10. M. Greco, G. Rossi, Nuovo Cimento **50A**, 168 (1967)
11. M. Greco, F. Palumbo, G. Pancheri, Y. Srivastava, Phys. Lett. **77B**, 282 (1978)
12. G. Curci, M. Greco, Phys. Lett. **79B**, 406 (1978)
13. S. Catani, M. Ciafaloni, G. Marchesini, Phys. Lett. B **168**, 284 (1986), (and references therein)
14. G. Parisi, Phys. Lett. **90B**, 295 (1980); G. Curci, M. Greco, Phys. Lett. **92B**, 175 (1980); G. Pancheri, Y. Srivastava, Phys. Rev. Lett. **43**, 11 (1979); G. Curci, M. Greco, Y. Srivastava, Phys.

Rev. Lett. **43**, 834 (1979); G. Parisi, R. Petronzio, Nucl. Phys. B **154**, 427 (1979); G. Curci, M. Greco, Y. Srivastava, Nucl. Phys. B **159**, 451 (1979); G. Altarelli, K. Ellis, M. Greco, G. Martinelli, Nucl. Phys. B **246**, 12 (1984)
15. R.B. Blumenthal, Phys. Rev. Lett. 14, et al., 660 and Phys. Rev. **144**(1966), 119 (1965)
16. M. Greco, B. Touschek, On the extension of the minimal coupling in QED, unpublished (1966)
17. C. Bernardini, High energy experiments in QED, Lectures at Boulder, Colorado, Frascati preprint LNF 68/42 (1968)
18. WAEP: B. Richter, Phys. Rev. Lett. **1**, 114 (1958); R.B. Blumenthal et al., Phys. Rev. **144**, 1199 (1966); E. Eislander et al., Phys. Rev. Lett. **18**, 425 (1967); S.C.C. Ting et al., Phys. Rev. 161, 1344 (1967)
19. J.J. Sakurai, Phys. Rev. Lett. **22**, 981 (1969). ((and references therein))
20. N.M. Kroll, T.D. Lee, B. Zumino, Phys. Rev. **157**, 1376 (1967)
21. A. Bramon, M. Greco, Lett. N. Cimento **152**, 739 (1971)
22. G. Veneziano, N. Cimento A **57**, 190 (1968)
23. N. Cabibbo, G. Parisi, M. Testa, Lett. N. Cimento **4**, 35 (1970)
24. B. Bartoli et al., N. Cimento A **70**, 615 (1970)
25. C. Bacci et al., Phys. Lett. B **38**, 551 (1972)
26. F. Ceradini et al., Phys. Lett. B **42**, 501 (1972)
27. V. Alles-Borelli et al., Phys. Lett. B **40**, 433 (1972)
28. A. Bramon, M. Greco, Lett. N. Cimento **3**, 693 (1972)
29. C. Bernardini, L. Paoluzi, in *2nd International Winter Meeting on Fundamental Physics* (1974)
30. A. Bramon, E. Etim, M. Greco, Phys. Lett. B **41**, 609 (1972)
31. J.J. Sakurai, Phys. Lett. B **46**, 207 (1973)
32. J.S. Bell, R. Bertlmann, Z. Phys, C 4, 11 (1980) and Nucl. Phys. B **177**, 218 (1981)
33. E. Etim, M. Greco, Lett. N. Cimento **12**, 91 (1975)
34. M.A. Shifman et al., Nucl. Phys. B **147**, 385 (1979)
35. J.J. Aubert et al., Phys. Rev. Lett. **33**, 1404 (1974)
36. J.E. Augustin et al., Phys. Rev. Lett. **33**, 1406 (1974)
37. C. Bacci et al., Phys. Rev. Lett. **33**, 1408 (1974)
38. A. De Rújula, Int. J. Mod. Phys. A **34**, 32 (2019)
39. T. Applelquist, H.D. Politzer, Phys. Rev. Lett. **34**, 43 (1975)
40. A. De Rújula, S.L. Glashow, Phys. Rev. Lett. **34**, 46 (1975)
41. S.L. Glasow, J. Iliopoulos, L. Maiani, Phys. Rev. D **2**, 1285 (1970)
42. G. Altarelli, N. Cabibbo, L. Maiani, G. Parisi, R. Petronzio, Letts. N. Cimento **11**, 14 (1974)
43. C. Dominguez, M. Greco, Lett. N. Cimento **12**, 439 (1975)
44. M. Greco, G. Pancheri, Y. Srivastava, Phys. Lett. B **56**, 367 and Nucl. Phys. B **101**(1975), 234 (1975)
45. D.N. Yennie, Phys. Rev. Lett. **34**, 239 (1975)
46. M. Greco, G. Pancheri, Y. Srivastava, Nucl. Phys. B **171**, 118 (1980)
47. M. Greco, Mod. Phys. Lett. A **30**, 39, 1530031 (2015). arXiv: 1503.05046; M. Greco, T. Han, Z. Liu, Phys. Lett. B **763**, 409 (2016)

Chapter 12
Accelerators at LNF: From AdA to EuPRAXIA

Andrea Ghigo

Abstract The accelerators realized, installed and operated in the Frascati National Laboratories (LNF) from ADONE to the present day are described together with the main characteristics necessary for the experiments that were carried out. The absolutely new elements that characterized all the accelerators realized in LNF and which were then used by other accelerators in the world are described: in particular Bruno Touschek's great contribution to accelerator physics is underlined. Present and future plans are also mentioned in the development of the new generation of accelerators in LNF.

12.1 Introduction

The scientific activity of the National Laboratories of Frascati (LNF) began with the realization of the first Italian high-energy accelerator: the Synchrotron, that produced an intense electron beam at a maximum energy of 1 GeV.

Giorgio Salvini, who was entrusted with the construction of the accelerator and its infrastructure, recruited young and brilliant physicists and engineers from the best Italian universities willing to move to Frascati to participate in this challenge.

The team that built the Synchrotron, an accelerator of formidable complexity, gave an example of how a truly complex system can be realized with study, commitment and perseverance, starting from scratch. The Synchrotron was completed, starting the operations, in 1959.

Bruno Touschek, working with the Synchrotron group, had the brilliant idea of accumulating and accelerating matter and antimatter in the same accelerator and then making them collide to create new elementary particles. Touschek proposed to inject a positron beam into the newly built synchrotron but Salvini was against because many experiments using the synchrotron beam were impatient to take data,

A. Ghigo (✉)
National Laboratories of INFN, Frascati, Italy
e-mail: andrea.ghigo@lnf.infn.it

© The Author(s) 2023
L. Bonolis et al. (eds.), *Bruno Touschek 100 Years*,
Springer Proceedings in Physics 287,
https://doi.org/10.1007/978-3-031-23042-4_12

then Giorgio Ghigo proposed to build a test accelerator. In only one year: AdA, Anello di Accumulazione was realized, in which electrons and positrons, produced with the Synchrotron beam, were injected and accumulated with a maximum energy of 500 meV. AdA was the world's first matter and antimatter storage ring and from AdA all the colliders built successively descend.

After the AdA great success, ADONE, an electron positron collider 100 m long with an energy up to 3 GeV, was realized in Frascati.

The story of the AdA and ADONE storage rings, the advantages of this collision scheme and the experiments in the particle physics field are presented in this book by my colleagues, showing how, starting from a brilliant intuition, a machine was born that no one thought could be made.

The example of this first generation of accelerators: Synchrotron, AdA and ADONE, has remained in the DNA of the laboratories.

12.2   ADONE Second Life

After the period of experiments on elementary particle physics, a second life was expected for ADONE. The use of storage rings for photon production gave to ADONE a new lifeblood; indeed, in the 1980s, synchrotron radiation lines were installed and ADONE was one of the first-generation synchrotron radiation sources for users.

The photon produced in bending magnet, wiggler and undulator, in the X and VUV range, have been used for many experiments in biological field and in material science.

Gamma rays were also generated in ADONE both using bremsstrahlung on gas jet and Compton backscattering of laser photons by electrons stored in the ring. These high energy photons were used in nuclear physics experiments, starting a generation of accelerators dedicated to these purposes.

At that time several accelerator projects have been proposed, without success, due to budget constraints, for the Frascati labs. INFN continued to participate in the construction of the large accelerators at CERN and of the equipment for high energy physics experiments.

12.3   LISA

While the ADONE accelerator was being used for photon experiments, the accelerator division of the laboratories continued a research and development program on new acceleration techniques. Sergio Tazzari proposed to develop new technologies for future accelerators such as superconducting accelerating cavities and low emittance beams for use in free electron lasers. A project of a low-energy, high repetition frequency linear accelerator LISA was funded, realized and installed in the Frascati Labs in the early '80. LISA was one of the first R&D activities in Europe that eventually gave birth to the European XFEL at DESY Hamburg.

The LISA infrastructure, consisting of an underground bunker and control room on the surface, is currently reused by the SPARC-LAB complex.

12.4   DAΦNE Φ-Factory

Touschek's attitude of looking forward reinvigorated the Frascati Labs in the '90s with the creation of a new class of accelerators: the very high luminosity electron–positron collider.

The cost and the size of the high energy accelerators became prohibitive, so the laboratories that have made collider history, such as LNF in Italy, SLAC in USA and KEK in Japan, decided to change the paradigm of the particle physics research with accelerators. Instead of chasing the limits of high energy, they conceived new medium–low energy accelerators with very high luminosity, called Factories, aimed at precision measurements, whose mission was to produce large quantities of particles to study rare events with high statistics.

All three laboratories reused the existing infrastructures to install their new machines in order to save on the budget and the time needed to build the new colliders. A collaborative competition pushed the three laboratories to impressive results in a very short time.

It was therefore decided to realize a Φ-Factory in Frascati, a collider just above 1 GeV in the center of mass, to produce and study the decays of Φ particles. SLAC and KEK, where they had longer tunnels, opted for colliders at 11.5 GeV to produce B particles. The primary purpose of these accelerators was to measure the violation of the charge-parity conservation theorem.

In the early '90s the INFN president Nicola Cabibbo set up a working group, chaired by Luciano Maiani, to draw up the possible experiments and define the parameters that the accelerator had to have in terms of luminosity.

INFN called back to Italy Gaetano Vignola, who, together with Mario Bassetti and all the accelerator Division staff, proposed a completely new concept collider: DAΦNE.

After the approval of the project, the group was formed, under the directorate of Enzo Iarocci, also with the recruitment of young people. In less than five years, from the end of ADONE's operations, the new collider was ready to begin testing (Fig. 12.1).

The basic idea behind the DAΦNE Project was to produce an amount of Φ-particles per year enough to measure the ratio between the direct (in the K-decay) and indirect (connected to the oscillation of K_L in K_S) components of CP violation. This ratio (ε'/ε) was expected of order 10^{-4}. On the resonance peak, 1020 MeV, the effective cross section for Φ production is approximately 2 μb. The collider goal

Fig. 12.1 DAΦNE: the Frascati Φ-Factory. Copyright INFN – LNF

was to produce 10^{10} Φs and 2×10^9 K_S per year. Therefore, a *luminosity* of 2.5 × 10^{32} cm^{-2} s^{-1} was initially requested.

The Luminosity formula can be written as simply as

$$L = f\, N_b\big(N_e N_p\big)/(4\pi\,\sigma_x\sigma_y)$$

where f is the revolution frequency of the machine and N_b the number of bunches stored in the rings (f N_b the collision frequency). N_e and N_p are the number of electrons and positrons in each bunch respectively and $\sigma_x\sigma_y$ the rms transverse dimensions at the interaction point.

The luminosity of 1×10^{30}, in single bunch, reached by the best 1 GeV collider at that time, VEPP2M, was taken as a reference for the proposal and it was decided to store up to 120 bunches in two rings 100 m long.

The max number of bunches is determined by the minimum distance between the bunches that avoid the simultaneous collision of two contiguous bunches in the detector.

Scaling all the parameters to get 10^{32}, the scary current of more than 2 A had to be stored in each ring with 10^{11} particles per bunch. In order to avoid parasitic collisions outside the interaction region proper, which contribute detrimentally to the beam-beam limit without *useful* luminosity, it was chosen to build two separate rings for electrons and positrons, intersecting at two interaction points (thus allowing the possibility of accommodating two distinct detectors, although not operating at the same time) where they collide at an angle. A frequency of the accelerating cavities of ~360 MHz has been chosen: since the revolution frequency is ~3 MHz. As many as 120 packets could be injected into 100 m long rings with a distance between them of 2.7 nsec.

In the interaction of two colliding bunches, the limit parameter was the so-called tune-shift. During the interaction of an electron bunch with its positron homologue, one beam, due to the electric charge, acts as a [de]focusing system on the other, with the consequence of producing a *tune shift*; if the tune shift is large the working point moves close to a resonance, with the consequence of widening the beam dimension, losing luminosity, or worse, making the beam completely unstable.

It was decided to make flat beams collide, i.e. they had a horizontal dimension of 2 mm and the vertical one 100 times smaller. This strong focusing had to be obtained at the center of the experimental setup with the latest magnetic lenses far enough away not to obscure the detector's field of view. The bunch length should not exceed 3 cm, otherwise the "*hour glass*" effect would have decreased the luminosity.

The first major concern in this interaction scheme was the synchro-betatron effect in the interaction angle, i.e. the longitudinal and transverse motion transfer from one plane to another with the risk of widening the beams, losing luminosity. The second was the multibunch instability in which each ring could induce longitudinal and transverse oscillations from one bunch to the subsequent ones, through the interaction with electromagnetic field due to the impedance of the components of the vacuum chamber, especially in the RF accelerating cavity, in which the particles traveled.

A great deal of attention has been paid to all other possible sources of limitation of the various parameters. Just to list a few:

- The current was so high that the desorption of the walls of the vacuum chamber required an impressive synchrotron pumping and light absorption capacity: vacuum pumps with enormous pumping capacity and mirror grade vacuum chamber (wall roughness below a micron).
- The very low impedance was obtained by designing all the components of the vacuum chamber with e.m. shielding system and tapered shape.
- The radiofrequency cavities were realized by suppressing the high order excitation modes produced by the beam field, that could act on the contiguous bunches, with innovative waveguide absorber systems.
- Clearing electrodes were installed in the rings to avoid *ion trapping* in the electron ring and to reduce the *electron cloud* effect in the positron one.
- To damp the longitudinal and transverse oscillation modes, a very effective bunch by bunch feedback system was realized, in collaboration with SLAC. This is one of the first examples of parallel data processing in which the position signals from beam detectors were sent to a digital signal processor to reach the calculation speed that allowed an immediate correction of the position of every single bunch.
- A series of wiggler magnets have been installed in order to increase synchrotron radiation emission to decrease the damping time of the injected bunches and to increase the emittance of the beam in order to store higher charge per bunch.
- A fast injection system at full energy in the ring has been realized to inject in top-up scheme. The injection system is composed by a e^+e^- linear accelerator working at 500 meV injection energy. An accumulator / damping ring has been placed between the linac and the main rings; this provides the injection of the charge per bunch with an emittance close to the main rings one, mainly for positrons.

- The *Crab Waist* collision scheme was proposed and implemented on DAΦNE from Pantaleo Raimondi to improve the luminosity.
- In the end the most important limitations of the storable and usable current in DAΦNE, and therefore of the maximum luminosity reached, were the lifetime of the particles which, due to the very high density in the individual bunches, was limited by the *Touschek effect* and the electron cloud.

Several particle physics experiments were installed on DAΦNE during 20 years of operation: first KLOE and KLOE2 which aimed to measure CP violation and make all the high statistical measurements of Kaon physics.

Subsequently the nuclear physics experiment FINUDA was installed in the other interaction region, which was proposed to produce hypernuclei with tags close to the interaction point. The Kaons produced, at the threshold energy by Φ decay, were allocated inside the nucleus, thus creating hypernuclei.

Another class of experiments: Dear, Siddhartha and Siddhartha2, aiming to study the Kaonic atoms, generated by the capture of the K, produced in the Φ decay, by light atoms of cryogenic targets, have been installed on the DAΦNE interaction region.

12.5 SPARC-Lab

A new type of accelerator was proposed in Frascati in the 2000s: a very low emittance injector.

The first experiments on Free Electron Lasers, FEL, to generate coherent synchrotron light in the ultraviolet and X region had been performed on ADONE with the strong limitation of high energy spread and emittance of the electron beam. To have a significant effect, intense beams of low-emittance electrons had to be produced.

SPARC was one of the first injectors in which the electron bunches were no longer generated by thermionic gun and static accelerating fields but by photo-emitters installed in radiofrequency cavities.

In SPARC the charges of the electron bunch and their temporal structures are generated by a laser pulse, of suitable wavelength, sent to the photocathode. The laser beam is easy to manipulate and, together with the possibility of changing amplitude and phase between the accelerating structures, different configurations of electron bunches can be experimented within the same RF bucket.

The cathode is installed in a radiofrequency cavity so that the accelerating electric field on its surface can reach values of the order of 100 MV/m. In this configuration the electrons reach ultra-relativistic speeds in a very short space, reducing the increase in emittance due to space charge. Furthermore, the laser pulse can be of short duration to generate pulses of high peak current (Fig. 12.2).

SPARC was the first radiofrequency photo injector in which the concept of emittance reduction was successfully employed by placing the first accelerating section on the maximum oscillation of the emittance at the gun exit, reducing the value

Fig. 12.2 SPARC seen from the FEL undulator side. Copyright INFN – LNF

below 1 mm·mrad. The method of compressing the length of the bunch through the *velocity bunching* technique passing the electron bunch in the off-crest radiofrequency oscillation has also been tested and verified. Different accelerating fields have been experienced between the head and the tail of the bunch to favor longitudinal compression with the achievement in the injector of the high peak current necessary to generate FEL radiation in magnetic undulators.

The SPARC scheme was subsequently adopted in all electron injectors dedicated to FEL or short electron pulse production.

With such intense beams obtained and with the possibility of generating multiple pulses with the laser in the same radiofrequency bucket, the experimentation of plasma acceleration based on the Wake Field Acceleration particle has begun. In this technique a plasma wave is formed by a driver bunch and a following bunch, witness, properly injected into the plasma wave, could be accelerated with accelerating gradients of the order of 10 GeV/m.

The SPARC injector flexibility has given rise to a series of important experiments.

– The two-color FEL was thus tested in Frascati, sending two bunches at slightly different energy to the undulator magnets. It has been successfully replicated at short wavelengths at FACET (SLAC) and in FERMI, the Trieste FEL.
– The train of pulses hundreds of fs short and spaced by 1 ps for efficient generation of THz radiation.
– The harmonics generation of the FEL radiation by injecting short wavelength photons, generated by lasers into a gas, together with the electron beam in the undulators to force their stimulated emission at shorter wavelengths.

- The X photons production by means of the backward diffusion due to the Compton effect of the pulses of the high power laser, FLAME, by the electron beam.
- Finally, the plasma acceleration of electron bunch. The production of FEL coherent radiation with an accelerated plasma beam was also measured at SPARC demonstrating that an accelerated plasma beam can maintains high quality emittance and energy spread characteristics.

On the basis of these experiments and those of the FEL, the new European user facility project EuPRAXIA was proposed and one of the project infrastructures will be built in Frascati.

12.6   EuPRAXIA

In the tradition of the Laboratories, the realization of new-concept particle accelerators, called EuPRAXIA, has been proposed to the European Commission involving the construction of a Free Electron Laser facility driven by a plasma accelerator.

The generation of the electron beam takes place with a high-brightness photoinjector followed by an innovative accelerator that uses radiofrequency in the X band. The beam thus generated is then accelerated by a pre-ionized plasma aiming to reach accelerating gradients greater than 10 GeV/m maintaining the excellent beam quality needed for FEL operations.

The EuPRAXIA project has been approved and funded by INFN and has entered the European road map of research infrastructures, ESFRI. The construction phase is starting with the civil infrastructure. INFN-LNF is also the headquarters of the European project.

The construction phase has begun with the design of the new building that will house the infrastructure and with the study of the accelerator components and the radiation beam lines for users.

12.7   Conclusion

The legacy that Touschek has left, since he first proposed to build AdA, has marked the 60 years of activity of the Frascati Laboratories. We have always projected ourselves into the realization of accelerators and experiments at the frontier of knowledge. In every accelerator that we have realized for research and development or for users there are concepts and elements of absolute novelty that have often been used by the world community of accelerators.

This is a source of pride and a sense of belonging for all those who have worked in the Labs, therefore we are grateful to Bruno Touschek for his teaching and inspiration.

Chapter 13
Accelerator Physics at IJCLab-ORSAY

Achille Stocchi

Abstract Infrastructures and Research activities at the Orsay Campus, IJCLab, are described.

Talk given at Bruno Touschek Memorial Symposium 1921–2021—Frascati 2–4 December 2021

13.1 Introduction. IJCLab a New European Laboratory

IJCLab is a joint research unit of the CNRS, the University of Paris Saclay and the University of Paris, located on the campus of the Faculty of Science in Orsay. It is named after Irène Joliot-Curie, an exceptional scientist who was behind the creation of the Orsay Science Campus and then the University. IJCLab was born from the merger of five physics laboratories: the Centre de sciences nucléaires et de sciences de la matière (CSNSM), the Imagerie et modélisation en neurobiologie et cancérologie (IMNC) laboratory, the Institut de physique nucléaire d'Orsay (IPNO), the Laboratoire de l'accélérateur linéaire (LAL) and the Laboratoire de physique théorique (LPT). These laboratories were all geographically and thematically close to each other on the Orsay campus and shared a common history, linked to the creation in 1956 and then to the development of the University Orsay Campus.

IJCLab brings together about 750 people and is one of the five largest laboratories in Europe in the field of "physics of the two infinities". Among the 750 members of IJCLab, there are about 230 researchers, 350 engineers and technicians, and 120 Ph.D. students. IJCLab's identity is centred on the field of physics of the two infinities and their applications, with all the richness of the themes that constitute this physics. This is reflected in the presence of strong historical poles, poles linked to emerging

A. Stocchi (✉)
Université Paris-Saclay, CNRS/IN2P3, IJCLab, Orsay, France
e-mail: achille.stocchi@ijclab.in2p3.fr

© The Author(s) 2023
L. Bonolis et al. (eds.), *Bruno Touschek 100 Years*,
Springer Proceedings in Physics 287,
https://doi.org/10.1007/978-3-031-23042-4_13

themes and activities at the interfaces. This laboratory has the capacity, vocation and ambition to have a strong global impact on a wide range of scientific and technical fields, by being the driving force behind major flagship projects at national and international level. It also encourages and helps to support projects at more local scales and faster cycles, which may emerge as a result of scientific developments and/or technical innovations.

The scientific activities of IJCLab are structured in 7 scientific poles: Astroparticles, Astrophysics and Cosmology; Accelerator Physics; High Energy Physics; Nuclear Physics; Theoretical Physics; Energy and Environment; Health. They are supported by an Engineering Pole which brings together highly expert technical services, strongly integrated into national technical network, and structured into four departments in the fields of electronics, computing, instrumentation and mechanics. The presence of other highly technical departments: cryogenics, power RF, optics completes the very wide technical range of IJCLab. This ensemble represents a unique potential for the design, development and use of the instruments necessary for the scientific challenges of the coming decades (accelerators and detectors), allowing IJCLab to be a "builder laboratory".

The presence of a vast set of research infrastructures and technological platforms (Andromède, ALTO, Laserix, SCALP, Supratech…), is also an essential feature of IJCLab and will be described in the following.

It is complemented by administrative, support and transversal services that vitally support all these scientific and technical activities.

Finally, this laboratory is located in the heart of a world-class scientific cluster in connection with two universities (Université Paris-Saclay and Université de Paris). This places IJCLab in an exceptionally favourable environment for teaching, training and knowledge dissemination activities.

13.2 Accelerators Research Physics at IJCLab

13.2.1 A Short Historical Overview

The history of Physics at Orsay is fully linked and associated with accelerators. In fact, the Orsay Campus was chosen by physicists based essentially in Paris to install particle accelerators and develop research in nuclear and particle physics. The leading person was Irène Joliot-Curie.

A Linear Electron Accelerator and a Synchrocyclotron were installed and put in operation in the late fifties at the newly created LAL and IPN laboratories (Fig. 13.1).

Figure 13.2 shows the impressive development of the accelerators constructed in Orsay and also the contribution of Orsay physicists and engineering for the accelerator machines worldwide.

Fig. 13.1 Top. The Orsay campus in 1956. Bottom left—LAL 1956: The linear electron accelerator. Bottom right—IPN 1958: the synchrocyclotron

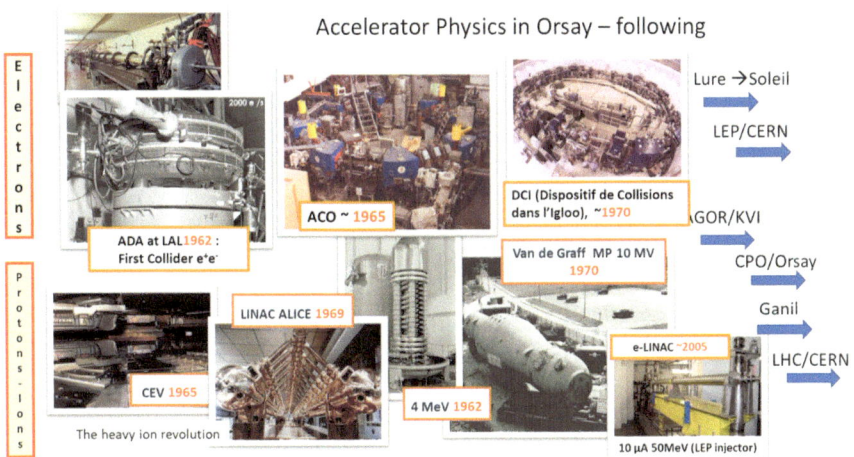

Fig. 13.2 A snapshot of the accelerator history of the accelerators at Orsay

A key moment in this history was the operation of AdA machine brought from Frascati and when the first e^+e^- collisions took place in a joint experiment with Orsay and Frascati teams led by Bruno Touschek. Please read the article of Jacques Haïssinski for a precise and detailed description of that period and of the close links between Frascati and Orsay and the leading role of Bruno Touschek.

13.2.2   IJCLab Accelerator Physics Today

The aim of the research pole in accelerator Physics is three-fold: to increase the R&D activities, to contribute to worldwide machines and to guarantee an efficient developing and operation of the complete set of local accelerator facilities of the new laboratory. Through its strengths and technical skills, this cluster is a global player, making a major contribution to the design and construction of large machines. This capacity to build large equipment is part of a national strategy, and it is proving to be a key tool in influencing the definition of scientific roadmaps for the research in nuclear and particle physics at international level.

Today the accelerator research department counts about 90 persons: physicists, engineers, technicians and about 10–15 Ph.D. students. In addition, several engineers and technician from the engineering departments (electronics, mechanics, instrumentation and informatics) contribute to the conception and the construction of the accelerators. All in all, more than 100 Full Time Equivalent Persons in a year work for accelerators Research, Development and Construction.

The research themes are: laser/plasma acceleration, Compton sources, and electron/laser interaction studies; the Physics of the instrumentation/diagnostic and beam manipulation for machine design and beam dynamics as well as conventional and "advanced" beam diagnostics; the design and implementation activities of the RF structures of an accelerator; the Vacuum, the studies on vacuum and ultra-dynamics vacuum and on materials/surfaces and layer deposition.

Over the last years, we have been playing a major international role in the field of high-intensity superconducting linear accelerators, participating to the construction of the Large Hadron Collider (LHC) at CERN, to Spiral-2 in GANIL, the X-FEL light source in DESY-Hamburg, the ESS neutron source in Sweden and now by participating to the PIP-II accelerator at Fermilab in the US for the neutrino factory facility. We have a remarkable expertise on electron and positron polarised beam production, intervening almost all the recently built electron/positron accelerator facilities (LEP, SLAC, KEK…). IJCLab has also a great expertise in laser/electron interaction with a recognized know-how on the improvement of laser finesse using a technology based on recirculating Fabry–Perot cavities (H1, KEK, ThomX…). In particular, ThomX is a new generation Compact Compton Source using 50 meV electrons to produce X-rays in the range of 50 keV and is actually under commissioning and has produced the first electrons.

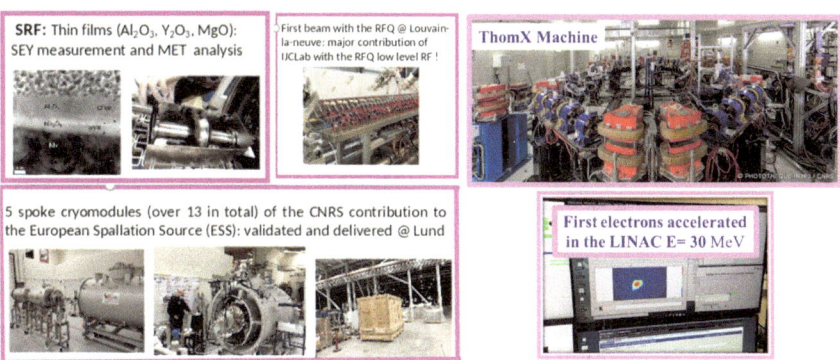

Fig. 13.3 A collection of few recent results and achievements in accelerators researches at IJCLab

The department is also involved in the PALLAS project that will implement a laser/plasma accelerator test facility consisting of a 150–200 meV electron Laser-Plasma Injector based on an existing laser driver (LaseriX).

IJCLab should host an ERL (Energy Recovery Linac) accelerator demonstrator as part of the particle physics roadmap. An intense 250 MeV (upgradable to 500 MeV) electron beam should be available with the PERLE project. The advent of PERLE would allow the design of a program to observe electron scattering from the interaction of the PERLE beam with a fixed self-confined target of 10^6 radioactive ions at relevant luminosities. This first step would fit perfectly as a prototyping and testing phase if the prospects of the nuclear physics community indicate a clear intention to focus on the electromagnetic probe for exotic nuclei.

A collection of few recent results and achievements at IJCLab are shown in Fig. 13.3.

The platforms at IJCLab are numerous and vary greatly in size and interaction with the outside world. Some are essential tools to develop scientific and technological activities of the laboratory, others, recognized "user facilities", constitute a major point of attraction for external collaborators. Here we described five of them.

The **ALTO** research platform gathers two accelerators unique in France: a 15 MV Tandem-type electrostatic accelerator for stable beams from proton to aggregates, and an electron linear accelerator for the production of radioactive beams by photofission. These machines serve a large variety of experimental devices on more than 10 beamlines for physics. The diversity of the beams produced allows one to carry out studies in nuclear physics, astrophysics and multidisciplinary studies (Fig. 13.4).

JANNuS-SCALP is an interdisciplinary research platform supporting many scientific fields ranging from materials sciences to astrophysics, including geology and nuclear physics. The platform consists of different equipments for ion irradiation/implantation (ARAMIS, IRMA and SIDONIE) and analysis (RBS, PIXE, PIGE, MET…). ARAMIS and IRMA are coupled with a Transmission Electron Microscope (TEM) allowing accelerating in situ inside the TEM (Fig. 13.5).

Fig. 13.4 The ALTO platform

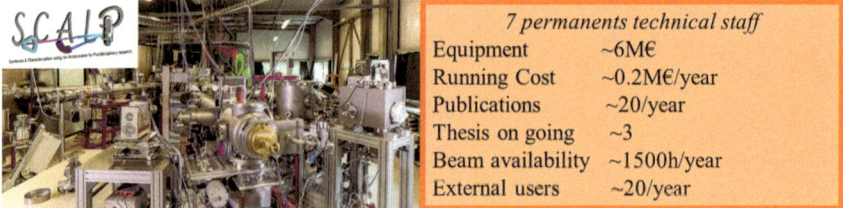

Fig. 13.5 The SCALP platform

LASERIX is a platform offering the scientific community to access to a complete range of coherent, intense and short (50 fs to 10 ps) sources in the near-infrared (800 nm) and EUV (30 to 90 eV) domains. LASERIX now extends its fields of application to the acceleration of electrons in a plasma wave created by a laser wake (Fig. 13.6).

Andromede is a multidisciplinary research platform, unique in the range of beams of several MeVs delivered: protons, multicharged atomic ions, gold molecules and nanoparticles. It is equipped with two beam lines. It allows studies related to ion/matter-aggregate interactions; measurements of nuclear sub-column fusion reactions of astrophysical interest; analyses of molecular fragmentation in the interstellar medium. It hosts experiments, for the surface analysis of samples essential for research in astrochemistry, biology, accelerator sciences and nanotechnology (Fig. 13.7).

A particular focus on **SUPRATECH** platform which is fully dedicated to R&D on the superconducting cavities of the future high-energy, high-power particle accelerators. It provides all the necessary equipment to prepare, package, assemble and test

Fig. 13.6 The LaseriX platform

Fig. 13.7 The Andromede platform

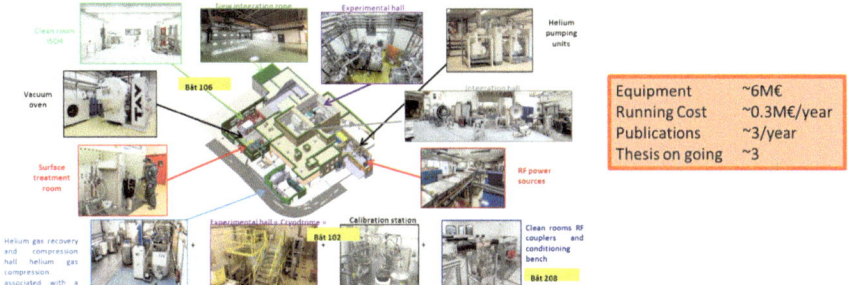

Fig. 13.8 The SupraTech platform

superconducting RF cavities for the projects in which IJCLab is involved. The platform equipment includes: a chemistry room, an ISO4 clean room, an assembly hall, an area dedicated to the integration of cryostats, two experiment halls, experimental areas equipped with vertical and horizontal cryostats for testing and validation of cavities with RF. In order to allow the optimal use of the previous infrastructures, the platform is equipped with: RF power sources at 88, 350 and 700 MHz, a helium facility comprising a helium liquefier and the associated recovery and compression facility; a 400 kW cooling system (HF sources) (Fig. 13.8).

13.3 Conclusions

The laboratories in the Campus were created to impulse fundamental physics in France and in Europe being able to contribute to the construction of large detectors and accelerators. Accelerators are the constitutive elements of the Orsay Campus (up to its urbanism!) since almost 70 years and the Orsay Campus of the University of Paris-Saclay is called the Valley of the Accelerators. IJCLab is continuing the tradition and continues to focus on fundamental science, also thanks to the researches, designs and contributions of accelerators.

Our links with Frascati are historical, strong and solid. We are twin laboratories and we have a common maestro, Bruno Touschek, whose memory I want to honour on the occasion of this memorial for the 100th anniversary of his birth in 1921.

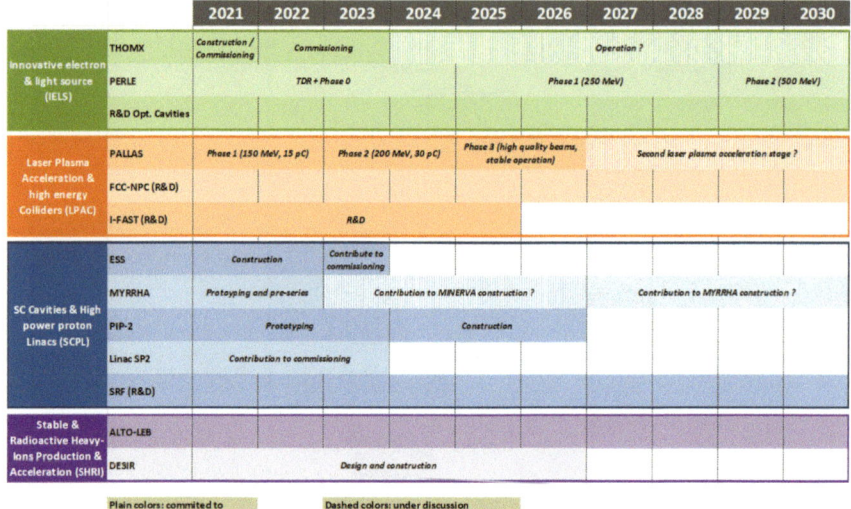

Fig. 13.9 The timeline of the accelerator researches, developments and constructions at IJCLab in the next 10 years

As a conclusion I would like to show in Fig. 13.9 the impressive future program in accelerator research and construction at IJCLab.

Chapter 14
Technical Challenges for Future Accelerators

Lucio Rossi

Accelerators have accompanied the development of nuclear and particle physics in the last ninety years. From first cyclotrons to the LHC and the discovery of the Higgs boson, throughout the collider concept demonstrated first by Bruno Toushek, accelerators have been instrumental for the discovery of new fundamental particles and mechanisms, thanks to an undeniable progress in performance supported by a continuous technical development. Now for the after-LHC era even larger challenges have to be faced, pushing existing technologies much beyond their present limits and pursuing until practical demonstration new technologies and concepts.

14.1 Introduction

The potential of accelerators as engines of discovery was clear since Lawrence, in collaboration with Livingstone, in Berkeley built the first accelerators capable to go well beyond a few MeV energy, i.e., the classical cyclotron for which he was credited with the Nobel Prize in 1939. Actually we like to cite the prophetic words used by Lord Rutherford in his opening speech at the 1927 Royal Society, in his capacity of President: "*The advance of science depends to a large extent on the development of new technical methods and their application... From the purely scientific point of view interest is mainly centred on the application of these high potentials to vacuum tubes in order to obtain a copious supply of high-speed electrons and high-speed atoms. This would open up an extraordinarily interesting field of investigation which*

L. Rossi (✉)
Dipartimento Di Fisica, LASA Laboratory, Università degli Studi di Milano, INFN-Sezione di Milano, Milan, Italy
e-mail: lucio.rossi@unimi.it

© The Author(s) 2023
L. Bonolis et al. (eds.), *Bruno Touschek 100 Years*,
Springer Proceedings in Physics 287,
https://doi.org/10.1007/978-3-031-23042-4_14

could not fail to give us information of great value, not only in the constitution of atomic nuclei but in many other directions" [1].

Already in the '40s and the '50s accelerators were a key tool for physics. But it was after two main breakthroughs: the proposal of the phase stability by Mc Millan and Veksler and the invention of the strong focusing by Christofilos, Courant, Livingstone and Snyder, applied to synchrotrons, that accelerators started rivaling with the cosmic rays for the discovery of new fundamental particles and mechanism. By providing copious flux of particles at the highest energy in a repeatable and predictable way, the increase in energy of the accelerated particle has accompanied the all new discoveries of fundamental particles from the sixties, as shown in the schematic of Fig. 14.1.

Accelerators are very complex instruments, with a variety of components, many of them having a strong influence on the performance of the accelerators. However, certainly the most significant components determining the performance of an accelerator are the accelerating structure, usually a cavity where an e.m. field with frequency ranging typically from the 30 MHz up to 30 GHz (for historical reason called radiofrequency, RF resonator) and the magnetic system providing the bending strength. Therefore in this paper we will discuss mainly the challenges for making progress in these two systems. However we will discuss also new accelerator schemes like the muon collider and the plasma acceleration.

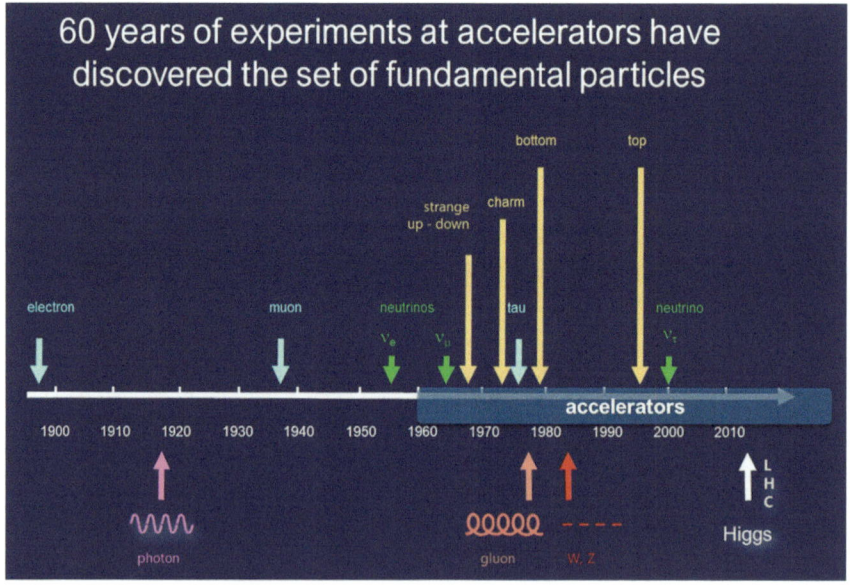

Fig. 14.1 Schematic of the timeline of discoveries of the fundamental particles with in evidence the ones discovered by accelerators (figure by L. Rossi)

14.2 The Accelerator Frontiers

14.2.1 The Energy Frontier

The first and most important parameters of an accelerator is the maximum kinetic energy attained by the particle beam. The energy gain is given almost invariably by the electric component of an electromagnetic (e.m.) field resonating in a cavity. The energy gain in a single cavity or in a gap between two electrodes of the cavity varies typically from about 100 keV to about 10 MeV. Considering here single charged particle, 1 MV voltage translates into a 1 MeV energy gain. We need to sum up the voltage of about ten thousands of cavities to reach the regime of hundreds GeV. Another way to express this is referring to the electric field. The highest performance RF cavities can provide, in pulsed operation mode, $E_{acc} \approx$ 30–50 MV/m if superconducting and $E_{acc} \approx$ 100–150 MV/m if high frequency normal conducting. Let's take the 100 MV/m, which is the baseline for the ee Clic collider, and clearly a length L = 10 km of RF field is necessary to reach 1 TeV. The Energy in a linac (linear accelerator) is simply given by: $E_{beam} = G \times L$ where G is the electric field (mostly referred as the potential gradient) and $L = \sum l_i$ with l_i being the length of a single cavity ($i =$ 1... N = total number of cavities). Of course the actual physical length of the whole accelerator needs to be multiplied by almost a factor two to account for all the space which is not covered by the accelerating field. To make just a 2 TeV c.o.m. collider with 100 MV/m cavity we need about a 2 × 20 km long infrastructure, considering that electron and positrons accelerators are independent and collide head-on. So the limit to the particle energy is feasibility and cost of the infrastructure, which scales with the accelerator length, and the technology limitation on the attainable electric field.

Another way to accelerate is of course based on the original idea that brought Lawrence to invent the cyclotron: recirculate the beam in the same cavity (ies) such as to use the same accelerating structure many times. To keep the particle in orbit a perpendicular magnetic field, that we call dipolar field, is needed. In theory we can pass the particles through the accelerating cavity as many times as we want and sum up energy without limitation. Of course this is not true since the increase in energy or momentum entails an increase of the magnetic field strength necessary to keep the particle in orbit; due to the barrier of the speed of light, speed is almost constant $v_{particle} \approx c$. So, as soon as we leave the favorable territory of the classical regime, where the magnetic force increases at constant field with velocity of the particle, the limit to the maximum attainable energy is the centripetal force we can provide through the magnetic field to keep the orbit. In relativistic approximation, for single charged particles, the relation turns to be quite simple: $E_{beam} \approx 0.3\ B \times \rho$ with E_{beam} in GeV, magnetic field B in tesla and ρ is the curvature radius inside the -uniform- magnetic field in meters. To reach the 7 TeV proton beam energy in the LHC in the 26.7 km long tunnel, considering that approximately only 2/3 of the tunnel length can be covered with dipoles, we need 8.3 T dipoles, a huge field that requires superconductivity. Therefore for circular accelerators, the maximum attainable energy is determined by cost and feasibility of infrastructure, i.e., the accelerator length or radius, and by the technological challenge, i.e., the maximum field intensity B we can generate perpendicularly to the particle trajectory.

In both cases, linear or circular, infrastructure and technology contribute with the same weight to the performance of the accelerator: technology is not all, after all! However, cost and size are of course limited, even for a community, like the one of high energy physics (HEP) that is used to large projects. Therefore pushing the technology is the way to secure progress of the so-called energy frontiers.

One can note that the equation $E_{beam} \approx 0.3 \, B \times \rho$ does not contains the rest mass, due to relativistic conditions, so it applies both to electrons and protons. However, beyond certain energy the electron dynamics in a synchrotron is dominated by the energy loss by radiation, due to centripetal acceleration, which is not called synchrotron radiation for nothing. This is the main reason why electron synchrotrons have an energy reach much less than for protons or other heavy particles, being the synchrotron energy loss proportional to $(E/E_0)^4$, where E_0 is the rest energy of the particle. It is clear that relativistic beam of electron radiates $\cong 10 \times 10^{12}$ times more power than protons with the same energy! Indeed the kinetic energy of the electron beam in the LHC tunnel, called LEP tunnel at the time, was limited at 100 GeV not by the magnetic field (just a very modest 0.2 T versus the 8 T used for the LHC) but the fact that at that beam energy the power loss by radiation was equal to the power transferred to the beam itself by the accelerating structure.

The complex rush toward high beam energy is depicted in Fig. 14.2, [2] which needs some explication. The graph, maximum beam energy versus time, is called Livingston plot, and is used to show the exponential increase of energy versus time. To make compatible in terms of physics reach the most recent accelerators, all of collider type, with the previous generation (fixed targets), the energy is reported as equivalent beam energy on a fixed target. Such a graph shows that the colliders, first realized by Touschek in Frascati, were essential to support the exponential growth, that otherwise would have stopped already in the sixties. Another factor that helped to support the exponential growth was the introduction of superconducting magnets for hadron colliders, in the late eighties. The graph of Fig. 14.2 shows clearly that even in the case the next projects would be timely realized: Fcc-ee and/or FCC-hh, ILC and/or CLIC, we are almost in a saturation regime. Since we cannot grow only by size, as mentioned before, this graph makes evident the urgency of a technology jump or of novel ideas in our field.

14.2.2   The Luminosity Frontier

For new discovery and studies, energy is not enough. Any energy increase must be accompanied by a consequent increase of the luminosity to compensate the decrease of cross section with energy. Luminosity is a fundamental parameter for measuring a collider performance, right after its beam energy, and is defined as: $L = \dot{n}/\sigma$ where $\dot{n}$ is the collision rate and σ the cross section of a particular event (or the

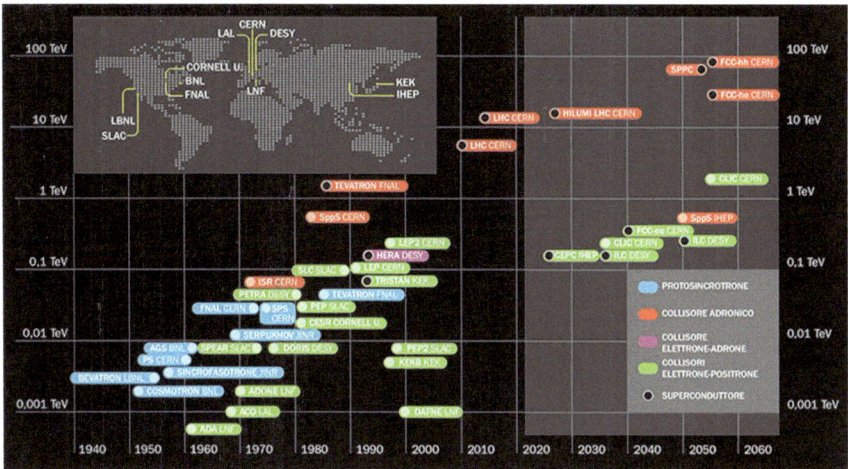

Fig. 14.2 Energy versus time for a compilation of accelerators (Livingstone plot). In grey background, on the right, the future accelerators under construction or under study (figure by L. Rossi published in Asimmetrie, INFN)

sum of cross sections). Even more interesting for physics reach is the concept of integrated luminosity over a period of time, that can be one year or the full span life of an accelerator: $L_{int} = n/\sigma$, where n is the total number of collisions over the time interval considered.

For a circular collider the luminosity depends: i) on the square of the bunch population, N^2 (we assume the colliding bunches have both the same population, N); ii) on the collision frequency, which means increase as much as possible the number of bunches; iii) on the inverse of the beam transverse size at collision point $\sigma_{xy} = \varepsilon \cdot \beta^*$, where ε is the transverse emittance and β^* is the optical function at the collision point. In many case a machine, after a first phase at maximum energy and at a certain luminosity, undergoes an important upgrade concerning mainly increasing the luminosity by a factor 3 to 10, with change in hardware limited to a few components and cost of a fraction, 10–25%, of the initial machine: a way to double the lifetime of a machine without spending too much, leveraging the initial investment.

In Fig. 14.3 the luminosity reached by various machines is plotted versus time, mixing lepton and hadron colliders. It is worth noticing that now SuperKEK overshadows the performance of any previous lepton (e^+e^-) while LHC the one of previous hadron (pp or pp_{bar}) colliders.

14.2.3 The Intensity Frontier

Another frontier is the one of beam intensity. Intense beams are necessary also in a collider for HEP, in order to increase the luminosity. However here we refer more

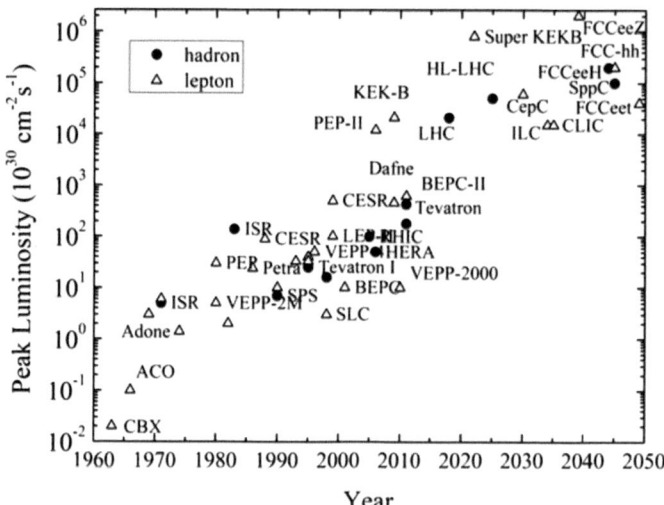

Fig. 14.3 Luminosity performance for a vast compilation of machines. HL-LHC us under construction, all other machine on the right are just in design study phase. Courtesy of V. Shiltev, FNAL

to beam used on targets. Demand of intense proton beams comes from neutron spallation sources, like SNS in the US, JPARC in Japan and ESS in Sweden. Intense proton beams are required also by neutrino sources, like JPARC, again, and FNAL in the US. In Fig. 14.4 we plot the beam intensity of the proton sources versus time.

Fig. 14.4 Beam power for proton accelerators versus time (compilation by V. Shiltev, FNAL)

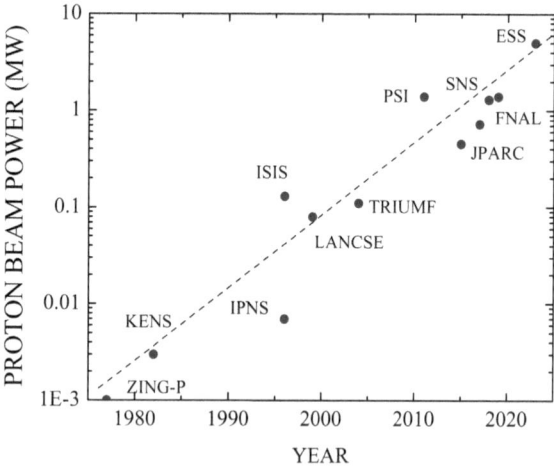

14.3   Technology Advance: Superconducting Magnets for h–h and RF Cavities for e-e

As mentioned before the progress of the energy frontier is linked to the progress of the superconducting magnet (SM) and of accelerating structures, both superconducting rf (SRF) or normal conducting. We will try to go over the main advance and the progress needed for the next generation colliders.

14.3.1   Superconducting Magnets

The LHC superconducting magnets are the summit of a 30 year-development for hadron colliders. The 8 T used in the LHC main dipole, see Fig. 14.5, is more or less the maximum that can be reached by using the superconducting Nb-Ti alloy.

To go beyond the limit of Nb-Ti new more advanced superconductors need to be employed, like the A3 compound Nb_3Sn. The situation is depicted in the graph of Fig. 14.6, reporting the performance of superconductors, in terms of engineering current density J_E, as a function of magnetic fields. As it can be seen, since magnets needs to operate in the region $J_E \approx 500$ A/mm^2, below 8–9 T Nb-Ti will be the

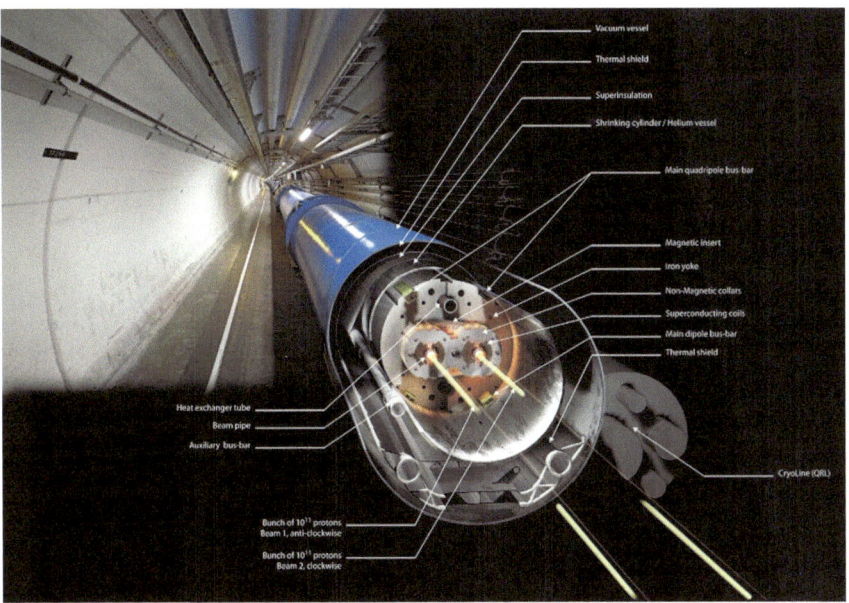

Fig. 14.5 The LHC main dipole: picture with opened cross-section superimposed. Yellow lines represent the two counter-circulating proton beams (not in scale). The superconducting coils is the crescent around the yellow beams (CERN Archive)

invariable choice, given its low cost and easy manufacturability, like the LHC magnets as indicated in Fig. 14.6; between 9 and 15 T Nb_3Sn is the suitable material, like it is used for HL-LHC (in the Fig. 14.6 a cross section of an HL-LHC dipole is shown for the Nb_3Sn regime); above 16 T the only choice is use of HTS (high temperature superconductor), either in form of YBCO (yttrium barium copper oxide) or bismuth based superconductors (Bi-2212 or Bi-2223). For various reasons REBCO (where RE stays for rare earth, since yttrium can be substitute partially or totally by gadolinium) is in this moment more favorite but the community has not yet made a clear decision. In Fig. 14.6 the use of HTS is depicted by a cross section of a dipole studied for the first idea of an HE-LHC [3].

Since a few years, and following the development of very high current density Nb_3Sn for accelerator magnets, there is a strong effort by CERN and by US laboratories (FNAL, LBNL, and BNL) to produce magnet suitable for accelerators. Magnets for colliders are the most difficult application of superconductivity because they need a very high current and a very precise field. The requirements translate into severe requirements on superconductors. The effort started some 20 years ago and is oriented to make magnets for the High Luminosity LHC project, of a level of 11–12 T [4]. It has been a huge effort, to overcome the difficulty due to the characteristics of Nb_3Sn: it requires a thermal treatment at 650–700 ºC of the whole coils, which poses technical challenges for the insulation, and the brittleness of Nb_3Sn when in superconducting state implies a complex mechanical structure with tight control of the tolerances. For the High Luminosity LHC (HL-LHC) project two types of high field superconducting magnets in Nb_3Sn are required: about 30 units of the inner

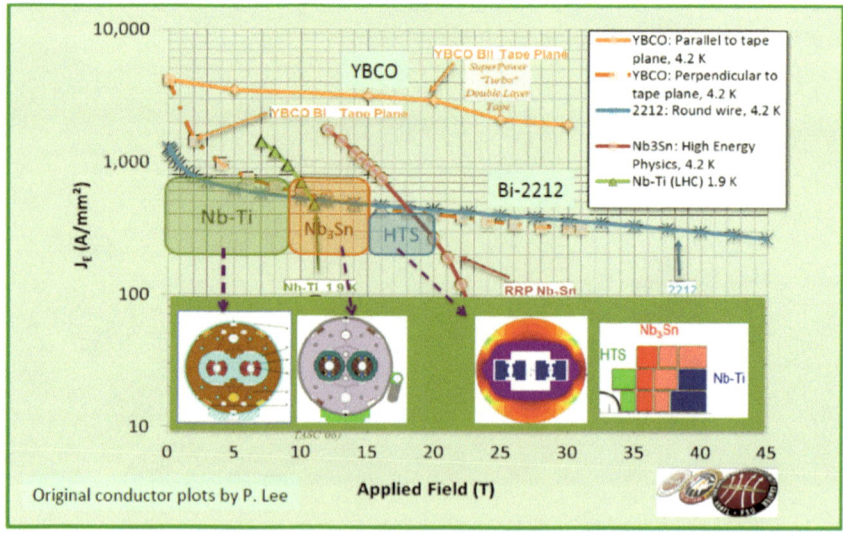

Fig. 14.6 Engineering critical current density for the practical superconductors and regime of application for various magnetic field level (figure by L. Rossi)

Fig. 14.7 Pictures of the two Nb3Sn magnets for the HL-LHC project at CERN (photo CERN Archive)

triplet (IT) quadrupole, with a very large aperture and 11.5 T peak field, 4.2 and 7.15 meters of length, and eight dipoles rated for 11 T and 5.5 m long. Despite the difficulties various short model magnets (1 m long) have been successfully manufactured and tested [5]. The passage to long magnets has been more painful than anticipated. However now there is a number of long magnets that have successfully reached the nominal field value, especially for the IT quadrupole that are the backbone of the project [6]. Their installation is foreseen in the period 2025–2027. In Fig. 14.7 the picture of the two among the first long prototypes for HL-LHC, an 11 T dipole and a IT quadrupole are shown.

In Fig. 14.8 the progress for magnetic field reached by magnets in various hadron colliders is shown. Blue dots refers to operating magnets in real accelerators. For Nb_3Sn this will happen at around 2027 (orange dot) at the commissioning of the HL-LHC magnets. The clouds of orange circles in the Nb_3Sn band of Fig. 14.8 indicates the number of R&D magnets for HL-LHC.

Beyond High Luminosity LHC we have the objective of the next hadron collider: FCC-hh, which is based on a tunnel of 100 km that would first host the lepton collider, FCC-ee and then the hadron collider, very much like the LEP/LHC tunnel. FCC-hh is designed for the CERN area, and the today baseline foresees the use of 16 T dipole magnets in Nb_3Sn. The possibility of going above, about 20 T is left open if HTS would be possible. The two green diamonds and the brown one in Fig. 14.8 refers to initial R&D for the FCC-hh: green refers to the baseline 16 T in Nb_3Sn and brown refers to a hybrid solution with HTS boosting the performance of a 15 T Nb_3Sn dipole.

In the period 2013–2018 a 16 T design was extensively investigated and about four magnet layouts were produced. First a classical $cos\vartheta$ design, by INFN Genova and Milano-LASA team, which has been chosen as baseline for the FCC-hh technical design report. Another design makes use of rectangular coil block, by CEA-Saclay, while the Spanish CIEMAT of Madrid ended up with a design called common coils, where the two apertures for the counter circulating beam are placed vertically, one on the top of the other, rather than side-by-side as in all other solutions. In Fig. 14.9 we report the cross section of these main dipoles proposed for FCC-hh.

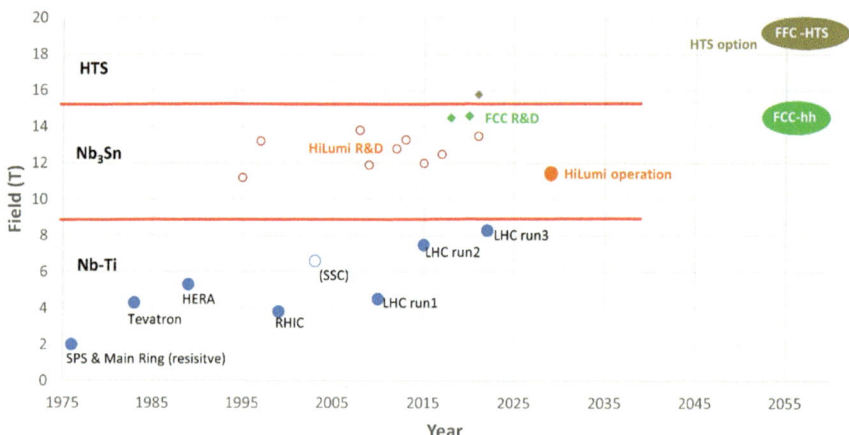

Fig. 14.8 Progress in time of the magnetic field for hadron collider magnets until HL-LHC (figure by L. Rossi)

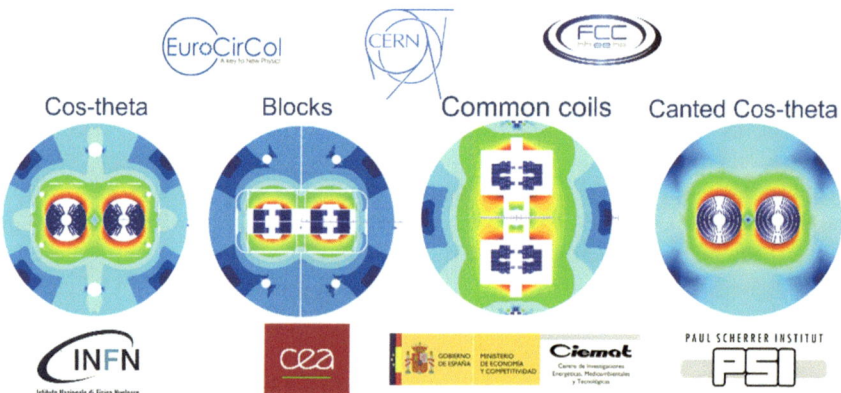

Fig. 14.9 Various magnet cross section studied in the H2020-Eurocircle project for FCC-hh by various Institutes under the guidance of CERN (figure by L. Rossi)

However, following the difficulties encountered in the HL-LHC for long magnets, now the community has made a step back and the idea is to produce magnets of 12 T with more robust characteristics with respect to the pioneering HL-LHC magnets. The jump from the 11–12 T level of HL-LHC, that has required about 20 year of development, see Fig. 14.8, up to the 16 T level seems now quite big, indeed. For FCC-hh almost 5000, 15 m long, magnets would be needed, versus the few tens of 4 to 7 m long magnets for HL-LHC. Therefore, demonstrating the manufacturability of 12 T long magnets in Nb3Sn seems a necessary intermediate step, which may take all the present decade. Then the solution is either to use it to built a 12 T based FCC, with an energy reach of 70–75 TeV c.o.m., or to continue an R&D for reaching the ultimate limit of Nb$_3$Sn. However, we are not sure that the limit of 16 T given by Nb_3Sn conductor performance can be actually attained: maybe mechanical degradation will impose a lower limit at 14 or 15 T. In such a case FCC-hh would need, if the final goal of 100 TeV performance remains important, the use of HTS superconductor.

As shown in the graph of Fig. 14.6, HTS may boost performance up to 20 T and beyond, being 25 T maybe the upper limit of such material. HTS materials, in particular REBCO, are mechanically robust: they can withstand stresses of up to 400 MPa versus the 150 MPa that can be applied to Nb3Sn. However they come in form of coated flat tape instead of multi-filamentary round wire, like the classical superconductors. This means that the field quality is today a serious issue for which there is no solution, yet. The second big issue with HTS is the difficulty in making a sound protection following a quench (quench being the sudden transition from the superconducting state to the normal conducting state). Because of the high transition temperature HTS have a huge stability margin, measurable in tens of joules rather than mJ typical for the classical superconductors. However, this stability margin entails that a quenched zone would propagate very slowly making a detection very difficult, with consequent possible irreversible damage of the coils. We have devised a strategy to limit this effect [7, 8], based on current redistribution, and demonstrated it in a small demo magnets of about 3–4 T. However a long R&D remains for introducing these features in a real accelerator magnet.

In Fig. 14.10 we show a compilation of the performance of the small HTS magnet built so far [9], with also a figure of the CERN small HTS dipole (35 mm free bore, 700 mm in length) holding the record dipole field of 4.5 T in standalone and having made a record of nearly 16 T when inserted in a high field facility (brown diamond in the plot of Fig. 14.8).

In conclusion we believe that for Nb_3Sn the 12 T level would be a sound field and to go beyond the 12 T is probably better to use HTS, if its cost decreases by at least a factor 3 to 5, since today is too expensive. Another possibility, also based in case of cost reduction, is to use HTS magnets for 12–16 T operating at higher temperature,

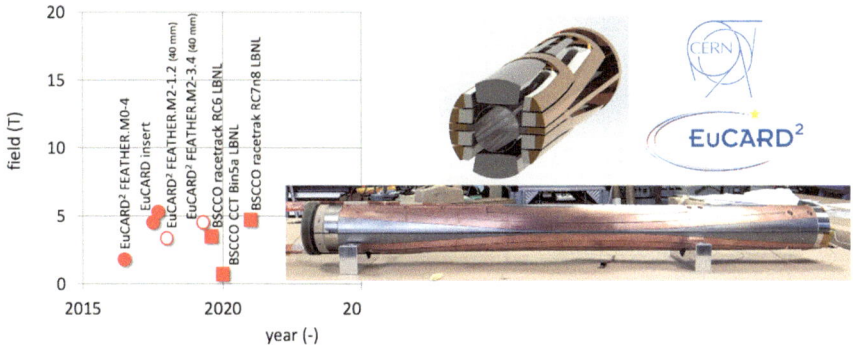

Fig. 14.10 Left: compilation of field result for various HTS racetrack coils (no bore) or dipole magnets (with an accessible bore); right: picture (bottom) and rendering (top) of the CERN dipole for Eucard2 program holding the record field for HTS dipoles (figure by L. Rossi)

e.g., 20 K instead of the 1.9 K necessary for Nb_3Sn. The 20 K operation temperature could save some 200 MW or more in the powering of the cryogenic plants, which is a very important goal both for environmental reasons and for cost reasons.

14.3.2 Superconducting RF Development

While for magnets the choice of superconductivity is very well established, the use of SRF or of normal conducting RF (NCRF) is not so straightforward. Indeed despite the better global energy efficiency in case of SRF the highest attainable gradients are still in the camp of NCRF, i.e., of copper cavity. Therefore the choice of using one or the other technology depends on the structure of the beam and on various considerations, not last also political ones.

In Table 14.1 we report a summary of the main points for the two technologies and in Fig. 14.11 is depicted the progress of SRF in the L-band (1.3 GHz, the most used for electron acceleration) both for single cell and for multicell cavity. In all cases we refer to Nb bulk cavities. As can be observed in Fig. 14.11 multicell performances have progressed steadily and have attained nearly 50 MV/m. Taking into account the inevitable contingency for operation in a cryomodule and for the fluctuation of a large production, which in case of X-FEL showed a variance of about 5 MV/m, we can say that the SRF technology is mature to go beyond the 30–35 MV/m of the ILC design, as indicated by the 45 MV/m indicated in parentheses in the RF performance table.

Normally linacs work with duty cycle. However the need of continuous (CW) beam is becoming high. For HEP machines in particular the FCC-ee, as well as its Chinese counterpart CepC, both e^+e^- machines at some 250–350 GeV c.o.m. collider would requires a large numbers of CW cavities. In such case the electric field is lower about 20 MV/m but the request on the Q_0 is very high, to limit the cryogenic losses. It is worth noticing the remarkable progress in Q_0 at moderate electric fields thanks to infusion and at higher gradient thanks to doping with nitrogen, as clearly shown in Fig. 14.12. This rather recent development shows that there is still a lot of room to improve in this technology [10].

A different interesting developing line is the study for using A15 compounds, namely Nb_3Sn as superconducting material for the SRF cavity: Nb_3Sn would allow operating the cavity in the 4.2 K or higher temperature, rather than the 2 K required by niobium bulk. This would result in a considerable saving on the electric bill and in increased cavity stability (heat capacity increases cubically at these temperatures so the same temperature margin gives eight time largest energy margin, very much like superconducting magnets). Recent result are very encouraging, [11] and alternative methods for coating like sputtering are being investigated, see Fig. 14.13. A similar line is pursuing the use of HTS thin films, opening the way to high temperature operation, even more than Nb_3Sn, as suggested in [12] on bulk Nb_3Sn cavity and an interesting development on Nb_3Sn coating on a copper cavity. If successful this last

Table 14.1 Comparison between normal and superconducting RF systems (*Source:* A. Yamamoto, KEK)

Parameters	Normal conducting (CLIC)	Superconducting (ILC)
Electric Field (MV/m)	70–100 Higher energy or shorter accelerator	30–35 (45) Higher efficiency, steady state beam power from RF input
RF frequency f (GHz)	12 High efficiency peak power Need precision alignment and stabilization for wake filed compensation	1.3 Large aperture → small wakefields
Quality factor Q0	$<10^5$ Resistive wall losses compensated by strong beam loading	$\cong 10^{10}$ Small losses Losses at cryogenic temperatures (250–500 factor)
Pulse structure	180 ns/50 Hz	700 μm 5 Hz
Fabrication issue	μm level mechanical tolerances	Material quality (purity) and complex clean room chemistry
Efficiency considerations	High-efficiency RF peak power production through long-pulse, low freq. klystrons and two-beam scheme	High-efficiency RF also from long-pulse, low-frequency klystrons

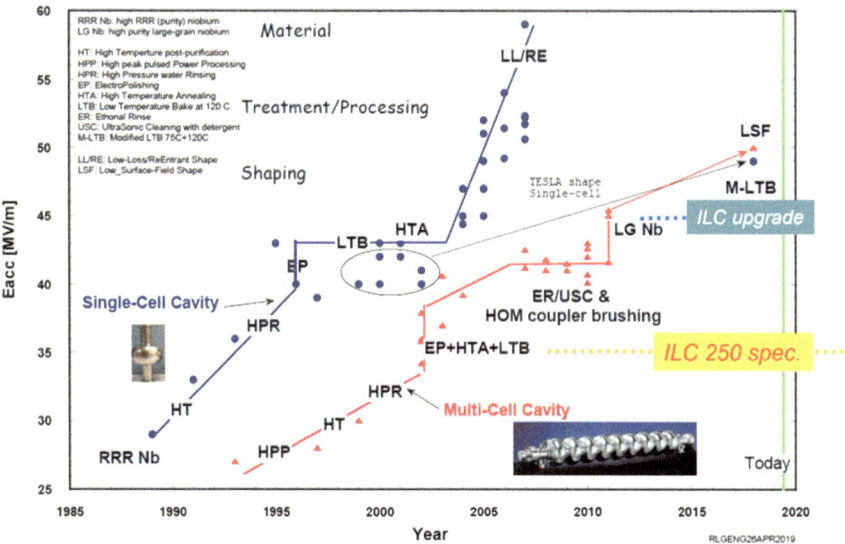

Fig. 14.11 Progress of the performance in SRF gradient for 1.3 GHz electron cavities (figure by CERN Courier, adapted by L. Rossi)

Fig. 14.12 Infusion & doping Q0 (Ref. [11])

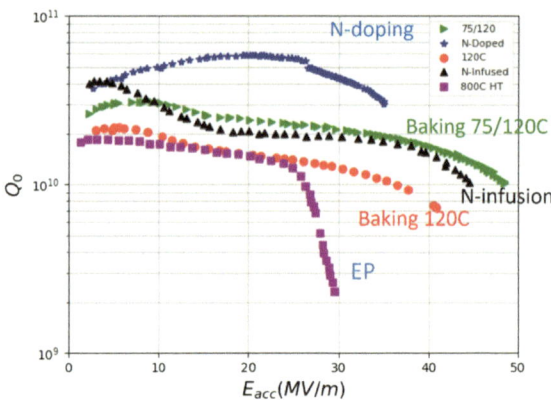

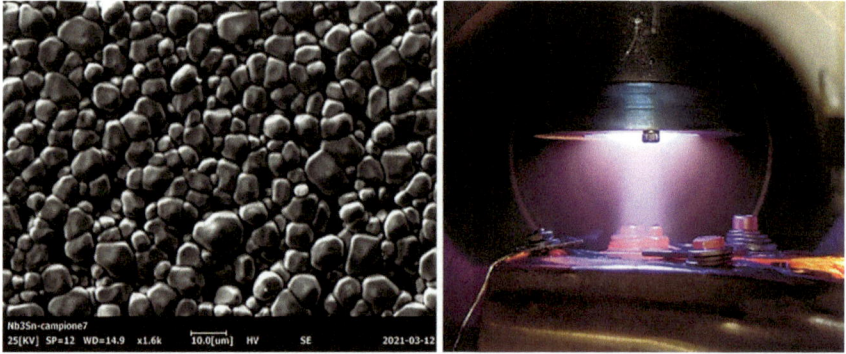

Fig. 14.13 Nb$_3$Sn thin film deposited on high conductivity Cu substrate for cavity R&D via sputtering (left) and picture of the sputtering process (right), Ref. [12]

development could open the way to higher gradient, with reduced cryogenic power and very high stability since the substrate would be high conductivity copper.

14.4 Beyond Superconducting RF Cavities and Superconducting Magnets

From what said in the previous sections, increasing basic performance, the fields in magnets or in cavities, takes a very long time constant: one can see a doubling each 20–25 years, and maybe more in the future.

An electric field two times larger than the one of SRF cavities is offered already now by CLIC technologies working at 12 GHz, as mentioned in the SRF section. We refer to specific paper on this [13].

Anyway, we think that apart from the incremental gain offered by continuous R&D along the routes previously described, we have two alternative routes:

The muon-collider, that holds the promise of accelerating leptons in a circular accelerator, where the luminosity is better controlled than in a linear collider, at energy of 3 TeV and then up to 10 TeV. At 10 TeV a lepton collider can claim a physics reach similar to the FCC-hh, but with an infrastructure size much smaller, comparable to the present LHC one. A muon-collider specificity is presented in [14].

Plasma acceleration. A lot is going on at present on this rather novel technology. This technology that holds the promise of reducing the infrastructure, for the same energy, by a factor of ten or so, thanks to electric field that in plasma can be as higher as tens of GV/m. While it looks like that for smaller size applications, like FEL, medicines, etc.... this technology may be a real game changer, the usefulness for colliders, also in view of the power consumption, is still to be demonstrated also in principle. But the challenge is one of the most interesting in all accelerator sectors. The plasma acceleration is discussed in Chap. 13 of these Proceedings and in R. Assman's contribution to the Symposium.

14.5 Conclusions

Accelerators have been a drivers of technology innovations. The use of NMR for spectroscopy and especially for MRI has been made possible by the development of the superconductors and superconducting magnets of the Tevatron. Now various machines for hadron therapy employ superconducting magnets. Use of superconducting magnets for cyclotrons and of superconducting cavities for CW linacs are essential for transmutation and efficiency improvement of nuclear power plants. Also CLIC technologies are being used for new type of flash-therapy with electron beams.

The request for new development in accelerator technology to face the challenges posed by the future HEP colliders will certainly profit the fundamental knowledge itself but will also certainly result in new interesting—maybe unexpected- societal applications.

References

1. Rutherford Radiation, http://www.aip.org/history/exhibits/rutherford/sections/rutherfords-rad iations.html
2. L. Rossi, *Megamacchine per il microcosmo*, on « Asimmetrie.it » INFN journal on line, https://www.asimmetrie.it/megamacchine-per-il-microcosmo
3. L. Rossi, E. Todesco, Conceptual design of 20 T dipoles for high-energy LHC, in *Proceedings of the "The High-Energy Large Hadron Collider"*, ed. by E. Todesco, F. Zimmermann. Malta, 14–16 October 2010, CERN-2011-003 and EuCARD-Conf-2011-001, 8 April 2011, pp. 13–19

4. L. Bottura, G. de Rijk, L. Rossi, E. Todesco, Advanced accelerator magnets for upgrading the LHC. IEEE Trans. Appl. Supercond. **22**(3), 8 (2012). https://doi.org/10.1109/TASC.2012.218 6109
5. E. Todeco et al., The HL-LHC interaction region magnets towards series production. Supercond. Sci. Technol. **34**(2021) 053001 (38 p) (2021). https://doi.org/10.1088/1361-6668/abdba4
6. O. Bruning, L. Rossi (eds.), The high luminosity large hadron collider, Adv. Ser. Dir. High Energy Phys. **24** (2015). World Scientific publisher; ISSN 1793-1339
7. L. Rossi et al., The EuCARD2 future magnets program for particle accelerator high-field dipoles: review of results and next steps IEEE Trans. Appl. Supercond. **28**(3), 10 (2018). https://doi.org/10.1109/TASC.2017.2784357
8. L. Rossi, C. Senatore, HTS accelerator magnet and conductor development in Europe. Instruments **5**(8), 33 (2021). Open access, https://doi.org/10.3390/instruments5010008
9. P. Vedrine et al., High-field magnets, in *The European Strategy on Particle Physics–Accelerator R&D Roadmap*, ed. by N. Mounet, CERN 2022-001
10. A. Grassellino et al., Accelerating fields up to 49 MV/m in TESLA-shape superconducting RF niobium cavities via 75C vacuum bake, arXiv:1806.09824v1 [physics.acc-ph]
11. S. Posen et al., Advances in Nb3Sn superconducting radiofrequency cavities towards first practical accelerator applications. Supercond. Sci. Technol. **34**, 025007 (10 p) (2021)
12. S. Bousson et al., High-gradient RF structures and systems, in *The European Strategy on Particle Physics–Accelerator R&D Roadmap*, ed. by N. Mounet, CERN 2022-001
13. S. Stapnes, CLIC, these proceedings
14. D. Schulte, Muon-Collider, these proceeding

Part III
Talks Given at Accademia Nazionale dei Lincei, Roma

Chapter 15
The Making of AdA: Bruno Touschek's Journey from Widerøe's Betatron to Storage Rings

Giulia Pancheri

Abstract In Italy, in 1960, Bruno Touschek conceived and built the first particle-antiparticle collider, called AdA. The different roads and pathways which led to the successful demonstration that it was possible to accumulate electrons and positrons in a single ring and make them collide, originated in different parts of Europe: Austria, Germany, Norway, UK, France and Italy, in parallel with similar developments in the US and USSR. AdA's success was due to Bruno Touschek's extraordinary formation as both an accelerator scientist and a theoretical physicist, and in the unique environment he found in the University of Rome and the Frascati Laboratories, where the post-war reconstruction of Italian physics was taking place, in parallel with the Europe-wide effort, that led to the creation of CERN.

15.1 Introduction

In 1972, Bruno Touschek was made a foreign member of the Accademia Nazionale dei Lincei, in recognition of his outstanding contribution to teaching and science. Three years before, ADONE, the beautiful machine in the Frascati National Laboratories, had successfully started operations. ADONE had started two beam operation in 1969, and new phenomena had appeared soon after electrons and positrons started circulating at a yet unsurpassed center of mass energy. Touschek had been envisioning and planning for it at least since November 1960, when AdA, the first ever electron-positron storage ring he had proposed to build in Frascati, could be seen to be well on its way. Since this first proposal, the world of particle physics had changed, and when Touschek was welcomed into the Academy, a number of particle-antiparticle colliders were in operation or in advanced planning and construction stage: ACO in France, VEPP-2 in the USSR, the ISR at CERN, SPEAR in the USA, DORIS in Germany. At the Academy, Touschek left one of his many legacies to Italian science and culture: the lectures he organized in the context of a project he called *Scienza*

G. Pancheri (✉)
Laboratori Nazionali di Frascati dell'INFN, Via Enrico Fermi, 54, Frascati 00044, Italy
e-mail: pancheri@lnf.infn.it

© The Author(s) 2023
L. Bonolis et al. (eds.), *Bruno Touschek 100 Years*,
Springer Proceedings in Physics 287,
https://doi.org/10.1007/978-3-031-23042-4_15

Fig. 15.1 Bruno Touschek at Accademia dei Lincei, on the occasion of his nomination, in 1972 (left panel), from family documents, courtesy of Francis Touschek, and (right panel) on April 15th, 1975, with (at center) Paul M. Dirac, the Accademia President Beniamino Segre and Marcello Conversi at right, courtesy of Sapienza University of Rome–Physics Department Archives, https:// archivisapienzasmfn.archiui.com, documents provided for purposes of study and research, all rights reserved

vivente, living science in English, where lectures on modern science were given by renowned scientists, such as Paul Dirac, Marcello Conversi, Rolf Widerøe, Edoardo Amaldi, himself, and others, Fig. 15.1.[1]

In addition to his contribution to particle physics, teaching and science outreach efforts [1], Touschek also left an important legacy to the Frascati National Laboratories, when he created a theoretical physics group, aimed at future planning and exploitation of ADONE's expected physics results. This legacy is found in this volume in the contributions from Paolo Di Vecchia, Mario Greco, Giancarlo Rossi, and in the present one, as I had the great privilege to be one of the young people Touschek gathered in Frascati, in what Fernando Amman, ADONE's director, called "the golden years of the Laboratory".[2]

I had graduated in physics from University of Rome in February 1966, with a thesis on a rather exotic process, *Coalescence and decay of photons on nuclei* under the supervision of Benedetto (Nino) De Tollis.

In 1966, returning to Rome after a vacation in the Dolomites which had followed my graduation, I learnt from Giancarlo Rossi and Paolo Di Vecchia that Touschek was starting a theoretical physics group in Frascati. After much hesitancy and fearful of rejection, I went to Touschek's office in the Physics Institute in Rome to inquire about the possibility of joining the new group. The answer which I received a fews days later was positive. Nino de Tollis had vouched on my behalf and in May I arrived in Frascati, with a post-graduation fellowship. This is when I started on the most important experience in my scientific life.

---

[1] The Lincei lectures were video-taped by Francis Touschek.

[2] Amman's letter to Edoardo Amaldi, 1978, Sapienza University of Rome–Physics Department Archives.

15.2   Who Was Bruno Touschek?

In this contribution I will present some highlights in the story of Bruno Touschek and the accelerators he built, as well as other legacies he left.

Touschek was a protagonist in the development of particle physics in Europe in the second half of the XXth century, and his life journey, seen in Fig. 15.2, mirrors the tragedy of World War II and the hopes of postwar reconstruction.

Born in Vienna in 1921, Bruno Touschek left Austria for Germany, when his studies in theoretical physics were interrupted by racial discrimination, because of his Jewish origin from the maternal side, Sect. 15.3.[3] Improbable as it may appear, once in Germany he was able to survive and study through the war, while moving between Hamburg and Berlin, protected by Arnold Sommerfeld's former colleagues and students. This is known to have happened in a context where some German scientists would employ their Jewish, or half-Jewish, friends in technical and scientific projects of interests to the military, in order to protect them from deportation to forced labor and ultimately death in the concentration camps [2]. Such destiny would have been Touschek's, but he survived through extraordinary circumstances, Sect. 15.3.1. In Germany he participated in a project to build a 15 MeV betatron, directed by the Norwegian scientists Rolf Widerøe, and financed by the *Reichsluftfahrtministerium*, (RLM), the Reich Ministry of Aviation. The war finished, his knowledge of particle accelerators became of interest to the Western Alliance, and he was first taken to Göttingen, where he obtained his Diploma in Physics, and then to the University of Glasgow to participate in the construction of a 300 MeV synchrotron and continue his studies for a doctorate, Sect. 15.4.3. Moving to Italy in 1952, he was a protagonist of the reconstruction of physics in and around the University of Rome. In the Frascati National Laboratories and, later, in France, Touschek catalyzed the energy of the scientists around him toward the construction of an "unthinkable" machine, which was named AdA, for Anello di Accumulazione. In AdA, for the first time ever, "particles which are not found in the world which surrounds us, were kept and stored for a long time" (in Touschek's own words), Sects. 15.4.4 and 15.4.5.

After he prematurely passed away on May 25th, 1978, his life and work were described by his mentor and friend Edoardo Amaldi, who gathered recollections and documents in a biographical portrait of still unsurpassed emotional and historical impact [1]. Not long after, two young historians of physics at the University of Rome, started preparing a catalogue of all the papers which Touschek had left in his office. They were alerted to the ongoing trashing of these papers by one of Bruno's young collaborators, Amilcare Bietti. Awed by the still vivid memory of Bruno's extraordinary accomplishments and personality in the Rome Physics Institute, Battimelli and De Maria rushed to Bruno's former office and literally extracted his papers from the large trash bin already on its way out. Their rescue efforts were published in a report with reproduction of unpublished notes, a full listing of Touschek's papers and a detailed guide to Touschek's archives, offering a vivid portrait of his personality [3]. The saving of Touschek's office papers signed the beginning of the extensive

---

[3] See Luisa Bonolis' contribution to these Proceedings.

Fig. 15.2 A cartoon showing Touschek's life journey through Europe, from his birth in Vienna in 1921, to his death in Innsbruck in 1978

collection of documents from other physicists, now existing in the Sapienza University of Rome–Physics Department Archives, https://archivisapienzasmfn.archiui. com. Thus, to a large extent, these archives, which are continuously enriched as the Rome institute professors retire, represent one more legacy coming from Bruno to the institution which welcomed him in 1953.

In those years, tributes to Bruno Touschek appeared in the context of the birth of electron-positron colliders, as from Fernando Amman [4], a protagonist in the development of the particle accelerator science in Italy. In 1987, the Bruno Touschek Memorial Lectures brought to Frascati National Laboratories a roster of the scientists from the international particle community, with novel contributions to Bruno Touschek's memory [5].[4] In the years to follow, Bruno's memory was kept alive by his close friend and collaborator Carlo Bernardini [6] through writings and public talks, while interest in Touschek's life increased alongside planning for new particle colliders [7]. Much more has since been published about Bruno Touschek, as presented in [8]. The present status of the narration of Bruno's life includes three

---

[4] A series of circumstances delayed the publication of the Lectures contributions until the year 2004.

docu-films[5] and readings from unpublished letters sent by Bruno to his family [9]. These letters were kept and chronologically ordered by Bruno's father, Franz Xaver Touschek. After his death in 1971, they were sent to Bruno, and, later still, preserved by Touschek's widow, Elspeth Yonge Touschek. Thanks to her and these letters, large parts of Bruno's life, unknown at the time of his death, became accessible, in particular shedding light on Bruno's role in the making of Rolf Widerøe's betatron.

Now, a hundred years after Bruno's birth, from these different sources a coherent description arises of the extraordinary circumstances which led to AdA's proposal and what its construction brought to particle physics [10].

15.3   Touschek from Vienna to Hamburg and Berlin

Bruno was born in Vienna, on February 3rd 1921. His mother, Camilla Weltmann, came from a well to do assimilated Jewish family. His father, Franz Xaver Touschek, was an officer in the Austrian Army, who had fought on the Italian front during World War I [1].

His early family life, with strong artistic and literary interests, was soon disrupted by a number of tragic losses. His mother died when he was only 9 years old, and a much admired maternal uncle, a doctor and a painter, committed suicide in 1934, probably following Hitler's accession to power, as it happened to many Jewish intellectuals. A small portrait of Oskar Weltmann was found among Bruno's office papers, a testimony of profound attachment to his maternal uncle.

A further disruption in Bruno's life took place in March 1938, with the annexation of Austria to Germany. In 1931, Bruno had started his high school studies at the Piaristengymnasium, one of the best schools in Vienna, but after the annexation, a reallocation of Jewish students to different institutions took place. Bruno was a *mischling*, a mixed race person, and already a rebel, and had to leave the Piaristen before he could obtain his *Maturazeugnis*, certifying that the student had passed the required examination for university admission. He transferred to the Schottengymnasium, a Benedectine school of high renown, and from there obtained his graduation certificate from the State gymnasium, in February 1939. During such a difficult time, the impending war and the ongoing tragedy of racial discrimination against Jews matured in him a decision to emigrate. But when he tried to go to England and study chemistry at the University of Manchester, it was already too late. After passing the *Matura*, he had visited his maternal aunt, Adele, nicknamed Ada, who lived in Rome, attending some lectures at the University, with more passion than profit, and waiting for the visa for England to arrive.[6] It did not happen, or, perhaps, hesitation to leave

---

[5] All the three movies were authored by Enrico Agapito, *Bruno Touschek and the art of physics* with Luisa Bonolis, 2005 ©INFN-LNF, *Bruno Touschek with AdA in Orsay* with Luisa Bonolis and Giulia Pancheri, 2013 © INFN-LNF, *Soixante années d'exploration de la matière avec des accélérateurs de particules*, with Giulia Pancheri, 2017 © IN2P3.

[6] Letters to father from Rome, during March 1939.

his family and a lack of money made him return to Austria. In September, he enrolled to study physics at the University of Vienna, gaining recognition and consideration for his extraordinary intellectual capacities. He came to know the theoretical physicist Hans Thirring, who remained his mentor and friend through most of Bruno's life. Through him, Bruno would later become a friend and collaborator of his son Walter [11], in Glasgow in 1951.

Bruno's dream of becoming a physicist was shattered in June 1940, when, after brilliantly passing his university exams, Bruno was told he could no more attend classes nor frequent the library. Unwilling to give up, he spent the following academic year studying at home, or with young assistant professor Paul Urban.[7] In Fall 1941, he reapplied to be admitted to the university, but the reply was negative and, in December, he was definitively expelled. By this time, it was clear that his future studies and life in Vienna were in danger. Paul Urban, a former student of Hans Thirring, with antinazi ideas and barely tolerated by university authorities, took upon himself to find a way out. In November, anticipating the university negative decision, Touschek and Urban had visited Sommerfeld in Munich. By February 1942 a plan was laid out for Bruno to go to Germany, first to Munich and from there to Hamburg, where one of Sommerfeld's former students, Günther Jobst, could employ Bruno in his electronic firm. The plan included the possibility for Bruno to unofficially attend physics classes and Seminars at the University of Hamburg.

15.3.1 Hamburg Days and a Journey to Berlin

After leaving Vienna, Bruno spent nine months in Hamburg, before moving to Berlin at the end of the year. He would later return to Hamburg under different circumstances and renewed confidence, but his first period in Germany was very difficult.

In Hamburg, Bruno started working at the Studiengesellschaft für Elektronengeräte, but it did not take him long to be unsatisfied with the work at the laboratory, and the poor pay. News from Vienna about his grandmother's deportation to Theresienstadt during the summer of 1942 are the probable cause of a depression which gripped him at that time. After the summer, he impulsively resigned from the position at the firm, forcing himself to give up whatever he was doing and start on a new road.[8]

The move which changed his life course, took place in November 1942. After resigning from his job, with his resources at the lowest, he decided to go to Berlin and claim his compensation for the referee work he was doing for the *Chemischen Zentralblatt*. This was an odd job, for which his friends from Vienna had recom-

---

[7] Paul Urban 1905–1995 was later to become Professor at University of Graz and, in 1962, founded the Schladming Winter School on Theoretical Physics.

[8] From his letters home during the war, Bruno worked for the Studiengesellschaft für Elektronengeräte from March until November 1942, in partial contradiction with [1], where it is said that he worked there for a "long time", pag. 4.

mended him, knowing both of his capacity to do it as well as his need for money. Thus he left for Berlin, embarking on what he later called an *extraordinary journey*, not knowing whether he would have the money to pay for his return to Hamburg.[9] When he went back on the following day, he had collected his dues and, through some rather chancy encounter with a girl he knew from Vienna, had even secured a position in Berlin at the firm Löwe Opta, whose director, Karl Egerer, was a man in the confidence of the German military and editor of the scientific journal *Archiv für Elekrotechnik*. Two months later, assisting Egerer in his editorial task, he would come across an article which changed his life.

15.4 Bruno Touschek's Legacy to Accelerator Physics

In the brief span of his life, Bruno Touschek built, or contributed to build, three particle accelerators; first came Widerøe's 15 MeV betatron during World War II (WWII), then AdA, an electron-positron storage ring, first in the world to observe collisions in 1964, and ADONE, an electron-positron collider which reached the world highest c.m. energy when it went into operation in 1969. Of these, the least known is his contribution to Widerøe's betatron [12].

15.4.1 Touschek and His First Accelerator: Widerøe's Betatron

Rolf Widerøe had proposed the betatron principle during his university years. Later, he had tried to built an electron accelerator built on this principle, but had not succeeded. He then turned to something easier to attain, and, for his Ph.D. thesis at University of Karlsruhe in 1928, built the first linear accelerator [13]. The article describing the construction of the linear accelerator included also the equation on which a betatron would work, and had an impact on the subsequent development of particle accelerators in the US. As a matter of fact, soon after its publication, the article reached Berkeley, catching the attention of Ernest Lawrence, who always acknowledged to have been inspired by Widerøe's article to conceive and build his cyclotron, the first circular electron accelerator, in 1933. The difficulties inherent in the cyclotron reaching higher energies, were overcome by the 1941 successful operation of the first betatron at University of Illinois, by Donald Kerst, who announced to have been able to accelerate electrons up to 2.3 MeV, using Widerøe's principle. Such energy had never yet been reached in the laboratory. When Widerøe came to know of this success, he saw the possibility to construct both a 15 MeV betatron and, in a later stage, to go up to 100 MeV in electron's energy and reaffirm his priority,

---

[9] Touschek's letter to his friend Peter on November 29th, 1942, from Franz Touschek's collection of his son's letters.

and, in September 1942, he submitted an article with this proposal to the *Archiv für Elekrotechnik*. This article caught Touschek's attention and started his long life interest in particle accelerators.

He discussed it with his boss Egerer, and, through him, Widerøe's proposal became of interest to the military, in search of a miracle weapon. The betatron, capable to produce high energy X-rays, would then be built in the context of on-going *death ray* projects [2]. Touschek had initially been critical of some theoretical aspects in Widerøe's proposal. After his objections had been taken in full consideration, he was asked to join Widerøe's project. Given his status as half-Jewish in Nazi Germany, he had no choice but to accept, and, in fall 1943, he was hired to participate in the construction of Widerøe's 15 MeV betatron, at the C.H. Müller factory in Hamburg. It is during these early times of the project, that Widerøe shared with Bruno Touschek a novel concept for increasing the collision center of mass energy, i.e. head-on-collisions [12]. As the story goes, Touschek was not impressed at the time, but later he would when he wrote "The first time I heard of head-on-collisions" was from Widerøe.[10]

In Hamburg, Touschek learnt the art of making an electron accelerator, under the leadership of Rolf Widerøe. Although this machine is usually referred to as Widerøe's betatron, a careful reading of sources, including Bruno's letters home during the years 1943–45, shows that Bruno was involved in the planning and construction of this betatron from its very beginning until its transfer to Wrist in March 1945, ordered by the German authorities in order to save it from the arriving Allied forces.

Proof of Touschek's impact on the functioning of the 15 MeV betatron comes from many sources. Apart from being hired in the project, the first evidence is that Touschek was able to avoid being drafted to forced labor, the first step to deportation and concentration camps. Touschek's letters to his family in 1944 and 1945 mention three such summons, all avoided through appeals to General Milch and Minister Speer, on the part of his co-workers in the betatron project. In all three cases, the only successful motivation could have been that his work was important for a project of interest to the war. This could have been a ruse invented by his friends, but the situation in Germany in 1944–45 was dire enough that friendship alone could not save many lives. Instead, in my opinion, this was exactly the truth, namely that without Touschek's theoretical knowledge and exceptional physics intuition, Widerøe's betatron would not have worked. The same statement, about Bruno's indispensable contribution, i.e. that the miracle device would not function without him, is found to be the reason of the more human treatment Bruno enjoyed during his imprisonment between March 17th and mid April 1945 [1, 9].

Touschek was well aware of how important his role was in Widerøe's project. In fall 1944, he wrote to his parents that he was finally engaged in a project he could call "his own", one that could establish a world record. It should also be noticed that during these months, 1944–45, Widerøe was often in Oslo, and the team went on

---

[10] Undated manuscript, Sapienza University of Rome–Physics Department Archives, https://archivisapienzasmfn.archiui.com.

without him. Touschek's contribution is clearly acknowledged by Rolf Widerøe, who wrote to Amaldi that Bruno did very many important calculations for the betatron group.

15.4.2 A Death-Ray Project in Hamburg and Touschek's Imprisonment

Of interest is also a mention of death-rays as 'decisive weapons' in the memories of Albert Speer, the Minister of Armament during the war, in Chap. 31 of [14]. In his reminiscences, Speer comments on wild notions flourishing as the enemy approached in the early days of Aril 1945, and a reference is made to the inventor's appeal having been rejected by the Ministry. If this sentence refers to Touschek, who was in prison in Fuhlsbüttel at the time, it would be a confirmation of his importance in the betatron project. No other death ray projects existed anymore, except for Widerøe's, and no other scientist working on a betatron project is known to have been applying for clemency or work, except for Bruno Touschek.

The appreciation of Touschek's contribution is present in the reports prepared by the US and UK occupation forces after the war [9, 15]. A direct consequence of such high opinion of Touschek's work, is that Bruno's future education was taken in charge by the British, interested in developing an accelerator program. As for the betatron, it was requisitioned and brought to Woolwich Arsenal near London for inspection and studies, its later whereabouts are unknown.

At the end of the war, Bruno was one of the few scientists in continental Europe who had a working knowledge of building a particle accelerator of an advanced type such as the betatron. Another one was clearly Widerøe, but he had gone back to Norway and was also entangled in an inquiry for collaborationism with the Germans [16]. By this time, the synchrotron principle had been discovered and new accelerators based on it were planned in the UK and, mostly, in the US. Touschek, who had been part of Widerøe's team, became of particular interest to the British in their postwar effort to develop particle accelerators. After earning his Diploma in Physics from the University of Göttingen in June 1946, and a six month period as Werner Heisenberg's assistant, Bruno was brought to the UK and enrolled in the PhD program at the University of Glasgow [17]. He was very actively participating in the British postwar accelerator program, which included building a number of synchrotrons, among them a 300 MeV in Glasgow itself, but not only. His contribution is glimpsed through his research report of the years 1947–48 and acknowledged through correspondence with Frank Goward,[11] and other notable scientists of the time.[12]

---

[11] F. Goward together with D.E. Barnes had demonstrated synchrotron acceleration for the first time in August 1946.

[12] See Touschek's papers in Sapienza University of Rome–Physics Department Archives, https://archivisapienzasmfn.archiui.com.

15.4.3 Bruno's Dream: To Become a Physicist

Touschek's success in proposing, and carrying through, the construction of the first electron-positron collider is not only due to the experience gained with Widerøe's betatron, but just as well on his theoretical physics capacities and insight.

In one of his letters home from Göttingen, Bruno, pressed by his father to return to Vienna, writes:

    I want to become a physicist.[13]

and so he did. Touschek's mentors in his formation as a theoretical physicist include some of the most illustrious theoreticians of last century physics: Hans Thirring, during Bruno's first—and only—year at the University of Vienna and through both the war and post-war years, Arnold Sommerfeld, who suggested and sponsored Bruno's moving to Germany, Werner Heisenberg who had Bruno as his Assistant in post-war Göttingen, Max Born during Touschek's years in Glasgow, and Wolfgang Pauli until his passing in 1958. Many other scientists influenced Bruno's thinking and were, in return, influenced by him, and the list can be glimpsed in [1].

After leaving Vienna, where Hans Thirring had been one of Bruno's teachers, Bruno remained in close contact with his former professor. Until 1945, Bruno travelled regularly from Berlin to Vienna, and in at least one occasion, his letters home mention physics discussion with him or other Vienna physicists. It is after one such discussion, probably on the occasion of a trip to Vienna to celebrate various family birthdays (his own, his father's, and his stepmother's) all occurring between January and February, that Touschek started thinking about the working of cyclotrons and the need to apply corrections when the electron's energy became close to be relativistic [9]. The last year of the war interrupted travel across Austria and Germany, and Bruno's visits to Austria resumed only during Bruno's Glasgow years. After his Ph.D. and becoming a Nuffield Lecturer, he could finally take a real vacation to his family favourite places, in Tyrol, Fig. 15.3. He visited the Thirring family in Kitzbühel, and became friends with Hans' son Walter, with whom he would write a paper which played an important role in Bruno's thinking about infrared radiative corrections [11, 18].

While still in Vienna, Bruno had also established a connection with Arnold Sommerfeld, which remained close until Sommerfeld's death in 1951. Thanks to Bruno's having approached him about some corrections to the second volume of his famous treaty *Atombau und Spektrallinien*, Bruno's high intellectual qualities and passion for physics became known to the great scientist. A scientific correspondence ensued, a rather surprising show of intellectual courage on the part of a twenty years old physics students and the father of atomic physics. Thanks to Sommerfeld, Bruno found a way out of Vienna where marginalization and discrimination were engulfing his hopes to become a physicist, threatening his life as well.

As Bruno moved into Germany, Sommerfeld's former friends or pupils befriended him, and he could attend, unofficially, lectures and seminars by Max von Laue and

---

[13] Ich will ein physiker warden, January 1947, letter to father.

Fig. 15.3 At left, a sketch found among Bruno Touschek papers, probably a memory of Bruno's walking with his father in Tirolean gear, during a summer vacation, courtesy of Sapienza University of Rome–Physics Department Archives, documents provided for purposes of study and research, all rights reserved. At right, a photograph of Bruno, seated with his father and stepmother, included in one of Bruno's letters to his father, probably from a summer 1950 vacation, Family Documents, © Francis Touschek, all rights reserved

Werner Heisenberg. A close relationship with Heisenberg was developed after the war, later in Göttingen, in 1946, during the first year of the reconstruction of German science under Heisenberg's leadership. Bruno was strongly influenced by Heisenberg's theoretical work on the observer as guiding principle in physics investigations, mired in a statistical approach. Such influence is present through the correspondence between Bruno and Heisenberg, in particular about ongoing questions of analyticity of the S-matrix, which continued through the years Bruno was in Glasgow.

Max Born, then Tait professor of Mathematical Physics at University of Edinburgh, was also to have influence on Bruno's interests. Bruno was introduced to him in May 1947 by Ian Sneddon, the only other theorist in Glasgow at the time, and soon became a regular attendee of the bi-weekly Seminar Born held in Edinburgh.[14] They exchanged letters and ideas about quantum mechanics, and Touschek collaborated to prepare the appendix on weak interactions of the second edition of Born's famous book *Atomic Physics*.

In 1952, when Bruno left the UK to accept a position with INFN at the University of Rome, he had developed into a brilliant theoretical physicist, aware of his genius and ready to take his own road and to exchange ideas, at level with no other than Wolfang Pauli, visiting Rome at the time of Bruno's arrival. Through the 1950s, Pauli became Bruno's intellectual companion, sparring ideas over a glass of wine in many occasions, often meeting at conference sites, inspiring Bruno's interest in the CPT theorem [19, 20]. After Pauli's death, in December 1958, the way was open for

---

[14] Letter to parents, May 3rd, 1947.

Bruno to be completely on his own, and start his greatest adventure, to explore the unknown with a new type of experiment, colliding matter against anti-matter. AdA, the first electron-positron collider in the world, was to be conceived in just over one year, and came to life not long after.

15.4.4 The Making of AdA

The official date of AdA's birth is March 7th, 1960, when the scientific council of the Frascati National Laboratories (LNF) approved its construction. Once approved by the laboratories, the project ran its course towards the national agencies, and, by the end of the month, orders had started to be placed, with Touschek in charge of the project. However, the March 7th meeting, where AdA's construction was approved, had not sprung out of nowhere.

15.4.4.1 Between Rome and Frascati

According to Nicola Cabibbo [21], Touschek's first proposal for an experiment to study electron-positron collisions came up in the discussion which followed a seminar held in Rome by Wolfgang Panofsky in late '59. Records kept in the Frascati National Laboratories (LNF) also show that, on October 26 1959, Panofsky held a seminar in Frascati, entitled *Sull'acceleratore lineare da due miglia*, 'About the two mile linear accelerator' in English. The seminar probably included what Panofsky had presented at the Kiev conference in July, in particular the ongoing electron-electron project at Stanford [22]. At the end of Panofsky's seminar, whether the one in Frascati or a similar one in Rome, Touschek launched the idea to make electrons collide against positrons.

From these records, late October 1959 may be considered the starting date of AdA's conception. A confirmation comes from the official permission for Touschek to enter the Frascati laboratories, dated as October 30th, 1960 [23]. Shortly after, a group of scientists from Frascati and the Rome physics institute started working on Touschek's idea. The interesting possibilities created by electron-positron annihilation in the study of the pion form factor were explored. The discussion involved some more senior theorists such as Raoul Gatto, Marcello Cini and the American visitor Laurie Brown, and younger ones such as Nicola Cabibbo and Francesco Calogero, who had graduated with Bruno Touschek in 1958.

In the months to follow, while the Rome theorists were calculating, Bruno Touschek's attention turned to the newly built electron synchrotron in the Frascati National Laboratories, and to its potential for physics experiments. The construction of a national laboratory had been approved by INFN in 1953, under Giorgio Salvini's direction, and the construction of the synchrotron had officially started in Frascati in 1957, beginning its operation on April 4th, 1959 [24]. This is how, by February 1960, everything was in place for Touschek's idea to become reality.

Fig. 15.4 From left: Carlo Bernardini, Giorgio Ghigo and Bruno Touschek, from [27]

Cabbibbo, Gatto, Brown and Calogero had finished and submitted their work in two separate articles to *The Physical Review Letters* [25, 26] while Touschek, pressed to become head of a future theoretical physics group in Frascati, remembered his years with Widerøe's betatron, and looked at the possibility of making the synchrotron into an electron-positron collider. A meeting of the laboratory council was held in Frascati on February 17th, 1960. Two conclusions were reached: the idea of using the synchrotron to make an electron-positron experiment was rejected, but, at the same time, a proposal to build a smaller, dedicated machine to study the feasibility of such an experiment was accepted and a mandate was given to the supporters of the idea, such as Carlo Bernardini, Giorgio Ghigo, and Bruno Touschek, to prepare a proposal, Fig. 15.4. On March 7th, the proposal was accepted and AdA's construction started in April.

15.4.4.2 AdA's Adventure in Orsay

To prove the feasibility of an electron-positron collider was not an easy task. It took almost four years, during which important effects affecting the operation of particle colliders were discovered and studied. The final measurements took place at the Laboratoire de l'Accélérateur Linéare, in Orsay. This is where AdA had been taken in 1962, to take advantage of the higher injection rate obtainable with the linear accelerator, which had started functioning around the same time as the Frascati synchrotron [27]. The idea to take AdA to Orsay had been put forward by Bernardini and Touschek during a visit by Pierre Marin and Georges Charpak to Frascati in July 1961 [28]. Following the exchange of letters ad visits between the two laboratories, the transportation from Frascati to Orsay was agreed. An almost epic trip took AdA and all its support system of vacuum pumps and power batteries, to the Laboratoire de l'Accélérateur Linéare, Fig. 15.5. The French team included a young doctoral student, Jacques Haïssiski, whose *Thèse d'État* gives the best description of how AdA reached

Fig. 15.5 At left, AdA in Orsay, placed in the 500 MeV hall, near the linear accelerator, and, at right, a 1967 picture of Pierre Marin and Jacques Haïssinki (in gray suit), whose 1965 thesis is the most complete description of AdA's operation in Orsay, photographs courtesy of Jacques Haïssinki

its success as proof-of-principle for electron-positron colliders to become the main tool in the exploration of high energy particle physics.

By February 1964, when the final runs were recorded in Orsay, the Touschek effect had been discovered [29] and new theoretical physics calculations had explored e^+e^- physics [30–32]. All over the world, in France with ACO, in the USSR with VEPP-2, at CERN with the ISR, in the USA with SPEAR, new particle-antiparticle accelerators had been designed and their construction approved.

15.4.5 ADONE: Touschek's Last Accelerator

ADONE, the better and more beautiful version of AdA, was Touschek's last accelerator. There exist a posthumous note by Touschek about stochastic cooling in proton-antiproton colliders, a contribution to hadron accelerators, discussed by C.Rubbia in these proceedings. It was of limited practical impact, but it highlights Touschek's ever lasting interest in matter-antimatter accelerators, [33].

Testimonies abound of Touschek's deep involvement in ADONE's beginnings, its construction and dedication to extract meaningful physics from it. Touschek had proposed the construction of ADONE in an handwritten note in November 1960, basically as soon as AdA's construction was sufficiently advanced that he could seriously suggest to build a machine with beam energy six times higher, 3 GeV in the c.m., and an eight times bigger radius. This preliminary note became a full fledged proposal, authored by Fernando Amman, Carlo Bernardini, Raoul Gatto, Giorgio Ghigo and Touschek [34].

According to Fernando Amman, director of the ADONE project, Touschek was integral part of ADONE's success. Still ADONE was a much bigger enterprise that AdA had been, and, as such, the responsibility, both merits and failures, belonged to many people, including political events on which Touschek had no control. Among the latter, between 1963, when ADONE's construction was formally approved by INFN, and 1969, when two beams circulated in ADONE, two interruptions occurred, both to leave long lasting consequences. The first is the so called Ippolito affaire, *il caso Ippolito*, taking place in 1964, and signalling a stop in the road to nuclear energy independence in Italy. Felice Ippolito was General Director of CNEN, the national agency in charge of funding nuclear energy related activities, in particular the Frascati laboratories. His arrest and inquiry on accusation of mismanagement and corruption almost stopped ongoing work in Frascati. The second interruption was the 1968 strike in the laboratory, following student and workers unrest, in the Italian universities and in the industrial sector. In the University of Rome, confrontations between students and the faculty led to slow down of scientific exchanges and affected Bruno's person as well,[15] as reflected in his many drawings of the time.

In Frascati, activities at ADONE restarted in 1969, and the many years work was rewarded by the first observation of abundant hadronic particle production, whose impact on particle physics is described by other contributions to these proceedings, by Mario Greco and Giorgio Parisi, in particular.

Unfortunately, the various problems which had marred ADONE's progress since 1967, delayed its operation. As the 1970s rolled in, other colliders had come into operation and made some of the discoveries which ADONE had been conceived to do. Still, ADONE's operation brought many interesting results in addition to first observing multihadron production: a new resonance, $\rho(1600)$, was studied and discovered, photon-photon collisions were observed just after similar observations with VEPP-2, and, in November 1974, ADONE gave an immediate confirmation to the American discovery of the J/Ψ resonance, with results published together with those from Brookhaven and Stanford [35]. Defying the odds of a late start, ADONE made Italy a member of the international particle accelerator community.

An indirect consequence of ADONE's construction is the rise of an important research field in theoretical particle physics, infrared resummation, which occupied Touschek's attention soon after ADONE's approval. Touschek saw that extraction of information for particle physics experiments at high energies such as those proposed for ADONE, needed to disentangle the process of interest from the radiation effects which always surround charged particle collisions. Thus, as early as 1963, he embarked on the problem of the "administration of the infrared radiative corrections" to ADONE's future experiments. His ideas influenced the young researchers in his Frascati group with a lasting effect on theoretical work in Frascati and Rome, and his approach to the problem may still be of interest.

---

[15] See also Ugo Amaldi's contribution to these proceedings.

15.5   Touschek's Way to the Infrared Catastrophe

Just like AdA's proposal had its roots in Bruno's past, his treatment of infrared corrections to ADONE's experiments can be traced back to his 1951 work with Walter Thirring, Heisenberg's influence on the role of the observer, and his own interest in radiation damping effects in electron's accelerators.

The problem of a divergence when photons of infinitely small energy are emitted by charged particles during their acceleration, was well known to Bruno Touschek, who had studied Bloch and Nordsieck [36] and formulated their work in covariant formalism during a 1950 visit by Walter Thirring [11]. Its solution had been found by Schwinger, who showed that the divergence arising from real photon emission was cured by cancelling a corresponding one from virtual photon exchanges, and hypothesized the exponentiation of a finite correction term. Further elaborated by Brown and Feynman, Lomon, Erikson, and fully treated to all orders in perturbation theory by Jauch and Rohrlich, Yennie, Frautschi and Suura, Schwinger's guess was confirmed, showing the exponentiation of a factor, which modifies the observed cross-section [37].

As always, Touschek's own way to the problem was very original. Two physics ideas join in Touschek's proposal to deal with the flood of soft photons which accompany a charged particle process, one is the Bloch and Nordsieck's result that the process of emission follows a Poisson distribution, the other that the observable quantity in reactions between charged particles is the energy-momentum loss due to soft photon emission. The emphasis on observable quantities shows the influence of Heisenberg's thinking about the concept of the observer as protagonist in physical observations. After his Diploma in June 1946 and until March 1947, Touschek had been Heisenberg's assistant in Göttingen for six months, and had continued to work on problems of Heisenberg's interest, such as analyticity properties of the S-matrix, after joining the University of Glasgow on April 1st 1947.[16]

Bloch and Nordsieck's fundamental theorem about the quantum theory of the emission of soft photons from a classical source had shown that the distribution of the number of soft photons is given by a Poisson distribution. Assuming a discrete momentum spectrum for the photons, corresponding to the quantization of the electromagnetic field in a finite conducting box, the probability that a scattering process among charged particles gives $n_{\mathbf{k}_1}$ photons with momentum $\mathbf{k}_1$, $n_{\mathbf{k}_2}$ photons with momentum $\mathbf{k}_2$, is given as

$$P(\{n_{\mathbf{k}}\}) = \Pi_{\mathbf{k}} \frac{[\bar{n}_{\mathbf{k}}]^{n_{\mathbf{k}}}}{n_{\mathbf{k}}!} e^{-\bar{n}_{\mathbf{k}}} \tag{15.1}$$

where $\bar{n}_{\mathbf{k}}$ the average value of number of photons of momentum $\mathbf{k}$.

To this result, with which he was utterly familiar, Touschek added the constraint of energy momentum conservation, via a four dimensional $\delta-$ function which would

---

[16] See Heisenberg-Touschek correspondence at Deutsches Museum Archives, in Munich, and Sapienza University of Rome–Physics Department Archives.

select the distributions $\{n_\mathbf{k}\}$ with the right energy-momentum loss K. With these two starting points, the probability of soft photon emission with overall energy-momentum K_μ in the infinitesimal interval between K_μ and $K_\mu + d^4K$ could be written as [18]:

$$d^4P(K) = \sum P(\{n_\mathbf{k}\}) d^4K \; \delta_4\left(K - \sum_\mathbf{k} kn_\mathbf{k}\right) = \sum \Pi_\mathbf{k} \frac{[\bar{n}_\mathbf{k}]^{n_\mathbf{k}}}{n_\mathbf{k}!} e^{-\bar{n}_\mathbf{k}} d^4K \; \delta_4\left(K - \sum_\mathbf{k} kn_\mathbf{k}\right)$$
(15.2)

where the sum $\sum$ is carried out over all the values of all the $n_\mathbf{k}$. By virtue of the δ-function, the sum and the product can be exchanged, and the result leads to the exponentiation of a regularized photon spectrum, namely

$$d^4P(K) = \frac{d^4K}{(2\pi)^4} \int d^4x \; e^{-iK \cdot x} exp\left\{-\sum_\mathbf{k} \bar{n}_\mathbf{k}[1 - e^{ik \cdot x}]\right\}$$
(15.3)

If the boundary conditions allow, as is the case in QED, one can take the continuum limit of (15.3). After integration over the unobserved variables, one can follow the calculation outlined in [18] and obtain the probability distribution for observing a total energy loss $K_0 = \omega$ as

$$N(\beta)dP(\omega) = \beta\frac{d\omega}{\omega}\left(\frac{\omega}{E}\right)^\beta \quad \text{with} \quad N(\beta) = \frac{\int_0^\infty dP(\omega)}{\int_0^E dP|(\omega)} = \gamma^\beta\Gamma(1 + \beta)$$
(15.4)

Touschek's derivation of (15.4) was based on positivity and analyticity of the energy distribution, starting from semi-classical considerations and statistical mechanics formalism, and confirmed results already well known since the 1950s. But Touschek's elegant treatment made it physically transparent. Then he added one of his jokes to the problem, calling *Bond factor* the quantity $\beta(E)$, whose numerical value at ADONE's energy was 0.07. This results was also obtained by Mario Greco and Giancarlo Rossi, using a coherent state approach, later extended to gauge theories.[17]

Interest in (15.3) did not stop at the energy distribution. Throughout 1966, Touschek, Etim and myself spent many months in trying to derive a closed form for the momentum distributions, obtainable from (15.3) after integration over the energy variable. We finally had to resort to an approximation, but this brought a long life to Touschek's thinking about the infrared problem. Indeed, his insistence to go beyond the energy distribution arose the interest in the Frascati group. The transverse momentum distribution, obtained from (15.3) as

$$d^2P(\mathbf{K}_\perp) = \frac{d^2\mathbf{K}_\perp}{(2\pi)^2} \int d^2\mathbf{b} \; e^{i\mathbf{K}_\perp \cdot \mathbf{b} - \int d^3\bar{n}(\mathbf{k})[1 - e^{-i\mathbf{k}_t \cdot \mathbf{b}}]}$$
(15.5)

was applied to study hadronic processes, as in the case of a constant coupling in the infrared limit [38]. In 1978, a landmark calculation by Giorgio Parisi and Roberto

---

[17] See Greco and Rossi's contributions to this volume.

Petronzio [39] obtained the transverse momentum distribution of Drell-Yan pairs arising from soft gluon emission using perturbative QCD and the asymptotic freedom expression for the strong coupling constant.[18] Other studies by members of the Frascati group, which had expanded to include theorists Fabrizio Palumbo and Calogero Natoli, followed, as did a calculation of the W-boson transverse momentum [40].

The potency of Touschek's way does not only rely on phenomenological applications. Standing mainly on its applicability to different types of interactions, it also has the possibility to extend it to the calculation of the zero energy mode in some theories [41]. The calculation addresses what happens in Abelian gauge theories in passing from the discrete to the continuum limit in (15.3) in the energy distribution case. The limit can be taken by first separating the zero energy mode from all the others, and then examining the zero mode in light of the boundary conditions in the theory under consideration. This leads to the overall energy distribution to be written as

$$dP(\omega) = \frac{d\omega}{2\pi} \int dt \, e^{i\omega t - h(t)}, \quad h(t) = h_0(t) + \bar{h}(t) \quad (15.6)$$

with $\bar{h}(t)$ having the usual expression

$$\bar{h}(t) = \int d^3 \bar{n}_k [1 - e^{-ik \cdot t}] \quad (15.7)$$

and

$$h_0(t) = \bar{n}_0 [1 - e^{-i\omega_0 t}] \approx i(\eta \tilde{\omega}) t \quad (15.8)$$

where η is a dimensionless parameter proportional to the coupling constant, while $\tilde{\omega}$ is energy and mass dependent (of the emitting particles), i.e.

$$\eta = \frac{4\pi e^2}{L^3 \mu^2 m} \qquad \tilde{\omega} = \frac{1}{2} m \left| \sum_i \epsilon_i \mathbf{v}_i \right|^2 \quad (15.9)$$

In this equation, m is rhe mass of the emitting particles, μ is a fictious photon mass used for the regularization procedure. The quantities L and μ depend on the way one passes from the discrete limit (in which one obtained the original resummation expression with classical statistical mechanics formalism) to the continuum, namely how the limits $L \to \infty$ (size L of the lattice), and $\mu \to 0$ are taken. In QED with the usual vanishing boundary conditions, the zero mode is killed by the measure of the integral, but it cannot be excluded that this treatment, directly derived from Touschek's approach to the infrared region, can be of relevance in other theories or in cosmology [42].

---

[18] Giorgio Parisi, 2021 Nobel Prize in Physics, was a member of the Frascati theory group from 1971 to 1981.

15.6 Conclusions

I have outlined Touschek's contribution to XXth century physics, through the three accelerators he built or helped building: the 15 MeV 1945 German betatron, the electron-positron colliders AdA, the first such machine ever in the world, and ADONE, where multihadron particle production first appeared. The extension of his legacy to theoretical physics in dealing with infrared phenomena was also outlined.

Acknowledgements I am grateful to Francis Touschek for his collaboration and for permissions to use his father's documents and family letters, photographed courtesy of Elspeth Yonge Touschek, at her home in 2009. I thank Luisa Bonolis for sharing documents about Touschek's life, her archival knowledge and historical expertise, Giovanni Battimelli, from University of Rome, for suggestions and advice concerning the Physics Department archives, their creation and content. From the Laboratoire de l'Accélérateur Linéare in Orsay, now part of the Joliot Curie Laboratoire, much is owed to JAcques Haïssinski, for advice and providing archival documents from LAL.

References

1. E. Amaldi, *The Bruno Touschek legacy (Vienna 1921 - Innsbruck 1978)*. No. 81-19 in CERN Yellow Reports: Monographs (CERN, Geneva, 1981). 10.5170/CERN-1981-019. https://cds.cern.ch/record/135949/files/CERN-81-19.pdf
2. P. Waloschek, *Death-Rays as Life-Savers in the Third Reich* (DESY, 2012). http://www-library.desy.de/preparch/books/death-rays.pdf
3. G. Battimelli, M. De Maria, G. Paoloni, *Le carte di Bruno Touschek* (Università La Sapienza, Rome, 1989)
4. F. Amman, Rivista di Storia della Scienza **2**, 130 (1985)
5. M. Greco, G. Pancheri (eds.). *1987 Bruno Touschek Memorial Lectures, Frascati Physics Series*, vol. XXXIII (INFN Frascati National Laboratories, Frascati, 2004). http://www.lnf.infn.it/sis/frascatiseries/Volume33/volume33.pdf
6. C. Bernardini, in *The Restructuring of Physical Sciences in Europe and the United States, 1945–1960*. ed. by M. De Maria, M. Grilli, F. Sebastiani (World Scientific, Singapore, 1989), p.444
7. V. Valente (ed.), *Adone, a milestone on the particle way, Frascati Physics Series*, vol. VIII (INFN, 1997)
8. L. Bonolis, Bruno Touschek Remembered. 1921-2021. Bibliography and Sources (2021)
9. L. Bonolis, G. Pancheri, European Physical Journal H **36**(1), 1 (2011)
10. G. Pancheri, *Bruno Touschek's Extraordinary Journey*, Springer Biographies (Springer Cham, 2022). https://doi.org/10.1007/978-3-031-03826-6
11. W.E. Thirring, B. Touschek, Philos. Mag. **42**(326), 244 (1951). https://doi.org/10.1080/14786445108561260
12. P. Waloschek, in *The Infancy of Particle Accelerators. Life and work of Rolf Widerøe* ed. by P. Waloschek (Friedr. Vieweg & Sons Verlagsgesellschaft (Braunschweig and Wiesbaden), Braunschweig, Germany, 1994). https://doi.org/10.1007/978-3-663-05244-9_7
13. R. Widerøe, Archiv für Elektrotechnik **21**(4), 387 (1928)
14. A. Speer, *Inside the Third Reich?: memoirs* (The MacMillan Company, New York, 1970)

15. L. Bonolis, G. Pancheri (2019). https://arxiv.org/abs/1910.09075
16. A. Sørheim, *Obsessed by a dream. The Physicist Rolf Widerøe – a Giant in the History of Accelerators* (Springer, Cham, 2020). 10.1007/978-3-030-26338-6. https://doi.org/10.1007/978-3-030-26338-6
17. G. Pancheri, L. Bonolis, (2020). https://arxiv.org/abs/2005.04942
18. G. Etim, G. Pancheri, B. Touschek, Il Nuovo Cimento B **51**(2), 276 (1967). http://inspirehep.net/record/1940376
19. G. Luders, Kong. Dan. Vid. Sel. Mat. Fys. Med. **28N5**(5), 1 (1954)
20. W. Pauli, *Exclusion Principle, Lorentz Group and Reflection of Space-Time and Charge* (McGraw-Hill, New York, 1955), pp.30–51
21. N. Cabibbo, in *Adone a Milestone on the Particle Way*, ed. by V. Valente (INFN Frascati National Laboratories, Frascati, 1997), Frascati Physics Series, p. 219
22. W.K.H. Panofsky, in *Proceedings, 9th International Conference on High Energy Physics, v.1-2 (ICHEP59): Kiev, USSR, Jul 15-25, 1959*, vol. 1, ed. by A. of Science USSR, I.U. of Pure, A. Physics (Moscow, 1960), vol. 1, pp. 378–409. http://inspirehep.net/record/44195/files/c59-07-15-p378.pdf
23. V. Valente, *Strada del Sincrotrone Km 12* (Istituto Nazionale di Fisica Nucleare, Frascati, 2007)
24. L. Bonolis, F. Bossi, G. Pancheri, Il Nuovo Saggiatore **37**, 47 (2021). https://www.ilnuovosaggiatore.sif.it/issue/65
25. N. Cabibbo, R. Gatto, Phys. Rev. Lett. **4**, 313 (1960). https://doi.org/10.1103/PhysRevLett.4.313
26. L.M. Brown, F. Calogero, Phys. Rev. Lett. **4**, 315 (1960)
27. G. Pancheri, L. Bonolis, arXiv:1910.09075 (2018). https://arxiv.org/abs/1812.11847
28. P. Marin, *Un demi-siècle d'accélérateurs de particules* (Éditions du Dauphin, Paris, 2009)
29. C. Bernardini, G.F. Corazza, G. Di Giugno, G. Ghigo, R. Querzoli, J. Haissinski, P. Marin, B. Touschek, Phys. Rev. Lett. **10**(9), 407 (1963). https://doi.org/10.1103/PhysRevLett.10.407
30. N. Cabibbo, R. Gatto, Phys. Rev. **124**, 1577 (1961). https://doi.org/10.1103/PhysRev.124.1577
31. V.N. Baier, Sov. Phys. Uspekhi **5**(6), 976 (1963). http://stacks.iop.org/0038-5670/5/i=6/a=R07
32. G. Altarelli, F. Buccella, Il Nuovo Cimento **34**(5), 1337 (1964). https://doi.org/10.1007/BF02748859
33. C. Rubbia, in *Bruno Touschek Memorial Lectures, Frascati Physics Series*, vol. 33, ed. by M. Greco, G. Pancheri (INFN-Laboratori Nazionali di Frascati, 2004), pp. 57–60. http://www.lnf.infn.it/sis/frascatiseries/Volume33/volume33.pdf
34. F. Amman, R. Andreani, M. Bassetti, C. Bernardini, A. Cattoni, R. Cerchia, V. Chimenti, G. Corazza, E. Ferlenghi, L. Mango, in *Proceedings, 4th International Conference on High-Energy Accelerators, HEACC 1963, v.1-3: Dubna, USSR, August 21 - August 27 1963*, ed. by A.A. Kolomenskij, A.B. Kuznetsov (NTIS, Oak Ridge, TN, 1965), pp. 309–327. http://inspirehep.net/record/918674/files/HEACC63_I_314-338.pdf
35. C. Bacci et al., Phys. Rev. Lett. **33**, 1408 (1974). https://doi.org/10.1103/PhysRevLett.33.1408, https://doi.org/10.1103/PhysRevLett.33.1649. [Erratum: Phys. Rev. Lett. **33**, 1649 (1974)]
36. F. Bloch, A. Nordsieck, Phys. Rev. **52**(2), 54 (1937). https://doi.org/10.1103/PhysRev.52.54
37. G. Pancheri, Y.N. Srivastava, (2020). https://doi.org/10.48550/arXiv.2011.05865
38. G. Pancheri-Srivastava, Y. Srivastava, Phys. Rev. D **15**, 2915 (1977). https://doi.org/10.1103/PhysRevD.15.2915
39. G. Parisi, R. Petronzio, Nucl. Phys. B **154**, 427 (1979). https://doi.org/10.1016/0550-3213(79)90040-3
40. G. Altarelli, R.K. Ellis, M. Greco, G. Martinelli, Nucl. Phys. B **246**, 12 (1984). https://doi.org/10.1016/0550-3213(84)90112-3
41. F. Palumbo, G. Pancheri, Phys. Lett. B **137**, 401 (1984). https://doi.org/10.1016/0370-2693(84)91742-8
42. S. Weinberg, Phys. Rev. **140**, B516 (1965). https://doi.org/10.1103/PhysRev.140.B516

Chapter 16
The Making of ADONE

Claudio Pellegrini

Abstract We review the history, physics challenges and final success of the electron–positron 1.5 GeV collider ADONE, from its inception as Bruno Touschek's brainchild to the beginning of elementary particle physics experiments. Many new problems were met along the road to the successful operation of ADONE, like the collective instability effects. These novel phenomena had to be understood and means to control them had to be found to reach ADONE design goals. Bruno contributed in many critical ways to a successful solution of these issues, leading the effort to make colliding beams the important particle physics instruments of discovery that they are today.

16.1 Introduction

High-energy electron and positron beams have been playing, for over half a century, a very important role in the exploration of the properties of matter at the molecular, atomic and subatomic levels. ADONE, and the following e^+–e^- colliders, explored the structure of subatomic matter starting in the 1960s, helping to establish the standard model of elementary particles. The development of electron storage rings spurred by e^+–e^- colliders, led to the many synchrotron radiation sources now in existence worldwide, exploring matter at the atomic/molecular level, giving critical contributions to biology, chemistry and physics. The theoretical and experimental study of electron beams collective instabilities, self-organization phenomena, necessary to bring the colliders luminosity to values useful for high energy physics experiments, together with the operation of electron–positron linear colliders, has been critically important to make free-electron lasers a reality, generating coherent X-ray beams that have opened, for the first time, the exploration of atomic/molecular processes at their characteristic length and time scales of 1 Ångstrom-1 femtosecond.

C. Pellegrini (✉)
SLAC National Accelerator Laboratory, Menlo Park, CA 94025, USA
e-mail: claudiop@slac.stanford.edu

© The Author(s) 2023
L. Bonolis et al. (eds.), *Bruno Touschek 100 Years*,
Springer Proceedings in Physics 287,
https://doi.org/10.1007/978-3-031-23042-4_16

ADONE was an important first step in these developments, that are continuing even today. ADONE was the brainchild of Bruno Touschek. He developed the initial concept and, together with a group of physicists and engineers at the Frascati National Laboratory, transformed his idea into a wonderful instrument to study high energy physics. In this paper I will reconstruct some of the early history of the ADONE project.

16.2 The Beginning of Electron–Positron Colliders

The starting point in the history of ADONE is the famous seminar given by Bruno Touschek at Frascati on March 7, 1960. He made a strong case for the scientific potential of electron–positron interaction and their annihilation in particle-antiparticle pairs for the study of elementary particle physics. He also discussed the kinematic advantage of colliding head–on electron and positron beams so that all their energy would be available for the creation of new particle pairs.

The fact that an accelerator system to collide electrons and their antiparticle, the positron, had never been built, did not decrease Touschek's enthusiasm for the physics to be explored and his enthusiasm was communicated to many of the other physicists at Frascati.

The impact of the seminar cannot be over emphasized. I like to use the words of Edoardo Amaldi and Burton Richter to describe it. Edoardo Amaldi in his paper "The Legacy of Bruno Touschek", [1] wrote:

"All of the arguments discussed by Touschek and their brilliant exposition [in the Frascati seminar], made a considerable impression on everyone present, including the then Director of the Laboratory Nazionali di Frascati, Giorgio Salvini, and Carlo Bernardini, Gianfranco Corazza and Giorgio Ghigo. During the same day, the three last mentioned persons began to work with Touschek on a project for the first e^+–e^- storage ring, essentially designed as a prototype for checking the feasibility of accelerators based on the ideas set forth by Touschek during the seminar."

Burton Richter in his paper "The Rise of Colliding Beams" [2] wrote: "The first step in the electron–positron direction was taken in Italy, and the key personality was Bruno Touschek. There is a seminal moment in this story that occurred at a seminar by Touschek at Frascati on March 7, 1960, in which Touschek outlined the scientific potential of electron–positron annihilation studies. Giorgio Salvini, then director of the Frascati laboratory, and the high-energy physics community in Italy were immediately convinced by Touschek's arguments and began to work to bring e^+–e^- colliders to life. The first machine was called AdA, and it was brought into operation less than a year after Touschek's seminar."

16.3 The ADONE Project

During the same year, 1960, a two prongs approach was started at Frascati. One was the construction and commissioning of AdA, to establish the feasibility of a collider. AdA (Anello di Accumulazione), shown in Fig. 16.1, was built in about one year. It had a 2 m diameter and a beam energy of 250 meV. The initial injector was the Frascati 1 GeV electron synchrotron. It was later moved to Orsay to use a linear accelerator as a more powerful injector.

The other, under the direction of Fernando Amman, was the design of ADONE (great AdA), conceived as a collider to investigate particle physics, extending the study of the processes generated in e^+–e^- collisions up to center-of-mass energies $W = 3$ GeV. A draft proposal for ADONE was already written by Touschek in November 1960 and served as the basis for the design that started in 1961. The proposal discussed in some details the physics that could be done with collider, as can be seen from Fig. 16.2.

The two prongs approach is well described by Touschek in a paper presented at a CERN conference in 1961 [3]. In the paper Touschek writes:

Fig. 16.1 AdA, the first electron–positron collider, with 2 m diameter, 250 meV beam energy, at the Laboratori Nazionali di Frascati

A D O N E - a Draft Proposal for a
Colliding Beam Experiment.

B.Touschek,
Rome, 9.Nov.60.

It is proposed to construct a synchrotron
like machine capable of accelerating simultaneously
electrons and positrons in identical orbits. The suggested maximum energy is 1.5 Gev for the electrons as well
as the positrons. This energy allows one to produce pairs
of all the so called 'elementary particles' so far known,
with the exception of the neutrino, which only becomes
accessible via a weak interaction channel.

It is assumed that experiments in which there
are only two particles in the final state are most easy
to interpret. There are 16 such reactions, namely:

(1) 2γ . This is the only reaction in which
the rest intermediate state is 'quasi real' and in which
therefore there should be no 'radiative corrections'.
This reaction should serve as a 'monitor'. The cross-
section is 2.6 10^{-31} cm^2.

(2) e^+,e^- . This reaction will show strong
angular variations and may require 'good geometry'. It
would give information on the breakdown of electrodynamics
at distances corresponding to about 1/3 the Comptonwave-
length of the proton.

(3) μ^+,μ^- . Test of electrodynamics in 'bad
geometry'. May also serve as an indication of the fundamental difference between electrons and muons.

(4) $\pi^+\pi^-$ reveals the interaction between
pions in odd parity states.

(5) $2\pi^0$: charge exchange interaction for pion-
pion scattering.

(6) K^+K^- : interaction of K-mesons in odd
parity states.

(7) $\overline{K}^0,K^0$: Charge exchange interaction between
K-mesons.

(8) $p,\overline{p}$: interaction of proton and antiproton
in even parity odd charge parity states.

(9) $n,\overline{n}$: same as (8) but for the charge
exchange reaction.

(10) through (15). Interactions simple or
with charge exchange of hyperons.

Fig. 16.2 The first page of Touschek's draft proposal for an electron positron collider

"Frascati is developing two storage rings. The first (code name AdA = anello di accumulazione = storage ring) designed for storing electrons and positrons of up to 250 meV is actually undergoing the first tests, the second (code name Adone) a storage ring for electrons and positrons of up to 1.5 GeV, is still being planned.

The AdA team consists of C. Bernardini, G.F. Corazza, G. Ghigo, R. Querzoli and myself. The magnet was planned by Dr. Sacerdoti and built in Terni, the radiofrequency by Dr. Puglisi.

Adone is a national effort. A design team headed by Dr. Amman has the task of arriving at a specific design proposal by the beginning of 1962. Simultaneously a committee is preparing the experiments to be carried out with the machine. If, by the beginning of 1962 it is found that the project has a reasonable chance of success from a technical point of view; it is expected that the machine should be working late in 1964."

The initial technical proposal for ADONE was written in January 1961 [4]. The paper starts with a discussion of the physics goals of the collider, establishes the beam energy needed to reach them and discusses the various technical options to design the accelerator/storage ring. The electron and positron beam energy was chosen to be 1.5 GeV, 3 GeV total in the center of mass system, large enough to produce pairs of all known particles from electron–positron annihilation. As written in ref. [4]: "Disponendo di 3 GeV nel baricentro si può pensare di ottenere dall'annichilamento e^+–e^- la produzione, in coppia, di tutte le masse conosciute." ["Having 3 GeV in the center of mass system it is possible to obtain the production, in pairs, of all known masses"].

The paper continues discussing all important elements of ADONE's design, including the choice of weak versus strong focusing of betatron oscillations, RF system, injection system, desired luminosity and corresponding beam current, effect of beam-beam interaction.

Most of the final choices for ADONE were not very different from those outlined in this first paper. The energy remained the same as well as many other parameters. One choice, however, remained still open at the time, what kind of magnetic focusing should be used in the ring, weak or strong focusing.

To obtain a large luminosity for a given current it was convenient to have strong focusing magnets and a smaller beam transverse cross-section area, instead of weak focusing and a larger area. However, in a strong focusing ring the emission of synchrotron radiation leads to exponential growth (anti-damping) of the betatron oscillation amplitude on a millisecond time scale, clearly too short for a storage ring [5]. A simple weak focusing system, that avoids this problem, could not give the desired luminosity. The solution was a new, never used until that time, separated function focusing system, where the trajectory bending is done by weak focusing magnets and the focusing by pairs of quadrupole magnets, as shown in Fig. 16.3.

Following ADONE, this has become the basic structure for e^+–e^- colliders and all synchrotron radiation storage rings.

It is important at this point to remember that Touschek, who was an active participant in all these choices, had previously worked on the construction of a betatron in Germany, in 1943–44, during the war, with Rolf Widerøe, who was also the first

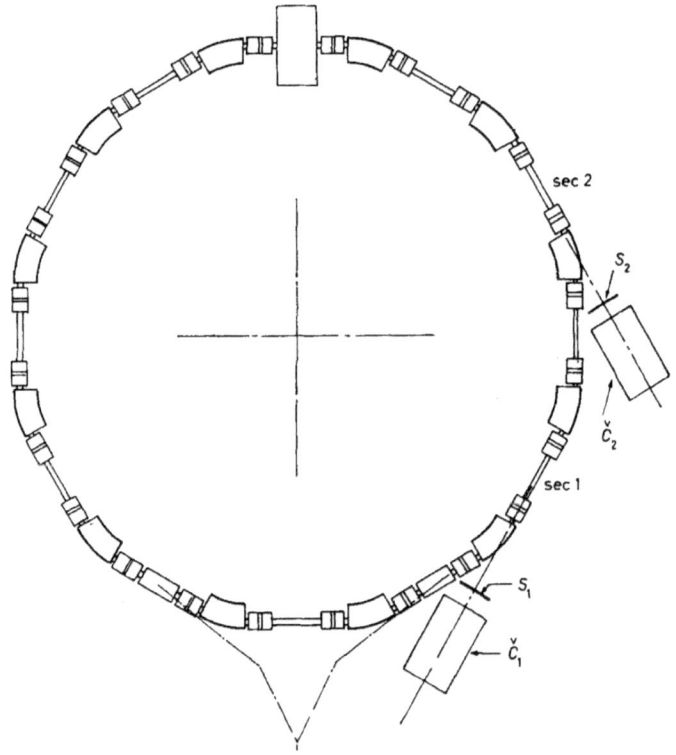

Fig. 16.3 ADONE magnetic structure. Twelve equal periods, each consisting of a long straight section, two quadrupole pairs and one weak focusing bending magnet. The boxes tangent to the ring are instruments to measure the luminosity, using small angle scattering or electron–positron going into electron–positron plus gammas

proponent of colliding particle beams and of linear accelerators. In 1946 at the University of Göttingen Touschek obtained the title of Diplomphysiker with a thesis on the theory of the betatron. When discussing AdA and ADONE he could and did use his previous theoretical and practical knowledge of particle accelerators to contribute to all aspects of their design. A more detailed biography of his life and work can be found in [1, 6].

16.4   High Intensity Effects in ADONE

I joined the ADONE group at the beginning of 1963. My initial task was to study single particle trajectories in the novel, separated functions, magnetic structure, including synchrotron radiation and radiation reaction effects in a strong, weak or separated functions accelerator. I worked mostly with Carlo Bernardini and Mario

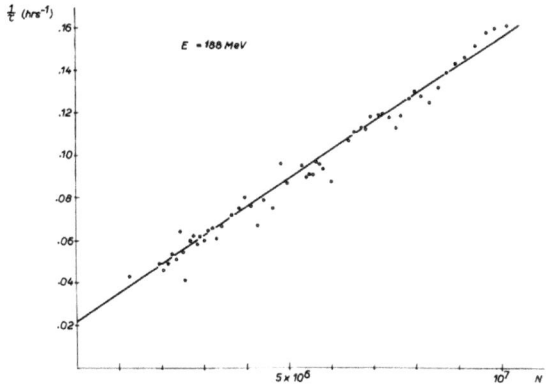

FIG. 1. Lifetime τ versus N, the number of stored particles in a beam, at the energy of 188 MeV.

Fig. 16.4 The plot showing the dependence of lifetime on the number of electrons in a bunch, from ref. [7]. The effect was observed on AdA and explained by Touschek in 1963

Bassetti. We also evaluated the beam lifetime and here it became obvious that the recent discovery on AdA of the Touscheck effect [7] required consideration of high intensity phenomena. The effect showed that some beam properties, in particular lifetime and transverse area, could change substantially when the beam current was raised. It also showed that scaling of these properties with the number of particles in the beam might be nonlinear and that thresholds for new phenomena might appear. The Touschek effect, a decrease in the beam lifetime when increasing the number of electrons in a bunch, as shown in Fig. 16.4, was discovered on AdA. Touschek explained the effect as due to the Coulomb scattering of electrons within the same bunch, generating a momentum transfer from the transverse to the longitudinal oscillation and leading to particle losses.

Another high intensity effect requiring attention was the resistive wall instability. In the period 1965–66 it was realized that the finite conductivity of the vacuum chamber, within which the beam or beams are moving, can be the source of a longitudinal and a transverse collective instability, limiting the maximum beam current. The theory was first developed for the case of a coasting beam [8, 9]. At Frascati the results were generalized to the case of interest to ADONE, two counter rotating bunched beams, by Touschek, Ferlenghi and Pellegrini [10]. The effect was analyzed, its dependence on the betatron frequency and other parameters was studied, what could be expected for ADONE, including possible luminosity limitations, was evaluated. The conclusion was that it should not be an obstacle to reaching the design goals. With the consideration and analysis of the Touschek effect and of the resistive wall instability we hoped to be ready for the start of the machine.

Here I would like to add a personal note. Working with Bruno Touschek on the resistive wall instability problem was, for a young person like me at that time, quite a learning experience, that influenced my work for the rest of my career. I have very clear memories of the time we spent together at Frascati, or sometime on a place near

Castel Gandolfo lake, working on the physics and the mathematics of the problem. What I learned from Bruno Touschek followed me through the rest of my career and helped me solve many other problems.

16.5 Commissioning ADONE

The collider, shown during assembly in Fig. 16.5, was completed in 1967–68 and its commissioning, with the active participation of Bruno Touschek, started at that time. He was very much present in the ADONE control room, always ready to give advice, discuss any problem, ready to help in any emergency. He was a reference point for all of us.

The first part of the commissioning, at low current, in the tens of μA range, was pretty good. ADONE behaved exactly as calculated, as far as orbits, synchrotron radiation effects, lifetime, beam size were concerned. But when we tried to increase the current to achieve the design current of 100 mA/beam, we encountered many unexpected effects generating sudden beam losses, limiting the current and the luminosity to values well below the design values and what was needed to do meaningful high energy physics experiments.

Fig. 16.5 ADONE in its building during assembly. The circumference is 100 m, the magnet bending radius 5 m and the energy 1.5 GeV. The injector is a 300 MeV electron linac built by Varian, with the capability of positron injection rate of 10 mA/minute. The design required 100 mA per beam to reach the design luminosity of about 3×10^{32} cm^{-2} h^{-1}

Amman discussed the situation in a paper he presented at 1969 Particle Accelerator Conference [11]. In this paper he summarized our experience with ADONE initial commissioning: "ADONE, after the first year of troubleshooting (talking of a storage ring it would be better to say instability-shooting), should start high energy physics experiments during 1969. It may seem strange that eight years after the initial operation of a storage ring, only one electron-electron (the Princeton-Stanford 550 MeV) and two e⁺–e⁻ rings, VEPP-II and ACO, have produced high energy physics results, and these are limited to experiments with very high cross section. I would like to remark that the first beam instabilities observed on the Princeton-Stanford ring, and interpreted as being due to the resistance of the walls, opened a new era in the accelerator field: it has been realized for the first time that the interaction of the beam with its environment makes a circular accelerator an essentially unstable system, that can become stable, in virtue of the Landau damping, when the beam density is not too high and the nonlinearities in the focusing forces give a frequency distribution of the particles large enough to compete with the instabilities. .... While a conventional accelerator operates usually at very low particle density, in an electron storage ring the radiation damping brings the density to very high values also when the current is in the mA range; a new set of theoretical and technical problems have therefore to be solved."

Longitudinal and transverse instabilities were observed in ADONE. The longitudinal instabilities were interpreted as due to the interaction with the RF cavity and were cured by separating the synchrotron frequency of the bunches and other techniques.

Particularly worrysome were the transverse instabilities. They could not be explained by the resistive wall effect and had a very low threshold current. In ADONE, at 300 MeV, the injection energy, with the natural beam dimensions, the threshold positron current was about 0.150 mA per bunch, to be compared with the value of 30 mA per bunch, 0.1 A per beam needed to reach the design luminosity of 10^{29}/cm²/s and start doing high energy physics.

16.6 Reaching ADONE Design Luminosity

Fortunately, the work done by the ADONE group, with the collaboration of Touschek and visitors like Matthew Sands from SLAC and Ralph Littauer from Cornell University, to understand and control the instabilities soon led to progress.

Quoting from a later paper by the ADONE group [12]: "The first electron beam was stored in ADON E in December 1967; parts of the ring still missing at that time have been installed during 1968, and the machine was completed in its present form by mid-1968. The experimental study of the single-beam instabilities has taken the major part of the ring operation until the beginning of 1969; the interpretation of the phenomena has allowed the development of suitable means of suppressing the instabilities. The multiple-bunch coherent-phase, longitudinal, oscillations have been cured by separating the synchrotron oscillation frequency of the bunches by means of a low-power radiofrequency cavity operating on a harmonic of the revolution frequency, but not of the main radio-frequency system [13].

Transverse betatron instabilities with very low current threshold (about 150 μ A/bunch at the injection energy, 300 MeV) were observed with a positron beam, or with an electron beam when the positive ions were swept out using transverse electric fields; these thresholds were much lower than those expected on the basis of current theories, and the dependence on the machine parameters indicated that the dynamics was not that of the resistive wall instability. It has been interpreted as being due to an interaction between the beam and rapidly decaying electromagnetic fields with frequencies extending in the GHz range induced by the beam in its environment; the theory has been found correct. [The interpretation of the effect has been first proposed by C. PELLEGRINI, M. SANDS and B. TOUSCHEK; a paper by C. PELLEGRINI on the subject is in course of publication*. Therefore, all the elements in the vacuum chamber should have been suitably terminated for frequencies in the GHz range, in order to reduce the forces acting on the beam. and to increase the rise time of the instability, while previous theories on beam instabilities were concerned with frequencies in the 10 MHz range and the machine was built accordingly."

The work done to control the beam instabilities led to an increase of the electron and positron beams currents and a corresponding increase in luminosity. In 1969 the luminosity reached a value near the design value, as shown in Fig. 16.6, and the high energy physics experiment could start. The paper reporting these results, [11], ended recognizing in the acknowledgements Touschek's contributions to ADONE

Fig. 16.6 ADONE Luminosity measurements with the scattering apparatus and three bunches per beam (full curve) at 1 GeV. The product of beam currents is also shown as a function of time (dashed curve). Errors are statistical, luminosity, left scale; i+–i−, right scale. The straight lines through the data are only indicative [11]

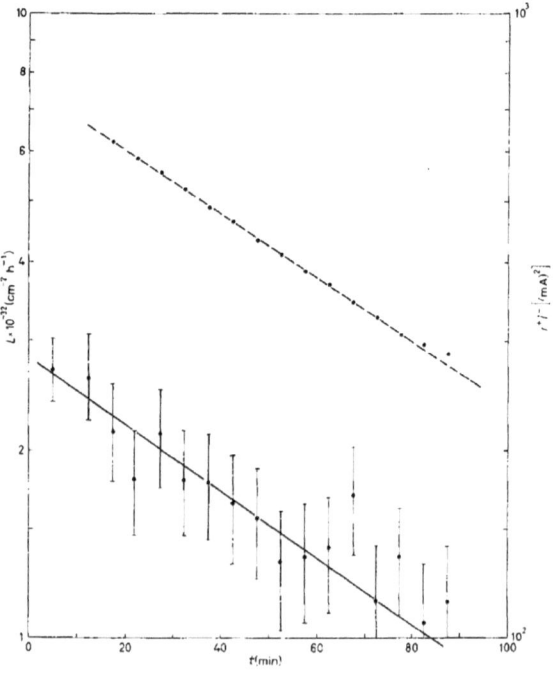

success: "We are grateful for helpful discussions with many physicists of the Frascati Laboratories and of other laboratories; we are especially grateful to Prof. C. BERNARDINI, whose contribution has been very important in the design stage, and to Prof. B. TOUSCHEK for his brilliant ideas and for suffering with us through the instability problems."

16.7   Conclusions

Bruno Touschek conceived the idea of electron–positron collider as a new way to explore high energy physics. He was a very active participant in the design of AdA and ADONE, in their commissioning and interpreting the many new physical effects and technical problems that had to be understood and solved before the colliders could contribute to elementary particle physics and the establishment of the standard model. He was also a mentor of young scientists in theoretical physics and in accelerator physics. I consider myself privileged to have been one of them.

References

1. E. Amaldi, The Bruno Touschek legacy. CERN Rep. 1–19 (1981)
2. B. Richter, The rise of colliding beams. SLAC-Pub-6023 (1992)
3. B. Touschek, The Frascati Storage Ring, LNF Report 61/45 (1961). Also presented at the Intern. Conf. of Theoretical Aspects of Very High Energy Phenomena, CERN, 61-22 (1961). p. 67
4. F. Amman, C. Bernardini, R. Gatto, G. Ghigo and B. Touschek, Anello di Accumulazione per elettroni e positroni (ADONE), Laboratori Nazionali di Frascati, LNF-61/5 (27.1.1961)
5. C. Pellegrini, Nuovo Cimento. Suppl. **22**, 603 (1962)
6. C. Bernardini, G. Pancheri, C. Pellegrini, Bruno Touschek: from betatrons to electron-positron colliders. Rev. Accel. Sci. Technol. **8**, 269 (2015)
7. C. Bernardini et al., Lifetime and beam size in a storage ring. Phys. Rev. Lett. **10**. 407 (1963)
8. V.K. Neil, A.M. Sessler, Longitudinal resistive instabilities of intense coasting beams in particle accelerators. Rev. Sci, Instr. **36**(429) (1965)
9. L.J. Laslett, V.K, Neil, A.M. Sessler, Transverse resistive instabilities of intense coasting beams in particle accelerators. Rev. Sci. Instr. **36**, 436 (1965)
10. E. Ferlenghi, C. Pellegrini, B. Touschek, The transverse resistive-wall instability of extremely relativistic beams of electrons and positrons. Il Nuovo Cimento, **XLIV**, 253 (1966)
11. F. Amman, Electron positron storage rings: status and present limitations. IEEE Trans. Nucl. Sci. **16**, 1073–1081 (1969)
12. F. Amman et al., Two-beam operation of the 1.5 Gev electron-positron storage ring ADONE. Lett. al Nuovo Cimento **15**, 729 (1969)
13. M. Bassetti, R. Littauer, C. Pellegrini, M. Sands, B. Touschek, ADONE longitudinal instabilities. Adone Int. Memoranda (1968) (un-published)

Chapter 17
Spontaneous Symmetry Breaking in Particle Physics

Giovanni Jona-Lasinio

Abstract I will review the appearance of spontaneous symmetry breaking (SSB) in particle physics at the end of the fifties and beginning of the sixties of the XXth century. I will recall Heisenberg non-linear spinor theory and the genesis of the first model (NJL) of fermion mass generation developed in collaboration with Yoichiro Nambu, based on the idea of spontaneous symmetry breaking. Both the non-linear spinor theory and the NJL model are invariant under a chiral transformation (γ_5—invariance) which was introduced by Bruno Touschek in 1957 and named by Heisenberg the Touschek transformation. Then I will briefly describe the subsequent evolution where the NJL model became an effective theory for low energy QCD and SSB was the key for the electroweak unification. Finally I will consider SSB in non-equilibrium which may be of interest in cosmology.

17.1 Introduction

The phenomenon of Spontaneous Symmetry Breaking (SSB) has been known for a long time even if it did not have a name. In a remarkable paper of 1759 "*Sur la force des colonnes*" Euler [1] discussed the following problem: "*Il s'agit de determiner le poids qu'une colonne peut soutenir, sans etre sujette à se plier*" and obtained a formula for the critical force necessary for bending a thin bar. After bending, the equilibrium configurations of the bar are degenerate as they lie in a plane which can have any orientation breaking the original rotational symmetry. The degeneracy of equilibrium states is a main feature of the phenomenon.

The concept of spontaneous breakdown of a symmetry is applicable to systems with infinitely many degrees of freedom and permeated the physics of condensed matter for a long time, magnetism is a prominent example. However its formalization and the recognition of its importance has been an achievement of the second half

G. Jona-Lasinio (✉)
Dipartimento di Fisica and INFN, Università di Roma "La Sapienza", Piazzale A. Moro 2, 00185 Roma, Italy
e-mail: gianni.jona@roma1.infn.it

© The Author(s) 2023
L. Bonolis et al. (eds.), *Bruno Touschek 100 Years*,
Springer Proceedings in Physics 287,
https://doi.org/10.1007/978-3-031-23042-4_17

of the XXth century and the name was adopted after the introduction in particle physics [2]. For the purpose of the present paper we can formulate this concept as follows:

SSB means that the lowest energy state of a system has a lower symmetry than the forces acting among its constituents or on the system as a whole.

The transfer of the idea of SSB from condensed matter to particle physics was an important case of *Cross Fertilization*. Heisenberg was probably the first to consider SSB as a possibly relevant concept in relativistic quantum field theory in the context of the comprehensive theory of elementary particles proposed by him and his collaborators [3, 4]. Mathematically the theory was based on a non-renormalizable non-linear spinor interaction. They tried to cope with the singularities of Quantum Field Theory by introducing an indefinite metric in the Hilbert space which made the approach very complicated, not transparent and comprehensible only to the intiated. Spontaneous symmetry breaking was really appreciated by the elementary particles community after Nambu and the present writer developed a specific model of relativistic field theory with a well defined physical interpretation [5, 6].

The NJL model was also non renormalizable and we simply introduced an invariant cut-off. Ideas like effective field theories [7] and asymptotic safety [8, 9], were not yet around. The model was formally close to Heisenberg theory in the sense that a non-linear spinor lagrangian was adopted and we considered the model with cut-off as a low energy theory of nucleons and mesons. I had been exposed more than once to the non-linear spinor theory, Heisenberg had visited Rome to explain it. Touschek was very interested in understanding the basic ideas and we had seminars on this subject.

We shared with Heisenberg symmetry properties because both approaches, considering the last version of his theory [4], were invariant under the following transformations

$$\psi \rightarrow e^{i\alpha}\psi, \qquad \bar{\psi} \rightarrow \bar{\psi}e^{-i\alpha}, \tag{17.1}$$

$$\psi \rightarrow e^{i\alpha\gamma_5}\psi, \qquad \bar{\psi} \rightarrow \bar{\psi}e^{i\alpha\gamma_5}. \tag{17.2}$$

The second transformation was named after Touschek who introduced it [10] to insure that the neutrino mass be equal to 0 in the theory of weak interactions.

To appreciate the innovative character of SSB in particle physics one may recall that one of the axioms of quantum field theory was that the vacuum, i.e. the state of lowest energy, must be invariant under the symmetries of the theory implemented by unitary operators. Therefore SSB in relativistic field theories represented a real turning point.

To understand the path leading to the NJL model we must start from an observation of Nambu. When I arrived in Chicago in September 1959 he was writing a short paper on the axial vector current conservation in weak interactions: in nature, the axial current is only approximately conserved and Nambu's hypothesis was that a small violation of axial current conservation gives a mass to the massless boson, which is then identified with the π meson, and renormalizes the axial vector part of the β-decay constant. So there must be a relation between these quantities. Under strict invariance under the Touschek transformation, γ_5-invariance, the structure of the axial vector current is

$$\Gamma_\mu^A(p', p) = \left(i\gamma_5\gamma_\mu - \frac{2m\gamma_5 q_\mu}{q^2}\right) F(q^2) \quad q = p' - p \qquad (17.3)$$

We see that it is compatible with a non-vanishing fermion mass provided there exists a zero mass pseudoscalar particle. Under Nambu's hypothesis, one can write

$$\Gamma_\mu^A(p', p) \simeq \left(i\gamma_5\gamma_\mu - \frac{2m\gamma_5 q_\mu}{q^2 + m_\pi^2}\right) F(q^2) \quad q = p' - p \qquad (17.4)$$

This expression implies a relationship between the pion nucleon coupling constant G_π, the pion decay coupling g_π and the axial current β-decay constant g_A

$$2mg_A \simeq \sqrt{2}G_\pi g_\pi \qquad (17.5)$$

This is the Goldberger–Treiman relation [11].

Nambu asked me to read a preliminary version of the paper. In order to support Nambu's idea one had to make some independent check. We did the following calculation: It was experimentally known that the ratio between the axial vector and vector β-decay constants $R = g_A/g_V$ was slightly greater than 1 and about 1.25. The following two hypotheses were then natural:

1. under strict axial current conservation there is no renormalization of g_A;
2. the violation of the conservation gives rise to the finite pion mass as well as to the ratio $R > 1$ so that there is some relation between these quantities.

Under these assumptions a perturbative calculation of the convergent difference of renormalization effects for $\mu_\pi \neq 0$ and $\mu_\pi = 0$ gives

$$R \simeq 1 + \Lambda(\mu_\pi) - \Lambda(0) \simeq 1 + \frac{G_\pi^2}{16\pi^2} \frac{\mu^2}{m^2} \ln \frac{m^2}{\mu^2} \simeq 1.24 \qquad (17.6)$$

where Λ is the contribution of the diagrams shown in the figure. In the second approximate equality we have retained the dominant logarithmic term.

Fig. 17.1 Typical graphs considered in the evaluation of $R = g_A/g_V$

We did this calculation independently obtaining at first very different results. Mine supported Nambu's conjecture while his was definitely against. The question was not entirely trivial as the result was the difference between two divergent expressions. We discussed for several days and finally Nambu agreed that my result was correct. It was a perturbative calculation with a large coupling constant so the numerical result close to the experimental value could not be taken too seriously, it showed however that the renormalization effects due to the pion mass went in the right direction.

The interpretation that Nambu suggested in the paper [12] was

> This situation may be understood by making an analogy to the theory of superconductivity
> originated by Bardeen, Cooper and Schrieffer [13] and refined by Bogoliubov [15].

In the case of supercoductivity the symmetry spontaneously broken is gauge invariance which in the present case is replaced by γ_5 invariance. Encouraged by the above calculation the construction of a relativistic model was mandatory and this is what we did following the analogy with superconductivity.

17.2   Spontaneous Symmetry Breaking in Superconductivity

A turning point for understanding microscopically the phenomenon of superconductivity was the theory of Bardeen, Cooper and Schrieffer [13], known with the acronym BCS theory. A different approach arriving at similar conclusions was independently proposed by Bogoliubov [14]. Further developments and refinements were due to Anderson [18], Ricayzen [19], Nambu [16]. See also the monograph by Bogoliubov, Tolmachev and Shirkov [15]. In this section I will follow [16] where a formalism close to quantum field theory is used.

Electrons near the Fermi surface due to the attractive phonon interaction [17] are paired (Cooper pairs) and described by the following equation

$$E\psi_{p,+} = \epsilon_p \psi_{p,+} + \phi\psi^\dagger_{-p,-} \tag{17.7}$$

$$E\psi^\dagger_{-p,-} = -\epsilon_p \psi^\dagger_{-p,-} + \phi\psi_{p,+} \tag{17.8}$$

with eigenvalues

$$E = \pm\sqrt{\epsilon_p^2 + \phi^2} \tag{17.9}$$

Here, $\psi_{p,+}$ and $\psi^\dagger_{-p,-}$ are the wavefunctions for an electron and a hole of momentum p and spin $+$, ϕ is the energy necessary to break a pair. The corresponding eigenstates are called quasi-particles.

Formally this is very similar to the Dirac equation which in the Weyl representation reads

$$E\psi_1 = \boldsymbol{\sigma} \cdot \boldsymbol{p}\psi_1 + m\psi_2 \tag{17.10}$$

$$E\psi_2 = -\boldsymbol{\sigma} \cdot \boldsymbol{p}\psi_2 + m\psi_1 \tag{17.11}$$

with eigenvalues

$$E = \pm\sqrt{p^2 + m^2} \tag{17.12}$$

Here, ψ_1 and ψ_2 are the eigenstates of the chirality operator γ_5 and the mass corresponds to the gap ϕ. However the quasi-particles are not eigenstates of the charge therefore there must be a "*backflow*" to recover the conservation of current.

Approximate expressions for the charge density and the current associated to a quasi-particle in a BCS superconductor are given by

$$\rho(x,t) \simeq \rho_0 + \frac{1}{\alpha^2}\partial_t f$$

$$\boldsymbol{j}(x,t) \simeq \boldsymbol{j}_0 - \boldsymbol{\nabla} f$$

where $\rho_0 = e\Psi^\dagger \sigma_3 Z\Psi$ and $\boldsymbol{j}_0 = e\Psi^\dagger(\boldsymbol{p}/m)Y\Psi$ with Y, Z and α constants while f satisfies the wave equation

$$\left(\nabla^2 - \frac{1}{\alpha^2}\partial_t{}^2\right) f \simeq -2e\Psi^\dagger \sigma_2 \phi\Psi$$

Here, $\Psi^\dagger = (\psi_1^\dagger, \psi_2)$

The Fourier transform of the wave equation for f gives

$$\tilde{f} \propto \frac{1}{q_0^2 - \alpha^2 q^2}$$

The pole at $q_0^2 = \alpha^2 q^2$ describes the excitation spectrum of a zero-mass boson. Due to the Coulomb force, the pole is cancelled [18] and the spectrum is shifted to the plasma frequency $e^2 n$, where n is the number of electrons per unit volume. In this way the electromagnetic field acquires a mass which lets the magnetic field penetrate only slightly in a superconductor (Meissner effect). This is the essence in a non-relativistic context of what will be known later as the Brout-Englert-Higgs-Guralnik-Hagen-Kibble mechanism. In the Landau-Ginzburg phenomenological theory of superconductivity [20] the vector potential $\mathbf{A}$ obeys the following equation

$$\nabla^2\mathbf{A} - e^2 n\mathbf{A} = 0 \tag{17.13}$$

that is the vector potential has a mass equal to the plasma frequency.

17.3 The NJL Model

The NJL model is a theory of nucleons and mesons based on a non-linear spinor lagrangian which however could be the limit of a theory with an intermediate particle of very high mass. The axial current is the analog of the electromagnetic current in BCS theory. In the hypothesis of exact conservation, as we already noted, the matrix elements of the axial current between nucleon states of four-momentum p and p' have the form of equation (17.3) and is compatible with a finite nucleon mass m provided there exists a massless pseudoscalar particle. Both superconductivity and the NJL model provide examples, of the following general statement, *Goldstone theorem* [36].

Whenever the original Lagrangian has a continuous symmetry group, which does not leave the ground state invariant, massless bosons appear in the spectrum of the theory.

The Lagrangian of the model is

$$L = -\bar{\psi}\gamma_\mu\partial_\mu\psi + g\left[(\bar{\psi}\psi)^2 - (\bar{\psi}\gamma_5\psi)^2\right] \tag{17.14}$$

It is invariant under the transformations (17.1) and (17.2)

By the Fierz transformation the non-linear term is equivalent to

$$-g\left[(\bar{\psi}\gamma_\mu\psi)^2 - (\bar{\psi}\gamma_\mu\gamma_5\psi)^2\right] \tag{17.15}$$

The simplest approximation we envisaged was a mean field approach.

$$m = -\frac{g_0 mi}{2\pi^4}\int\frac{d^4p}{p^2 - m^2 - i\varepsilon}\,F(p, \Lambda)$$

or

$$\frac{2\pi^2}{g\Lambda^2} = 1 - \frac{m^2}{\Lambda^2}\ln\left(1 + \frac{\Lambda^2}{m^2}\right)$$

where Λ is the invariant cut-off. The model exhibits an interesting spectrum of bound states as shown in the table.

Fig. 17.2 The self-consistent equation for the mass of the fermion in diagrammatic form

The NJL model had a considerable follow up. Its structure was generalized in [21, 22] and shown to be equivalent, as far as the calculation of the S-matrix is concerned, to a more conventional renormalizable theory.

The NJL model has been mainly reinterpreted as an effective theory of low energy QCD where the nucleons of the original model are interpreted as quarks. The literature on the subject is rather extensive and we refer e.g. to the following reviews [23–25]. One is interested in the low energy degrees of freedom on a scale smaller than some cut-off $\Lambda \sim 1$ Gev. In [23] the short distance dynamics above Λ is dictated by perturbative QCD and is treated as a small perturbation. Confinement is also treated as a small perturbation. The total Lagrangian is then

Nucleon number	Mass μ	Spin-parity	Spectroscopic notation
0	0	0^-	$^1S_0$
0	$2\,m$	0^+	$^3P_0$
0	$\mu^2 > \frac{8}{3}m^2$	1^-	$^3P_1$
± 2	$\mu^2 > 2m^2$	0^+	$^1S_0$

$$L_{QCD} \simeq L_{NJL} + L_{KMT} + \varepsilon\,(L_{conf} + L_{OGE})$$

where the Kobayashi–Maskawa–'t Hooft term

$$L_{KMT} = g_D \det_{i,j}\left[\bar{q}_i(1 - \gamma_5)q_j + \text{h.c.}\right]$$

mimics the axial anomaly and L_{OGE} is the one gluon exchange potential. Applications in particle and nuclear physics of the NJL model are still quite frequent.

The argument showing that SSB actually takes place in the NJL model was based on a self-consistent field approximation and a formulation independent of any approximation was desirable. The similarities of the formalisms of quantum field theory and statistical mechanics is part of the common wisdom. This is emphasized for instance in the book of Bogoliubov and Shirkov [26]. In statistical mechanics both classical and quantum there are variational principles determining the equilibrium states of a system. In the quantum case variational principles have been introduced by Lee and Yang [27] followed by Balian, Bloch and De Dominicis [28] and generalized by De Dominicis and Martin [29]. The independent variables appearing in these principles are quantum averages of operators, that is c-numbers.

In a similar vein I found natural to characterize the vacuum and therefore SSB in terms of a variational principle for an effective action [30], a c-number action functional which turned out to be the generating functional of one-particle irreducible amplitudes. The independent functional variable is the vacuum expectation value of the field. It generalizes the effective-potential introduced by Goldstone whose theorem becomes very simple in this formalism. The effective action differs from a classical action as it is non-local in space and time and involves the whole history of the system.

Many years later I learnt that the effective action in a semi-classical context had appeared in a paper by Heisenberg and Euler in the thirties [31] where they calculated quantum corrections to Maxwell's equations. However the effective action was fully appreciated after its use to describe SSB. It became a standard approach to SSB in textbooks of quantum field theory [32, 33] to which the reader is referred. See also [34, 35].

17.4 SSB in Gauge Theories and the Electroweak Unification

We have seen in the case of superconductivity of charged fermions that the long range Coulomb interaction eliminates in the excitation spectrum the massless collective mode which becomes the plasmon and the electromagnetic vector potential acquires a mass. Several people observed that one can take advantage of this mechanism to eliminate unwanted zero mass Goldstone bosons and give a mass to vector mesons in gauge invariant theories. See the articles by Anderson [37], Brout and Englert [38], Higgs [39], Guralnik et al. [40, 41].

To illustrate the mechanism in relativistic field theories we consider the following simple example [38]. Consider a complex scalar field $\varphi = (\varphi_1 + i\varphi_2)/\sqrt{2}$ interacting with an abelian gauge field A_μ

$$H_{\text{int}} = ieA_\mu\varphi^\dagger \overset{\leftrightarrow}{\partial_\mu} \varphi - e^2\varphi^\dagger\varphi A_\mu A_\mu$$

If the vacuum expectation value of φ is $\neq 0$, e.g. $\langle\varphi\rangle = \langle\varphi_1\rangle/\sqrt{2}$, the polarization loop $\Pi_{\mu\nu}$ for the field A_μ in lowest order perturbation theory is

$$\Pi_{\mu\nu}(q) = (2\pi)^4 ie^2\langle\varphi_1\rangle^2 \left[g_{\mu\nu} - \left(q_\mu q_\nu/q^2\right)\right]$$

Therefore the A_μ field acquires a mass $\mu^2 = e^2\langle\varphi_1\rangle^2$ and gauge invariance is preserved, $q_\mu\Pi_{\mu\nu} = 0$.

The discovery of how this could be used for the electroweak unification is due to Weinberg [42] and Salam [43] building on previous work by Glashow [44]. A very clear introduction to the path leading to the model presented in [42] is Weinberg's Nobel lecture [46]. We shall not describe in detail the electro-weak unification as there are comprehensive expositions in books, see e.g. [45]. The following quotation from [46] shows that the application of SSB to the electroweak unification was far from obvious.

At some point in the fall of 1967, I think while driving to my office at MIT, it occurred to me that I had been applying the right ideas to the wrong problem [the strong interactions]. It is not the ρ meson that is massless: it is the photon. And its partner is not the A1, but the massive intermediate bosons, which since the time of Yukawa had been suspected to be the mediators of the weak interactions. The weak and electromagnetic interactions could then be described in a unified way in terms of an exact but spontaneously broken gauge symmetry.

17.5   SSB in Non-equilibrium

After my collaboration with Nambu I progressively shifted to many-body physics, still under the spell of BCS and Bogoliubov theories, and more generally to statistical mechanics where I continued to explore the analogies with field theory, in particular in critical phenomena and non-equilibrium states.

SSB has been studied so far mainly as an equilibrium phenomenon. It was discovered however that out of equilibrium SSB can take place through mechanisms not available in equilibrium: currents are flowing through the system and their dynamics is crucial.

Stationary states are the obvious generalization of equilibrium states but the conditions under which SSB takes place in nonequilibrium are different from equilibrium. In stationary nonequilibrium states SSB may be possible even when it is not permitted in equilibrium.

To illustrate this statement let us consider the following toy model [47]: during a time interval dt three types of exchange events can take place between two adjacent sites, see Fig. 17.3

$$+0 \to 0+ , \quad 0- \to -0 , \quad +- \to -+ , \tag{17.16}$$

with probability dt. The last one takes place only on the bridge. At the left of the access lane of plus particles we have

$$0 \to + , \tag{17.17}$$

with probability αdt. At the right end of the exit lane of plus particles

$$+ \to 0 , \tag{17.18}$$

with probability βdt, and similarly for minus particles after reflection.

The model is clearly invariant under the discrete CP transformation. The authors have shown that that there is a phase transition spontaneously breaking CP invariance that manifests itself by blocking one of the access lanes so that in the stationary state charges of one sign prevail. On the other hand there is no phase transition in equilibrium. Unfortunately the study of SSB in non-equilibrium is considerably more difficult due to the lack of a general theory of non-equilibrium states.

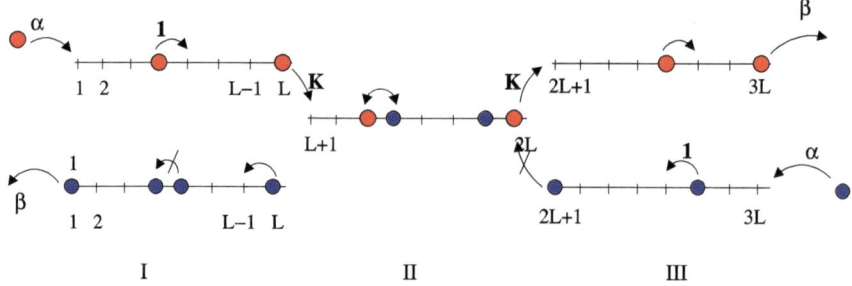

Fig. 17.3 The bridge model with two junctions from [47]. Positively (negatively) charged particles hop to the right (left). The model is invariant with respect to left-right reflection and charge inversion. Section 17.2 is the bridge. It contains positive and negative particles and holes. Sections 17.1 and 17.3 comprise parallel segments each containing pluses and holes or minuses and holes

The question naturally arises of how non-equilibrium SSB could be relevant in particle physics. At the cosmic scale matter is widespread and we do not see regions with antimatter. Explanations have been proposed invoking initial small asymmetries which are amplified over a long nonequilibrium evolution, See for example the following quotation [48].

> Baryogenesis gives a possible answer to the following question: Why there is no antimatter in the Universe? A (qualitative) solution to this problem is known already for quite some time: the Universe is charge asymmetric because it is expanding (the existence of arrow of time, in Sakharov's wording), baryon number is not conserved and the discrete CP-symmetry is broken. If all these three conditions are satisfied, it is guaranteed that some excess of baryons over anti-baryons will be generated in the course of the Universe evolution. However, to get the sign and the magnitude of the baryon asymmetry of the Universe (BAU) one has to understand the precise mechanism of baryon (B) and lepton (L) number non-conservation, to know exactly how the arrow of time is realized and what is the relevant source of CP-violation.

This type of explanation seems to shift the problem back in time. Non-equilibrium is considered after Sakharov [49] a precondition for explaining the matter-antimatter asymmetry in our universe. Phase transitions due to non-equilibrium spontaneous symmetry breaking may be relevant in cosmology. If the conditions are such that an approximately stationary state is an SSB phase, depending on the initial conditions a system will relax to one of the degenerate states avoiding the difficulties of reconstructing a history with many uncertainties. This may require a departure from the prevailing big-bang view of the origin of the universe. See, for an alternative, the recent theories of cyclic universes [50, 51].

References

1. L. Euler, Memoires de l'Academie des Sciences de Berlin **13**, 252 (1759)
2. M. Baker, S. Glashow, Phys. Rev. **128**, 2462 (1962)
3. W. Heisenberg, Rev. Mod. Phys. **29**, 269 (1957)
4. W. Heisenberg, W.H. Dürr, H. Mitter, S. Schlieder, K. Yamazaki, Zeit. f. Naturf. **14**, 441 (1959); W. Heisenberg, Proceedings of the 1960 Rochester Conference, p. 851
5. Y. Nambu, G. Jona-Lasinio, Phys. Rev. **122**, 345 (1961)
6. Y. Nambu, G. Jona-Lasinio, Phys. Rev. **124**, 246 (1961)
7. S. Weinberg, Physica A **96**, 327 (1979); S. Weinberg, *Effective field theory: past and future*, opening talk at the 6th International Workshop on chiral dynamics, Bern 2009, reproduced in the Memorial Volume for Y. Nambu, world Scientific 2016
8. S. Weinberg, *Critical Phenomena for Field Theorists in Understanding the Fundamental Constituents of Matter*, ed. by A. Zichichi (Plenum Press, New York, 1977)
9. G. Parisi, *On Non-renormalizable Interactions in New Developments in Quantum Field Theory and Statistical Mechanics*, ed. by M. Levy, P. Mitter (Plenum Press, New York, 1977)
10. B. Touschek, Nuovo Cimento **5**, 299 (1957)
11. M.L. Goldberger, S.B. Treiman, Phys. Rev. **110**, 1478 (1958)
12. Y. Nambu, Phys. Rev. Lett. **4**, 380 (1960)
13. J. Bardeen, L.N. Cooper, J.R. Schrieffer, Phys. Rev. **106**, 162 (1957)
14. N. N. Bogoliubov, J. Exptl. Theoret. Phys. USSR **34**, 65 (1958), translation Soviet Phys. JEPT **34 (7)**, 41 (1958)
15. N.N. Bogoliubov, V.V. Tolmachev, D.V. Shirkov, *A New Method in the Theory of Superconductivity* (Academy of Sciences of USSR, Moscow, 1958)
16. Y. Nambu, Phys. Rev. **117**, 648 (1960)
17. L. Cooper, Phys. Rev. **104**, 1189 (1956)
18. P.W. Anderson, Phys. Rev. **112**, 1900 (1958)
19. G. Ricayzen, Phys. Rev. **115**, 795 (1959)
20. V. L. Ginzburg, L. D. Landau, J. Exptl. Theoret. Phys. USSR **20**, 1064 (1950)
21. T. Eguchi, H. Sugawara, Phys. Rev. D **10**, 4257 (1974)
22. T. Eguchi, Phys. Rev. D **14**, 2755 (1976)
23. T. Hatsuda, T. Kunihiro, Phys. Rept. **247**, 221 (1994)
24. J. Bijnens, Phys. Rept. **265**, 369 (1996)
25. M.K. Volkov, A.E. Radzhabov, Phys. Usp. **49**, 551–561 (2006). (arXiv:hep-ph/0508263)
26. N.N. Bogoliubov, D.V. Shirkov, *Introduction to the Theory of Quantized Fields* (Interscience Publishers Inc., NY, 1959)
27. C.N. Yang, T.D. Lee, Phys. Rev. **113**, 1165 (1959); T.D. Lee, C.N. Yang, Phys. Rev. **117**, 22 (1960)
28. R. Balian, C. Bloch, C. De Dominicis, Nucl. Phys. **25**, 529 (1961); Nucl. Phys. **27**, 294 (1961)
29. C. De Dominicis, P.C. Martin, J. Math. Phys. **5**, 14 (1964)
30. G. Jona-Lasinio, Nuovo. Cimento **34**, 1790 (1964)
31. W. Heisenberg, H. Euler, Z. Phys. **98**, 711 (1936)
32. C. Itzykson, J-B. Zuber, *Quantum Field Theory* (McGraw-Hill, 1980)
33. M. Peskin, D. Scroeder, *An Introduction to Quantum Field Theory* (CRC Press, 1995)
34. S. Coleman, E. Weinberg, Phys. Rev. D **7**, 1388 (1973)
35. S. Coleman, Secret symmetry: an introduction to spontaneous symmetry breakdown and gauge fields, in *Aspects of Symmetry: Selected Erice Lectures* (Cambridge University Press, 1985)
36. J. Goldstone, Nuovo Cimento **19**, 154 (1961); J. Goldstone, A. Salam, S. Weinberg, Phys. Rev. **127**, 965 (1962)
37. P.W. Anderson, Phys. Rev. **130**, 439 (1963)
38. F. Englert, R. Brout, Phys. Rev. Lett. **13**, 321 (1964)
39. P.W. Higgs, Phys. Rev. Lett. **13**, 508 (1964); P.W. Higgs, Phys. Rev . **145**, 1156 (1966)
40. G.S. Guralnik, C.R. Hagen, T.W.B. Kibble, Phys. Rev. Lett. **13**, 585 (1964)

41. T.W.B. Kibble, Phys. Rev. **155**, 1554 (1967)
42. S. Weinberg, Phys. Rev. Lett. **19**, 1264 (1967)
43. A. Salam, in *Proceedings of the Eighth Nobel Symposium*, ed. by N. Svartholm, Stokholm, p. 367 (1968)
44. S.L. Glashow, Nucl. Phys. **22**, 579 (1961)
45. T.D. Lee, *Particle Physics and Introduction to Field Theory*, Chap. 22 (Harwood Academic Publishers, 1981)
46. S. Weinberg, Rev. Mod. Phys. **52**, 515 (1980)
47. V. Popkov, M.R. Evans, D. Mukamel, J. Phys. A: Math. Theor. **41** (2008)
48. M. Shaposhnikov, J. Phys.: Conf. Ser. **171**, 012005 (2009)
49. A.D. Sakharov, JEPT Lett. **5**, 24 (1967)
50. A. Ijjas, P. Steinhardt, Phys. Lett. B **795**, 666 (2019)
51. Anna Ijjas, Paul J. Steinhardt, Phys. Lett. B **824**, 136823 (2022)

Chapter 18
String Theory

Paolo Di Vecchia

Abstract I start describing my interaction with Bruno during my thesis and then in his group in Frascati in connection with the calculation of the total cross-section of double bremsstrahlung that, at that time, was considered a good candidate as a monitor reaction for Adone. Then I discuss my transition to S-matrix theory and to the work that brought from the Dual Resonance Model to String Theory. I conclude describing the main results of String Theory in a way that could be followed by non-experts in the field.

18.1 Bruno and Me

In my third year of physics in 1964 I decided to follow the course on Statistical Mechanics held by Prof. Bruno Touschek and I was immediately fascinated by his personality. I liked a lot his informal personality, his way of doing and explaining physics and his free spirit. Therefore, at the beginning of 1965, I went to him asking for a thesis. He was very positive and told me to follow his course at the Scuola di Perfezionamento on QED where he was discussing first the Bloch-Nordsieck method and then he went on to discuss the quantisation of the electromagnetic field. When he finished with the Bloch-Nordsieck method [1, 2] he told me to use it for computing the cross-section of the double bremsstrahlung, corresponding to the process $e^+ + e^- \to e^+ + e^- + 2\gamma$, that, at that time, he was thinking to use as a monitor process for the luminosity of Adone. After his lectures it was easy to compute such a cross-section obtaining the leading soft term. Bruno helped me to write a letter on this result [3]. Adding to my thesis also the calculation of the forward amplitude I managed to get my Laurea in Physics in February 1966.

P. Di Vecchia (✉)
Niels Bohr Institute, University of Copenhagen, Blegadamsvej 17, 2100, Copenhagen, Denmark
e-mail: divecchia@nbi.dk; paolo.divecchia@su.se

Nordita, KTH Royal Institute of Technology, University of Stockholm, Hannes Alfvéns väg 12, 11419 Stockholm, Sweden

L. Bonolis et al. (eds.), *Bruno Touschek 100 Years*,
Springer Proceedings in Physics 287,
https://doi.org/10.1007/978-3-031-23042-4_18

Immediately after I got a fellowship to work in Bruno's group in the Laboratori Nazionali di Frascati (LNF). There I met Mario Greco who was also trying to compute the total cross-section for the double bremsstrahlung at high energy for any frequency of the two photons. We joined the forces and we published a paper after many months of work [4]. We computed the cross-section in two different ways finding agreement with [5] but not with [6]. After one year of fellowship I received a permanent position and continued to work in Bruno's group in Frascati.

After having finished the calculation of the total cross-section for double bremsstrahlung I did not want to go into the computation of loops and into the study of infrared divergences relevant for the experiments with Adone, as the rest of the group was doing. Instead I decided to move into S-matrix theory that was being developed in the sixties under the influence of the Berkeley school. For my work on the finite energy sum rules I got the possibility of visiting Caltech for three months where I discussed various issues of S-matrix theory with Frautschi and followed the wonderful lectures on particle physics given by Feynman. It was an incredible experience that made me to apply for a NATO fellowship to be able to go back to the US. In 1969 I got a NATO fellowship and I decided to use it at MIT to work with Sergio Fubini on the newly found Veneziano model. I had leave of absence from LNF for one year, but, when I asked to continue it for a second year, I received a negative answer from the Director of LNF. At the end of 1970 I resigned from LNF and stayed one more year at MIT.

This is the beginning of my peregrinations that brought me first to Cern for two years, then to Nordita in Copenhagen for four and half years, then to Cern again for one year. In 1979 I got a permanent position at the Freie Universität in Berlin and in 1980 I moved to a better position at the Bergische Universität Gesamthochschule Wuppertal where I stayed until the end of January 1986. From February 1986 I started to work again at Nordita and now I divide my time between the Niels Bohr Institute in Copenhagen and Nordita in Stockholm.

The Bloch-Nordsieck method and, more in general, the study of the infrared divergences in QED was the main activity in the theoretical group led by Bruno at the LNF and this activity has been very important for the experiments done with Adone. In the last couple of years I went back to these methods applying them to gravity [7] in connection with the experiments done at Ligo/Virgo where gravitational waves coming from the merging of two black holes have been observed.

18.2   From Dual Resonance Model to String Theory

The big successes obtained in the forties and fifties in the computation of the electro-magnetic processes starting from the QED Lagrangian did not seem to be possible for strong interactions because of the large value of the pion-nucleon coupling constant $\frac{g_{\pi NN}^2}{4\pi} \sim 14$ that did not allow perturbative calculations.

Therefore, in the sixties, many people, led by Chew and Mandelstam in Berkeley, gave up the idea of writing a Lagrangian for strong interactions and pushed instead the idea of computing directly the S-matrix of the strong processes by implementing its basic properties as analyticity, Regge behaviour, crossing symmetry and unitarity by means of a not well specified bootstrap approach.

The most successful result of these new ideas was the Veneziano model [8], originally constructed for the process $\pi\pi \to \pi\omega$ and immediately after extended to the scattering of four scalar particles:

$$B_4 \sim \frac{\Gamma(-\alpha(s))\Gamma(-\alpha(t))}{\Gamma(-\alpha(s) - \alpha(t))} \sim \Gamma(-\alpha(t))(-\alpha's)^{\alpha(t)} \qquad (18.1)$$

It contains an infinite number of narrow width resonances lying on a linearly rising Regge trajectory $\alpha(t) = \alpha_0 + \alpha't$ and on its daughter trajectories spaced by integers and, for $s \gg |t|$, has Regge behaviour as shown in the right-hand-side of the previous equation. Also the external scalar particle lies on the leading Regge trajectory and has a mass given by $\alpha_0 + \alpha'm^2 = 0$ in terms of the intercept α_0 and the slope α' of the leading Regge trajectory.

Shortly after the Veneziano model the previous amplitude has been extended to the scattering of N scalar particles [9]

$$B_N \sim \int_{-\infty}^{\infty} \frac{\prod_1^N dz_i \theta(z_i - z_{i+1})}{dV_{abc}} \prod_{i=1}^{N} \left[(z_i - z_{i+1})^{\alpha_0-1}\right] \prod_{i<j} (z_i - z_j)^{2\alpha' p_i \cdot p_j} \qquad (18.2)$$

and this N-point amplitude was called Dual Resonance Model (DRM). It satisfies all the axioms of S-matrix theory, except unitarity, with an infinite number of zero-width resonances lying on linearly rising Regge trajectories.

Having found the S-matrix the next questions were: is it a consistent S-matrix? What is the underlying theory?

The first step in this direction was taken by Fubini, Gordon and Veneziano [10] and by Nambu [11] and Susskind [12] who rewrote it in terms of an infinite number of harmonic oscillators spaced by integers satisfying the commutation relation: $[a_{n\mu}, a_{m\nu}^{\dagger}] = \delta_{nm}\eta_{\mu\nu}$ with $n, m = 1, 2\ldots$ and with the Lorentz metric given by $\eta_{\mu\nu} = (-1, 1, 1, 1)$. In particular, in [10] it was written as follows:

$$B_N = \int \frac{\prod_{i=1}^{N} dz_i}{dV_{abc}} \langle 0| \prod_{i=1}^{N} V(z_i, p_i)|0\rangle \qquad (18.3)$$

in terms of the vertex operator of the external scalar particle $V(z, p)$ and $Q^{\mu}(z)$:

$$V(z, p) =: e^{ipQ(z)} :$$

$$Q^{\mu}(z) = \hat{q}^{\mu} - 2i\alpha'\hat{p}^{\mu} \log z + i\sqrt{2\alpha'} \sum_{n=1}^{\infty} \sqrt{n}\left(a_n^{\mu}z^{-n} - a_n^{\dagger\mu}z^n\right) \qquad (18.4)$$

where $\hat{q}_0$ and $\hat{p}_0$ satisfy the commutation relation $[\hat{q}_0, \hat{p}_0] = i\eta^{\mu\nu}$.

Factorising the amplitude at the pole for $s_{ij} \sim M^2 = \frac{N-1}{\alpha'}$ (for simplicity in the case with $\alpha_0 = 1$) for $N = 0, 1 \ldots$ one gets the states $|\lambda\rangle$ with mass M^2 that contribute to its residue:

$$N|\lambda\rangle = \sum_{n=1}^{\infty} n a_{n\mu}^{\dagger} a_{n\nu} \eta^{\mu\nu} |\lambda\rangle \; ; \quad \eta_{\mu\nu} = (-1, 1, 1, 1) \tag{18.5}$$

For $N = 0$ the previous equation is satisfied by the vacuum $|0\rangle$, for $N = 1$ by the vector state $a_1^{\dagger\mu}|0\rangle$ and so on.

For a generic value of α_0 it turns out that the DRM contains states with negative norm [13, 14] violating tree-level unitarity.

Virasoro [15] found that for $\alpha_0 = 1$ there are extra conditions (Virasoro conditions) that possibly eliminate ghosts.

For $\alpha_0 = 1$, together with Del Giudice [16], we found that the on-shell physical states are characterised by the following conditions

$$L_n|Phys.\rangle = (L_0 - 1)|Phys.\rangle = 0 \; ; \quad n > 0 \tag{18.6}$$

that generalise the Fermi condition in QED: $\partial^\mu A_\mu^{(+)}|Phys.\rangle = 0$.

Fubini and Veneziano [17] showed that the Virasoro generators L_n satisfy the conformal algebra in two space-time dimensions called Virasoro algebra:

$$[L_n, L_m] = (n - m)L_{n+m} + \frac{D}{24}n(n^2 - 1)\delta_{n+m,0} \tag{18.7}$$

where the central charge was obtained by Weis [18], using the expression of L_n in terms of the oscillators.

Campagna et al. [19] generalised the amplitude to include any physical state (not just the ground state). For $\alpha_0 = 1$ the lowest state is a tachyon with mass $m^2 = -p^2 = -\frac{1}{\alpha'}$ and the next state is a massless photon with vertex operators [19, 20]

$$V(p, z) =: e^{ipQ(z)} : \; ; \quad V_i(k, z) = \left(\epsilon_i \frac{dQ(z)}{dz}\right) e^{ikQ(z)} \tag{18.8}$$

Those vertex operators and, more in general, the vertex operators of any physical state are conformal fields with dimension $\Delta = 1$ that satisfy the following commutation relation with the conformal generators L_n:

$$[L_n, V_\alpha(z, p)] = \frac{d}{dz}\left(z^{n+1}V_\alpha(z, p)\right) \tag{18.9}$$

The N-point amplitude involving N physical states can be written in terms of their vertex operators [19]:

$$A_N = \int \frac{\prod_{i=1}^{N} dz_i}{dV_{abc}} \langle 0| \prod_{i=1}^{N} V_{\alpha_i}(z_i, p_i)|0\rangle \tag{18.10}$$

realising the idea [21] that there is a complete democracy among the physical states and, as a consequence, there is no physical state more fundamental than the others. All of them lie on Regge trajectories.

The photon vertex operator was then used to construct the $(D - 2)$-dimensional DDF operators [22]:

$$A_{n,i} = \frac{i}{\sqrt{2\alpha'}} \oint_0 dz (\epsilon_i \frac{dQ(z)}{dz}) e^{ikQ(z)} ; \ k^2 = \epsilon k = 0 ; \ [L_m, A_{n,i}] = 0 \tag{18.11}$$

They satisfy the algebra of the harmonic oscillators:

$$[A_{n,i}, A_{m,j}] = n\delta_{ij}\delta_{n+m;0} ; \ A_{-m,i} \equiv A^{\dagger}_{m,i} ; \ n > 0 \tag{18.12}$$

and generate an infinite number of physical states with positive norm (no ghosts): but not all of them for arbitrary D (only for $D = 26$).

Already in 1969 Nambu [11, 23], Nielsen [24] and Susskind [12] suggested that the structure underlying the DRM was that of a string theory. In particular, in the Nambu formulation, the Virasoro algebra appeared to be a classification algebra as in Conformal Field Theory, while in the DRM was a gauge algebra needed to eliminate ghost states. It took a while to understand how to eliminate this discrepancy and this delayed the connection of the DRM with string theory.

The Nambu-Goto [23, 25] action was written down in 1970 as a generalisation of the one for a point particle

$$L_{part.} = -m \int \sqrt{-dx_\mu dx^\mu} ; \ L_{string} = -T \int \sqrt{-d\sigma_{\mu\nu} d\sigma^{\mu\nu}} \tag{18.13}$$

where m is the mass of the particle and $T = \frac{1}{2\pi\alpha'}$ is the string tension. But it took a while to understand how to use it. In terms of the world-sheet coordinates σ and τ one gets

$$L_{String} = -T \int_{\tau_i}^{\tau_f} d\tau \int_0^\pi d\sigma \sqrt{(\dot{x}x')^2 - \dot{x}^2(x')^2} \tag{18.14}$$

where $x^\mu(\tau, \sigma)$ is the string coordinate, $\dot{x}^\mu = \frac{\partial x^\mu}{\partial \tau}$ and $(x')^\mu = \frac{\partial x^\mu}{\partial \sigma}$. It is invariant under any choice of coordinates σ and τ.

It is more convenient to use a simpler and classically equivalent action:

$$S(x, g) = -\frac{T}{2} \int_{\tau_i}^{\tau_f} d\tau \int_0^\pi d\sigma \sqrt{-g} g^{ab} \partial_a x^\mu \partial_b x^\nu \eta_{\mu\nu} \tag{18.15}$$

where g^{ab} is the metric of the two-dim world-sheet (σ, τ). The equation of motion for g_{ab} implies the vanishing of the world-sheet energy-momentum tensor:

$$T_{ab} = \partial_a x \partial_b x - \frac{1}{2} g_{ab} g^{cd} \partial_c x \partial_d x = 0 \implies \dot{x}^2 + (x')^2 = \dot{x} x' = 0 \quad (18.16)$$

Choosing the conformal gauge

$$g_{ab} = \rho(\sigma, \tau) \eta_{ab} \quad (18.17)$$

we get a free action with the constraint of the vanishing of T_{ab}. The gauge is not completely fixed because we can still perform conformal transformations remaining in this gauge. To fix the gauge completely we can go to the light-cone gauge imposing an extra condition:

$$x^+ = 2\alpha' p^+ \tau \; ; \; x^\pm = \frac{x^0 \pm x^{D-1}}{\sqrt{2}} \quad (18.18)$$

Using the light-cone gauge condition and the vanishing of the two-dimensional energy momentum tensor, one can determine x^- in terms of the components $x^\perp$ (orthogonal to $x^\pm$). The only independent components are the $D - 2$ transverse $x^\perp$. This analysis was performed by Goddard et al. [26]. They checked that the D-dimensional Lorentz generators, written only in the terms of the $D - 2$ transverse $x^\perp$, satisfy the Lorentz algebra only if

$$\alpha_0 = 1 \; ; \; D = 26 \quad (18.19)$$

For $D = 26$ the DDF operators generate a complete set of physical states implying that the bosonic string is ghost free. This finally shows that the spectrum of physical states of the DRM for $D = 26$ is identical to the spectrum of string theory described by the Nambu-Goto action. Concerning the interaction Cremmer and Gervais [27] and Mandelstam [28] showed that the on shell three-point amplitude computed in string theory was identical to that of three arbitrary DDF states [29]. The equivalence of the DRM and string theory for higher N-point amplitudes was shown in [28, 30].

Even before finding its connection with string theory, the DRM, with the infinite set of zero width resonances, was considered a tree diagram of a unitary theory. At tree level, unitarity requires absence of ghosts and this property was satisfied for $\alpha_0 = 1$. On the other hand, loop diagrams were necessary in order to have the total widths of the resonances Γ_T to be equal to the sum of all partial widths $\sum_i \Gamma_i$. In order to implement this property one-loop and even multiloop amplitudes were constructed [31–33].[1]

---

[1] A complete expression for the multiloop amplitude in the bosonic string was only possible in the eigthies [35, 36] after the discovery of BRST symmetry.

Lovelace [34] showed that the non-planar loop had cuts violating unitarity unless $D = 26$. If $D = 26$ those cuts become poles that later turned out to be the states of a closed string that also lie on linearly rising Regge trajectories:

$$\alpha_{open}(s) = 1 + \alpha's \ ; \ \ \alpha_{closed}(s) = 2 + \frac{\alpha'}{2}s \qquad (18.20)$$

This means that unitarity requires that open strings always include closed strings. Closed strings require open strings but only at non-perturbative level as we will see later. As a consequence, Gauge Theories always include Gravity and vice-versa.

18.3   The Dual Pion Model

A N-point amplitude for pions was proposed by Neveu and Schwarz [37]. Unlike the one for the ground state particle of the bosonic string, it has the nice property that it vanishes for an odd number of pions, consistently with the fact that the pion has G-parity equal to -1. All previous analysis done for the DRM that turned out, for $\alpha_0 = 1$, to be equivalent to the bosonic string has been repeated for the NS model finding that it corresponds to a spinning string, i.e. a string with also spin degrees of freedom along it. It turned also out that it has no ghost if

$$\alpha_0 = 1 \ ; \ \ D = 10 \qquad (18.21)$$

Actually, if $\alpha_0 = \frac{1}{2}$, as for the ρ Regge trajectory (for $m_\pi = 0$), one gets the four-point Lovelace-Shapiro model [38]:

$$A(s,t) \sim \frac{\Gamma(1 - \alpha_\rho(s))\Gamma(1 - \alpha_\rho(t))}{\Gamma(1 - \alpha_\rho(s) - \alpha_\rho(t))} \ ; \ \ \alpha_\rho(s) = \frac{1}{2} + \alpha's \ ; \ \ \alpha' = \frac{1}{2m_\rho^2} \qquad (18.22)$$

with Adler zeroes as expected for pions ($A(s = 0, t = 0) = 0$).

The N-point generalisation of the LS model is discussed in a recent paper with Bianchi and Consoli [39]:

$$A_N = \int \frac{\prod_{1=1}^{N} d\theta_i dz_i}{dV_{abc}} \prod_{i<j} (Z_i - Z_j)^{2\alpha'k_ik_j} \prod_{i=1}^{N} (Z_i - Z_{i+1})^{-\frac{1}{2} - \alpha'm_\pi^2} \qquad (18.23)$$

using a super-conformal formalism, where $Z_i = (z_i, \theta_i)$. It has the correct Adler zeroes. It reduces to the non-linear σ model when $\alpha' \to 0$. But it has negative norm states: ghosts.

The reason is that, while the NS model is super-conformal invariant, the integrand of the amplitude in (18.23) is only super-projective invariant. This means that there are not enough conditions to decouple the ghosts. The conclusion is that it seems

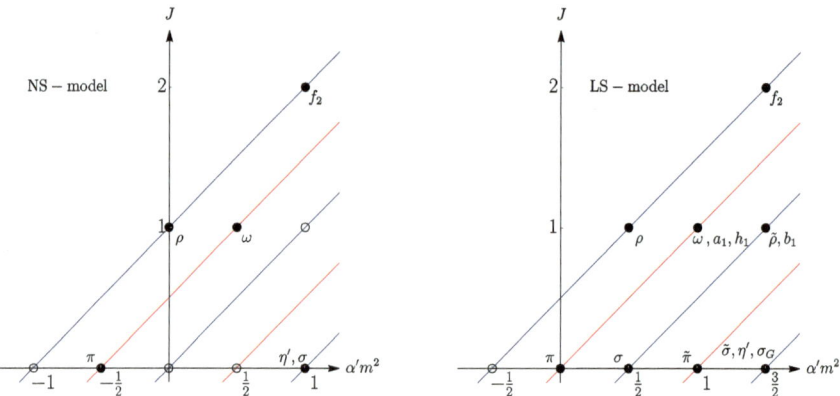

Fig. 18.1 Spectrum of NS (left) and LS (right) model in four dimensions, Regge trajectories in blue (red) have G-parity +1 (−1). Bullets represent 'physical' states, open circles represent 'missing' states

impossible to write a consistent N-pion amplitude with exact linearly rising Regge trajectories [40] (Fig. 18.1).

The NS model contains Regge trajectories with both integer and half-integer intercept. The particles lying on those with integer intercepts have G-parity $+1$, while the particles lying on those with half-integer intercepts have G-parity -1. As a consequence, the amplitudes with an odd number of particles lying on the Regge trajectories with half-integer intercepts are vanishing.

In order to extend the DRM to fermions Ramond [41] constructed the Ramond model and, later on, it turned out that the NS and the R model are part of the same model called RNS model. In addition to the bosonic oscillators of the bosonic string, it also contains an infinite set of fermionic oscillators with half-integer labels in the NS model and integer labels in the R model.

18.4 Unification of Gauge Theories and Gravity

In hadron physics only the pion is massless in the chiral limit. The consistent string theories that we have discussed do not allow for a massless pion, but contain instead massless gauge fields in the open string sector and a massless graviton in the closed string sector. This implies that the string theories that we have discussed cannot describe hadron physics as it was intended in the beginning with the Veneziano model.

In 1973 QCD was formulated and those interested in hadron physics left string theory and joined QCD.

In 1974 Scherk and Schwarz [42] proposed to use string theory, not for hadrons, but as a theory consistently unifying gauge theories with gravity. A very important property of string theory is that Gauge Theories and GR are not put by hand, but emerge together as an unavoidable part of the theory, as gauge invariance and invariance under diffeomorphisms are necessary ingredients for a consistent description of massless spin 1 and 2, respectively.

Unlike in field theory, in string theory we have a single interaction: that among strings. The only free parameter is the Regge slope α' related to the string tension. The string coupling constant g_s that enters in the loop expansion is not a free parameter but is given by the vacuum expectation value of a particular state of the closed string, the dilaton $g_s \sim e^\phi$ and it is fixed by the minimum of the dilaton potential.

We must finally stress that String Theory is an extension of Field Theory: field theory amplitudes are recovered in the limit of $T \to \infty$ or $\alpha' \to 0$ [43, 44].[2]

In conclusion, the softness of string at high energy that was a problem for hadrons becomes now a virtue providing a finite theory of gravity.

18.5 From the RNS Model to Superstring

The RNS model contains two sectors: one with ten-dimensional fermions and another with ten-dimensional bosons. The spectrum of states in the bosonic sector is given by

$$\alpha' m_B^2 = N - \frac{1}{2} \;\; ; \;\; N|\lambda\rangle = \sum_{n=1}^{\infty} \left(n a_n^\dagger a_n + \left(n - \frac{1}{2} \right) \psi_{n-\frac{1}{2}}^\dagger \psi_{n-\frac{1}{2}} \right) |\lambda\rangle \quad (18.24)$$

where N can be integer and half-integer. The lowest state is still a tachyon $|0\rangle$ and the next state is a massless gauge field $\psi_{\frac{1}{2};\mu}^\dagger |0\rangle$.

One can define a world-sheet fermion number:

$$(-1)^F \;\; ; \;\; F = \sum_{n=1}^{\infty} \psi_{n-\frac{1}{2}}^\dagger \psi_{n-\frac{1}{2}} - 1 \quad (18.25)$$

The states with an even (odd) number of fermionic oscillators have eigenvalue -1 $(+1)$ under the action of $(-1)^F$. $(-1)^F$ corresponds to the G-parity that we discussed in the Dual Pion Model.

The spectrum of states in the fermionic sector is given by:

$$\alpha' m_F^2 = N \;\; ; \;\; N|\lambda\rangle = \sum_{n=1}^{\infty} \left(n a_n^\dagger a_n + n \psi_n^\dagger \psi_n \right) |\lambda\rangle \quad (18.26)$$

---

[2] See also [45].

where N is an integer. The lowest state is a massless ten-dimensional fermion. There is a fermionic zero mode that satisfies the same algebra as that of the Dirac Γ-matrices:

$$\{\psi_0^\mu, \psi_0^\nu\} = \eta^{\mu\nu} \tag{18.27}$$

This means that the ground state $|0, A\rangle$ has a ten-dimensional Dirac spinor index A. Also in this case we have a fermion number operator:

$$(-1)^F = \Gamma_{11}(-1)^{F_R} \;;\; F_R = \sum_{n=1}^{\infty} \psi_n^\dagger \psi_n \tag{18.28}$$

In 1976 Gliozzi et al. [46] proposed to truncate the spectrum of the RNS model keeping only the states that are even under the action of the fermion number operator:

$$(-1)^F |\psi\rangle = |\psi\rangle \tag{18.29}$$

It is called GSO projection.

The GSO projection eliminates the states in the NS sector that lie on half-integer Regge trajectories and in R sector imposes to the ground state to be a Weyl-Majorana fermion.

The two lowest states are a gauge field in the bosonic and a massless Weyl-Majorana fermion in the fermionic sector. They have the same number of physical degrees of freedom: 8 in $D = 10$. It turns out that, after the GSO projection, we get at each level of the spectrum the same number of bosons and fermions. One gets the spectrum of the open type I string theory that is supersymmetric in $D = 10$. Type I contains also a supersymmetric closed string sector with gravitons, gravitinos and other massless states. This is the first string theory without a tachyon in the spectrum that has been constructed.

18.6 Superstring Theories: Type IIA, Type IIB, Type I

After 1976 and before 1985 two closed superstring theories were constructed by Green and Schwarz [47]. They contain two bosonic sectors, called NS-NS and R-R, and two fermionic sectors, called R-NS and NS-R.

Type IIB theory is a chiral closed superstring theory that, in the massless NS-NS sector, contains a graviton, a dilaton and a Kalb-Ramond field described by an antisymmetric tensor $B_{\mu\nu}$ and in the R-R sector the potentials C_0, $C_{2\mu\nu}$, $C_{4\mu\nu\rho\sigma}$. The two fermionic sectors contain two gravitinos and two dilatinos with the same chirality.

Type IIA is instead a non-chiral closed superstring theory that has the same massless NS-NS sector as type IIB, while the R-R sector contains the potentials $C_{1\mu}$, $C_{3\mu\nu\rho}$.

The two fermionic sectors contain two gravitinos and two dilatinos with opposite chirality.

We conclude this section with Type I whose massless open string sector we have already discussed and we have seen that it contains a gauge boson and a gaugino. The closed string sector contains instead a graviton, a dilaton, a $C_{2\mu\nu}$ potential and one gravitino and one dilatino. Furthermore, in order to cancel gauge and gravitational anomalies the gauge group must be $SO(32)$.

Those are the three superstring theories that were constructed before string theory became popular again around 1985. The developments of string theory from the origin to 1985 are described in a book edited together with Cappelli et al. [48].

18.7 D(irichlet)p-Branes

In the previous section we have seen that the type I and type II theories contain potentials with more than one index. They are a generalisation of the electromagnetic potential A_μ and, as the electromagnetic potential is coupled to point-like particles, they are instead coupled to p-dimensional objects through the following generalisation of the electromagnetic coupling:

$$\int A_\mu dx^\mu \implies \int A_{\mu_1\mu_2\ldots\mu_{p+1}} d\sigma^{\mu_1\mu_2\ldots\mu_{p+1}} \tag{18.30}$$

It turns out that there exist classical solutions of the low-energy string effective action that are coupled to the metric, the dilaton and are charged with respect to one of these RR fields [49]. For them we get the following asymptotic behaviour

$$C_{01\ldots p} \sim \frac{1}{r^{D-3-p}} \iff C_0 \sim \frac{1}{r} \ if \ D = 4, p = 0 \tag{18.31}$$

that reduces to that of the electromagnetic vector potential for $p = 0$. They correspond to non-perturbative states of string theory with tension and RR charge given by:

$$\tau_p = \frac{Mass}{p - volume} = \frac{(2\pi\sqrt{\alpha'})^{1-p}}{2\pi\alpha' g_s} \ ; \ \mu_p = \sqrt{2\pi}(2\pi\sqrt{\alpha'})^{3-p} \tag{18.32}$$

where g_s is the string coupling constant.

In 1994 Polchinski [50] showed that, in string theory, these objects are required by T-duality that, in the case of a closed string exchanges Kaluza-Klein modes with winding modes, while, in the case of an open string, changes Neumann with Dirichlet boundary conditions. For this reason they are called D(irichlet)p-branes. Open strings satisfy Neumann boundary conditions along the direction of the world-volume of the Dp-brane and Dirichlet boundary conditions along the directions orthogonal to the

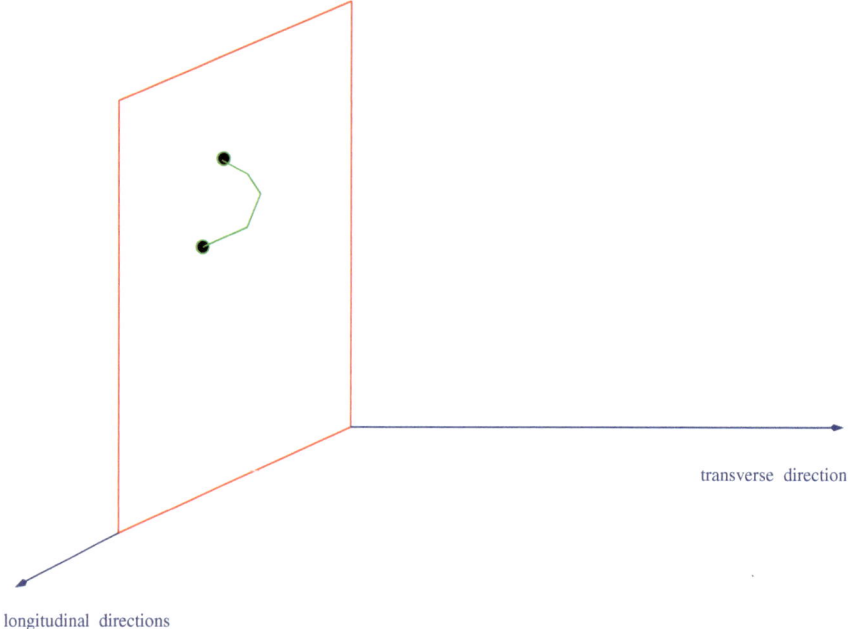

transverse direction

longitudinal directions

Fig. 18.2 The endpoints of an open string can move freely inside the world-volume of the Dp-brane, but they cannot move along the directions orthogonal to the world-volume of the Dp-brane

world-volume of a Dp-brane, as shown in Fig. 18.2. Besides the perturbative states, type I and II string theories contain also the Dp-branes that are non-perturbative states of type I and II string theories and are characterised by the fact of having open strings attached to their world-volume.

It follows that open strings and the corresponding gauge theories live in the (p+1)-dim. world-volume of a Dp-brane, while closed strings (gravity) live in the entire ten dimensional space.

If we have a stack of N parallel coincident Dp-branes, then we have N^2 open strings having their endpoints on the D branes, corresponding to the degrees of freedom of the adjoint representation of $U(N)$. The massless bosonic states correspond to the gauge fields of $U(N)$, while the massless fermionic states correspond to their supersymmetric partners, called gauginos.

The gauge theory living on N maximally supersymmetric D3-branes is the maximally supersymmetric $\mathcal{N} = 4$ super-Yang-Mills with $U(N)$ gauge group containing one gluon, 6 scalars and 4 Majorana fermions, all transforming according to the adjoint representation of the gauge group. It is conformal invariant with vanishing β-function. Maldacena [51] conjectured that this theory is equivalent to 10-dimensional string theory on $AdS_5 \otimes S_5$. By now there is a lot of evidence for it and a lot of applications have been made both for hadrons and condensed matter systems.

What remains to understand is how to extend the previous exact duality to non-conformal gauge theories as QCD and what is the string theory appearing in the 't Hooft large N expansion of QCD [52].

18.8 M-theory

Up to now we have discussed the three superstring theories that were constructed before 1984. After that, two more fully consistent string theories were constructed. They are the two heterotic strings. They are closed string theories, but, unlike type II theories, they contain a gauge theory: one with gauge group $SO(32)$ and the other with gauge group $E_8 \times E_8$. These two gauge groups are required in order not to have gauge and gravitational anomalies.

The five superstring theories that we have discussed are all consistent string theories in ten-dimensional Minkowski space-time and, at the perturbative level, they are all independent from each other.

This has generated a puzzle for many years: If string theory is a unique theory why do we have five theories instead of just one?

It turns out that, if we also include their non-perturbative behaviour, they are related to each other through a web of weak-strong dualities [53, 54] and they are all part of a unique 11-dimensional theory, called M-theory that, at low energy, reduces to the unique 11-dimensional supergravity. Starting from M-theory, in which two directions are compactified on $S^1 \times S^1$, one recovers type IIA and type IIB theories, in which one of the ten directions is compactified on S^1, that are T-dual to each other. If we instead compactify two directions of M-theory on $S^1 \times \frac{S^1}{Z_2}$ one recovers the two heterotic strings and type I theory [55]. In particular, type I and heterotic with gauge group $SO(32)$ are related by weak-strong duality [56].

The unification of all consistent string theories in ten dimensions in a unique 11-dimensional theory is a very beautiful result, but we should not forget that we live in four and not eleven dimensions. This means that eight directions of M-theory must live in a compact manifold that must be small enough in order not to contradict experiments. Unfortunately this compactification can be done in too many consistent ways and, at the moment, it seems impossible to use M-theory or string theory to predict the low energy physics that we see in experiments at present energies. This is the Landscape Problem that unfortunately is still with us at present.

References

1. F. Bloch, A. Nordsieck, Note on the radiation field of the electron. Phys. Rev. **52**, 54–59 (1937)
2. W. Thirring, B. Touschek, A covariant formulation of the Bloch-Nordsieck method. Phil. Mag. Ser. **7**(42), 244–249 (1951)
3. P. Di Vecchia, A note on double bremsstrahlung. Nuovo Cimento **45**, 249–251 (1966)

4. P. Di Vecchia, M. Greco, Double photon emission in $e^{\pm}e^-$ collisions. Nuovo Cimento 50, 319–332 (1967)
5. V.N. Bayer, V.M. Galitsky, Double bremsstrahlung in electron collisions. JETP Lett. 2, 165–167 (1965)
6. M. Bander, SLAC-TN-64-93 (1965)
7. P. Di Vecchia, C. Heissenberg, R. Russo, G. Veneziano, Radiation reaction from soft theorems. Phys. Lett. B 818, 136379–136388 (2021). (arXiv:2101.05772 [hep-th])
8. G. Veneziano, Construction of a crossing-symmetric, Regge-behaved amplitude for linearly rising trajectories. Nuovo Cimento A 57, 190–197 (1968)
9. K. Bardakçi, H. Ruegg, Reggeized resonance model for arbitrary production processes. Phys. Rev. 181, 1884–1889 (1969); C.J. Goebel, B. Sakita, Extension of the Veneziano form to N-particle amplitude, Phys. Rev. Lett. 22, 257–260 (1969); Chan Hong-Mo, T.S. Tsun, Explicit construction of the N-point function in generalized Veneziano model, Phys. Lett. B 28, 485–488 (1969) Z. Koba, H.B. Nielsen, Reaction amplitude for n mesons: a generalisation of the Bardacki-Ruegg-Virasoro model, Nucl. Phys. B 10, 633-655 (1969)
10. S. Fubini, D. Gordon, G. Veneziano, A general treatment of factorisation in dual resonance models. Phys. Lett. B 29, 679–682 (1969)
11. Y. Nambu, in *Proceedings of the International Conference on Symmetries and Quark Models*, Wayne State University, 1969 (Gordon and Breach, 1970) p. 269
12. L. Susskind, Dual-symmetric theory of hadrons. 1., Nuovo Cimento A 69, 457–496 (1970) and Harmonic-oscillator analogy for the Veneziano model. Phys. Rev. Lett. 23, 545–547 (1969)
13. S. Fubini, G. Veneziano, Level structure of dual-resonance models. Nuovo Cimento A 64, 811–840 (1969)
14. Bardakçi, S. Mandelstam, Analytic solution of the linear-trajectory bootstrap. Phys. Rev. 184 1640–1644 (1969)
15. M.A. Virasoro, Subsidiary conditions and ghosts in dual resonance models. Phys. Rev. D 177, 2933–2936 (1969)
16. E. Del Giudice, P. Di Vecchia, Characterization of physical states in dual-resonance models. Nuovo Cimento A 70, 579–591 (1970)
17. S. Fubini, G. Veneziano, Algebraic treatment of subsidiary conditions in dual resonance models. Ann. Phys. 63, 12–27 (1971)
18. J. Weis, Priv. Commun. (1970)
19. S. Campagna, E. Fubini, E. Napolitano, S. Sciuto, Amplitude for n nonspurious excited particles in dual resonance models. Nuovo Cimento A 2, 911–928 (1971)
20. L. Clavelli, P. Ramond, New class of dual vertices. Phys. Rev. D 4, 3098 (1971)
21. G. Chew, *"The Analytic S Matrix", a Basis for Nuclear Democracy* (1966)
22. E. Del Giudice, P. Di Vecchia, S. Fubini, General properties of the dual resonance model. Ann. Phys. 70, 378–398 (1972)
23. Y. Nambu, *Lectures at the Copenhagen Symposium* (1970). Unpublished
24. H.B. Nielsen, in *Paper Submitted to the 15th International Conference on High Energy Physics*, Kiev, 1970 and Nordita preprint (1969)
25. T. Goto, Relativistic quantum mechanics of one-dimensional mechanical continuum and subsidiary conditions in dual resonance model. Progr. Theor. Phys. 46, 1560–1569 (1971)
26. P. Goddard, J. Goldstone, C. Rebbi, C. Thorn, Quantum dynamics of a quantum relativistic string. Nucl. Phys. B 56, 109–135 (1973)
27. E. Cremmer, J.L. Gervais, Combining and splitting relativistic strings. Nucl. Phys. B 76, 209–230 (1974)
28. S. Mandelstam, Interacting string picture of dual resonance models. Nucl. Phys. B 64, 205–235 (1973)
29. M. Ademollo, E. Del Giudice, P. Di Vecchia, S. Fubini, Coupling of three excited particles in the dual-resonance model. Nuovo Cimento A 19, 181–203 (1974)
30. M. Ademollo, A. D'Adda, R. D'Auria, E. Napolitano, S. Sciuto, P. Di Vecchia, F. Gliozzi, R. Musto, F. Nicodemi, Theory of an interacting string and dual resonance model. Nuovo Cimento A 21, 77–145 (1974)

31. V. Alessandrini, D. Amati, M. Le Bellac, D. Olive, The operator approach to dual multiparticle theory. Phys. Rep. **1**, 269–346 (1971)
32. C. Lovelace, M-loop generalized veneziano amplitude. Phys. Lett. B **32**, 703–708 (1970)
33. V. Alessandrini, D. Amati, A general approach to dual multiloop diagrams. Nuovo Cimento A **2**, 321–352 (1971)
34. C. Lovelace, Pomeron form factors and dual Regge cuts. Phys. Lett. B **34**, 500–506 (1971)
35. P. Di Vecchia, M. Frau, A. Lerda, S. Sciuto, A simple expression for the multiloop amplitude in the bosonic string. Phys. Lett. B **199**, 49–56 (1987)
36. S. Mandelstam, The n-loop string amplitude: explicit formulas, finiteness and absence of ambiguities. Phys. Lett. B **277**, 82–88 (1992)
37. A. Neveu, J.H. Schwarz, Factorizable dual models of pions. Nucl. Phys. B **31**, 86–112 (1971) and Quark model of dual pions. Phys. Rev. D **4**, 1109–1111 (1971)
38. C. Lovelace, A novel application of Regge trajectories. Phys. Lett. B **28**, 264–268 (1968); J. Shapiro, Narrow resonance model with Regge behaviour for pi pi scattering. Phys. Rev. **179**, 1345–1353 (1969)
39. M. Bianchi, D. Consoli, P. Di Vecchia, On the N-pion extension of the Lovelace-Shapiro model. JHEP **03**, 119 (2021). (arXiv:2002.05419 [hep-th])
40. S. Caron-Huot, Z. Komargodski, A. Sever, A. Zhiboedov, Strings from massive higher spins: the asymptotic uniqueness of the Veneziano amplitude. JHEP **10**, 026 (2017). (arXiv:1607.04253 [hep-th])
41. P. Ramond, Dual theories for free fermions. Phys. Rev. D **3**, 2415–2418 (1971)
42. J. Scherk, J.H. Schwarz, Dual models for nonhadrons. Nucl. Phys. B **81**, 118–144 (1974)
43. J. Scherk, Zero slope limit of the dual resonance model. Nucl. Phys. B **31**, 222–234 (1971). A. Neveu, J. Scherk, Connection between Yang-Mills fields and dual models. Nucl. Phys. B **36**, 155–161 (1973)
44. T. Yoneya, Connection of dual models to electrodynamics and gravidynamics. Prog. Theor. Phys. **51**, 1907–1920 (1974)
45. P. Di Vecchia, L. Magnea, A. Lerda, R. Russo, R. Marotta, String techniques for the calculation of the renormalization constants in field theory. Nucl. Phys. B **469**, 235–286 (1996). hep-th/9601143 [hep-th]; P. Di Vecchia, L. Magnea, A. Lerda, R. Marotta, R. Russo, Two loop scalar diagrams from string theory. Phys. Lett. B **388**, 65–76 (1996). hep-th/9607141 [hepth]
46. F. Gliozzi, J. Scherk, D. Olive, Supergravity and the spinor dual model. Phys. Lett. B **65**, 282–286 (1976); Supersymmetry, supergravity theories and the dual spinor model. Nucl. Phys. B **122**, 253–290 (1977)
47. M.B. Green, J.H. Schwarz, Supersymmetrical string theories. Phys. Lett. B **109**, 444–448 (1982)
48. A. Cappelli, E. Castellani, F. Colomo, P. Di Vecchia, *The Birth of String Theory* (Cambridge University Press, 2012)
49. M.J. Duff, R.R Khuri, J.X. Lu, String solitons. Phys. Rep. **259**, 213–326 (1995), hep-th/9412184; K.S. Stelle, *Contribution to the ICTP Summer School in High-Energy Physics and Cosmology*, hep-th/9803116; R. Argurio, *Brane Physics in M Theory*, hep-th/9807171
50. J. Polchinski, Dirichlet branes and Ramond-Ramond charges. Phys. Rev. Lett. **75**, 4724–4727 (1995). (hep-th/9510017)
51. J. Maldacena, The large N limit of superconformal field theories and supergravity. Adv. Theor. Math. Phys. **2**, 231–252 (1998). (hep-th/9711200)
52. G. 't Hooft, A planar diagram theory for strong interactions. Nucl. Phys. B **72** 461–473 (1973)
53. C. Hull, P. Townsend, Unity of superstring dualities. Nucl. Phys. B **438**, 109–137 (1995). (hep-th/9410167)
54. E. Witten, String theory dynamics in various dimensions. Nucl. Phys. B **443**, 85–126 (1995). (hep-th/9503124)
55. P. Horava, E. Witten, Heterotic and type I string dynamics from eleven dimensions. Nucl. Phys. B **460**, 506–524 (1996). (hep-th/9510209)
56. J. Polchinski, E. Witten, Evidence for heterotic-type I string duality. Nucl. Phys. B **460**, 525–540 (1996). (hep-th/9510169)

Chapter 19
Multi-messenger Astronomy

Marica Branchesi

Abstract On 2015 September 14, the first observation of gravitational-waves by the Advanced Laser Interferometer Gravitational-wave Observatory detectors concluded a long scientific quest, which began 100 years before with Einstein's prediction of their existence. This detection opened a new exploration of the Universe making it possible to access the properties of space-time at extreme regime, to probe the properties of compact objects (binary systems of neutron stars and stellar-mass black holes), and investigate their formation and evolution. On August 17, 2017, the first observation of gravitational waves from the inspiral and merger of a binary neutron-star system by the Advanced LIGO and Virgo network, followed 1.7 s later by a weak short gamma-ray burst detected by the Fermi and INTEGRAL satellites initiated the most extensive world-wide observing campaign which led to the detection of multi-wavelength electromagnetic counterparts. Multi-messenger discoveries are unveiling the rich physics of most energetic transient phenomena in the sky, probing relativistic astrophysics, nuclear physics, nucleosynthesis, and cosmology. Here, we give an overview of the recent gravitational-wave and multi-messenger discoveries, and the perspectives for the future.

19.1 Introduction

The multimessenger astronomy is based on observations of astrophysical objects through different cosmic messengers (electromagnetic radiation, gravitational waves, neutrinos and cosmic rays) which can provide a complementary and complete view of astrophysical sources and their environment. Its onset resides in the discovery of neutrinos from a supernova exploded in the Large Magellanic Cloud in 1987, SN1987A [29, 44]; the neutrinos arrived a few hours before the optical emission and

M. Branchesi (✉)
Gran Sasso Science Institute, INFN, 67100 L'Aquila, Italy
e-mail: marica.branchesi@gssi.it

Laboratori Nazionali del Gran Sasso, 67100 Assergi, Italy

© The Author(s) 2023
L. Bonolis et al. (eds.), *Bruno Touschek 100 Years*,
Springer Proceedings in Physics 287,
https://doi.org/10.1007/978-3-031-23042-4_19

were detected by Kamiokande II, the Irvine-Michigan-Brookhaven detector, and the Baksan Neutrino Observatory.

Another recent discovery marked the history of multi-messenger observations, giving a huge boost to the field and showing the tremendous potential of combining multi-messenger observations to probe the physics of the most energetic events of the Universe: GW170817. On 2017 August 17, the merger of a binary neutron-star system has been observed through gravitational waves (GW170817) [13], and multi-wavelength photons from gamma rays (GRB 170817A), X-ray, ultraviolet-optical-near infrared (AT2017gfo), to radio [7]. The multi-messenger signals associated with this spectacular event represent the first strong observational evidence that binary neutron-star mergers power short gamma-ray bursts [6, 40, 62] and kilonovae [33], unveiling properties of relativistic jets [38, 54] and showing that binary neutron-star mergers are one of the major channels of the formation of heavy (r-process) elements in the Universe [56]. Neutron stars are unique laboratories to probe matter in extreme conditions, and the multi-messenger observations can constrain the neutron-star equation of state [51, see e.g.]. The distance estimated from the gravitational-wave signal combined with the recessional velocity of the host galaxies enable to evaluate the Universe expansion rate, showing a new way to make cosmology [5].

This paper covers the major discoveries related to the gravitational-wave astronomy since 2015 (Sect. 19.2), the multi-messenger observations of GW170817 and their scientific return (Sect. 19.3), the perspectives of the future multi-messenger astronomy (Sect. 19.4).

19.2 Gravitational-Wave Astronomy

The LIGO-Hanford, LIGO-Livingstone [1] and Virgo [25] interferometers observing the sky as a network made it possible to observe gravitational-waves. They have performed three run of observations; the first observational run lasted from September 2015 to January 2016, the second run from November 2016 to the end of August 2017, the third run from April 2019 to the end of March 2020. During the intervals between runs, technological upgrades increased the sensitivity of the detectors, making larger and larger volumes of the Universe accessible through gravitational wave observations. Rare events such as the coalescences of binary systems of neutron stars, neutron stars and black holes, and black holes have begun to be observed, and the frequency of their observations has increased significantly from the first run to the following ones. The first run led to the detection by the LIGO interferometers of three gravitational-wave signals from the coalescence of a binary system of stellar-mass black holes [11, 12]. These events showed us that black-holes exist in binary systems, that they can merge within the Hubble time, and that stellar-mass black-hole can be more massive ($> 30 M_{sun}$) than expected before [10]. The second run increased the detected events to eleven, including binary black-hole coalescences and a binary neutron star coalescence, GW170817 [16]. During the third gravitational waves have been detected with a rate of about 1.5 detections per week. Seventy-nine candidate gravitational-wave events have been added to the 11 confident detections of the first and second observation runs. The majority of the signals are classified as binary black hole coalescences, but they also include another binary-neutron-star and two confident binary neutron-star black-hole coalescences [4, 18]. In a few years from a few merging binary black-holes, a significant number of detections was accumulated making possible population studies [2, 15, 19]. We have now direct measurements of binary black-hole properties, such us mass and spin distributions, and their frequency of merging. This had a huge impact on our knowledge of formation and evolution of these astrophysical systems, and indirectly also on their progenitors, the death of massive stars [10]. These events also provide unique access to the properties of space-time at extreme conditions under the strong-field and high-velocity regime. They enable us to define stringent constraints on testing general relativity [3, 17, 20]. Among the detections, some events were particularly interesting. GW190412 is a signal from a highly asymmetric mass binary black-hole system, component masses of $30M_{sun}$ and $8M_{sun}$. This signal made it possible to find for the first time strong evidence for gravitational radiation beyond the leading quadrupolar order, in complete consistency with the Einstein's general theory of relativity [21]. GW190814 is a signal from the coalescence of a black hole of $23M_{sun}$ with a compact object of mass $2.6M_{sun}$. Its unequal mass ratio and its secondary component consistent either with the lightest black hole or the heaviest neutron star ever discovered in a binary compact-object system are unprecedented, and challenges all current models of the formation of compact-object binaries [23]. GW190425 is the second detected signal from a binary neutron-star merger after GW170817. The total mass of the system, $3.4M_{sun}$, is significantly larger than those of any other known binary neutron-star system [8]. GW190521 is a signal from the coalescence of a highest mass binary black-hole system $(66 - 85M_{sun})$ forming a final black hole of $142M_{sun}$. This is the firm evidence of the existence of intermediate-mass black holes $(100 - 1000M_{sun})$ [22, 24].

19.3 Multi-messenger Astronomy Including Gravitational-Waves

The first detection of gravitational waves from a binary system of neutron stars by the Virgo and LIGO network, GW170817 is an epochal discovery which represents a landmark for multi-messenger astrophysics including gravitational waves [13]. The relatively small sky-localization of the signal enabled the most extensive electromagnetic observational campaign in human history, which led to the observation of the gravitational-wave source in all electromagnetic wavelengths (X-rays, ultraviolet, optical, infrared, and radio) [7]. In the following we summarize the different observations (see Fig. 19.1) and the implications of the revealed signals in the astrophysical knowledge of the source.

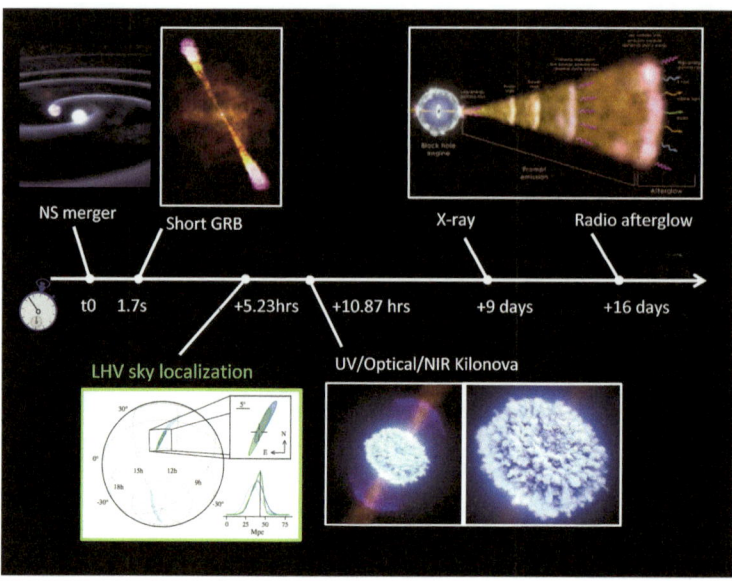

Fig. 19.1 Timeline of the discovery of the gravitational-wave signal (GW170817), the gamma-ray burst (GRB 170817A), the release of the gravitational-wave sky localization, the discovery of the optical emission from the kilonova (AT 2017gfo), and the discoveries of the X-ray and radio emission from the GRB relativistic jet

The gravitational wave signal enabled to infer the component masses of the binary system in the range $0.86 - 2.26M_{sun}$ which is consistent with the masses of known neutron stars in our Galaxy [13]. It also enabled to constrain the neutron star tidal deformability (each neutron star in the binary system is tidally deformed when under the influence of tidal field of the other). This macroscopic observable can be used to study neutron star interiors, in particular to infer the neutron star equation of state (EoS). The measurement obtained for GW170817 favors larger tidal deformability values, and thus softer EoSs are preferred with respect to stiffer ones [14]. The amount of tidally ejected mass which gives origin to the baryon mass powering the electromagnetic emission depends on the EoSs with stiffer EoSs producing larger amount of tidally ejected mass than the softer ones (see also below for EoS constraints from the electromagnetic observations).

The Fermi and INTEGRAL satellites independently detected a short gamma-ray burst (GRB) with a time delay of 1.7 s from the merger time. The time delay and the source distance made it possible to measure the propagation speed of gravitational waves; gravitational waves propagate at the speed of light to within $1:10^{15}$ [6]. This ruled out several classes of modified gravity models. Nine and sixteen days after the merger an X-ray signal [64] and a radio signal [42] were discovered. The following observations showed a slow non-thermal emission flux-rise in the radio, optical, and X-rays for about 150 days [49, 52] and then a slow decay [27, 34, 36, 41]. These multi-wavelength observations are consistent with both a slightly relativistic isotropic outflow (choked jet) and a successful structured jet (with energy and velocity decreasing with angular distance from the jet axis) observed off-axis. The magnificent resolution of Very Long Baseline Interferometry observations enabled to measure a superluminal proper motion of the radio counterpart [54] and to constrain the apparent size of the source [38], demonstrating that the later scenario, i.e. a relativistic jet successfully emerged from the neutron star merger, has occurred. GW170817 and observations from the gamma to the radio, have provided the first firm observational evidence that binary neutron-star mergers power short GRBs.

Neutron star mergers represent the perfect event for producing heavy elements, the temperature ($T > 10^9$K) and high neutron start density (10^{22} cm^{-3}) of the merger dynamical ejecta make neutron capture much faster than the β-decay. The formed heavy nuclei radioactively decay heating the material around and powering an ultraviolet (UV), optical and infrared (IR) transient, known as kilonova. While in the tidal tail ejecta the nucleosynthesis produce heavy elements up to lanthanides and actinides, whose opacity makes the spectral peak of the emission in the near-infrared and the peak of the light curve on one week timescale, in other components of the ejecta, such as the shock-heated ejecta and the accretion disc wind outflow, weak interactions (neutrino absorption, electron/positron capture) prevent the production of the heavier elements. This gives rise to smaller opacity and a bluer kilonova component peaking on day timescale (for a complete review see [53]). Eleven hours after the merger, optical transient emission was discovered from a galaxy, NGC 4993, at the same distance as the one evaluated from the gravitational-wave signal, pinpointing the location of the merger [33]. The observations from the near infrared to the ultraviolet taken for about ten days showed a transient thermal emission with a blue component fading within two days and a red component evolving in one week [65, e.g.]. The spectra revealed signatures of the radioactive decay of r-process nucleosynthesis [56, 63, 66], showing that binary neutron star mergers are one of the major channels of formation of heavy elements in the Universe. The brightness and evolution of the UV/optical/infrared data enabled to constrain the ejected masses providing a lower bound on the tidal deformability, and ruling out extremely soft equations of state. Joint kilonova and gravitational-wave observations are thus complementary, and rule out EoS in different directions [58]. The identification of the host galaxy through the kilonova detection enabled to use the recessional velocity of the host galaxies together with distance estimated from the gravitational-wave signal to evaluate the Hubble constant [5]. Figure 19.2 shows a summary of the major implications in astrophysics of the multi-messenger discovery of GW170817.

In the next run of observations of the current gravitational-wave detectors, currently planned to start at the end of 2022/early 2023, a few to ten binary neutron star mergers are expected to be detected [9]. The search of the electromagnetic counterparts will be more difficult due to the larger distances accessible by the upgrades of the LIGO, Virgo and KAGRA detectors. However, improved sensitivity observatories are expected to operate in synergy with the gravitational-wave detectors, for example the James Webb Space Telescope [46, JWST], the Vera C. Rubin Observa-

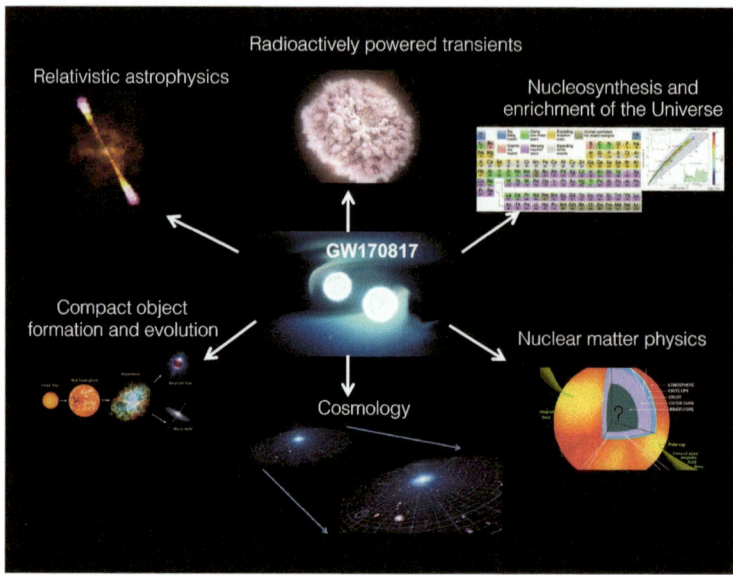

Fig. 19.2 Summary of the major astrophysical fields impacted by the multi-messenger discovery of GW170817

tory [45], GECAM [47], the Space-based multi-band astronomical Variable Objects Monitor ECLAIRs [67, SVOM-ECLAIRs], and Einstein Probe [68] to mention a few.

19.4 The Future of Gravitational-Wave and Multi-messenger Astronomy

Despite the enormous impact of LIGO-Virgo discoveries on many research fields, from fundamental physics and astrophysics to nuclear physics and cosmology, we are only at the dawn of this new exploration of the Universe. A new generation of more sensitive detectors is needed to address fundamental questions of gravitational-wave astro(physics) and cosmology which require to make precise measurements of the source parameters, to observe the evolution of the sources along the cosmic history, and to reach and explore the early Universe. The Einstein Telescope is the European ground-based gravitational-wave detector, evolution of second-generation detectors, which was recently included in the European Strategy Forum on Research Infrastructures (ESFRI) roadmap.

It will feature a system of triangular shape nested detectors, where the arm length is increased to 10 km (compared to 3 km for Virgo and 4 km for LIGO). The larger size and the implementation of new technologies will enable to achieve an improved sensitivity by at least a factor of ten compared to the second generation instruments. ET will be built a few hundred meters underground, reducing terrestrial gravity noise and seismic noise and thus extending the sensitivity toward low frequencies. The ET extraordinary sensitivity and wide frequency band will make it possible to access the entire population of stellar mass black-holes up to the early Universe, to detect primordial black holes, and to unveil intermediate mass black-holes (up to $1000M_{sun}$) enabling us to understand their origin, evolution, and demography. It will probe the physics near the black-hole horizon enabling unprecedented general-relativity test. It will help understanding the nature of dark energy and possible modifications of general relativity at cosmological scales. ET will make gravitational waves powerful tools for comprehending fundamental forces in extreme regimes such as in the interiors of neutron stars, revealing the nature of compact objects and the properties of nuclear matter [50]. New gravitational-wave sources are expected to be detected including core-collapse supernovae, isolated neutron stars, stochastic backgrounds of astrophysical and cosmological origin, and cosmic strings. ET will operate in synergy with a new generation of innovative electromagnetic observatories, such as the Cherenkov Telescope Array [26, CTA], Athena [55, 57], the Vera Rubin Observatory, JWST, the European Southern Observatory Extremely Large Telescope [39, ELT], the Square Kilometre Array [35, SKA] and the mission concepts THESEUS [28, 32] and TAP [31]. Multi-messenger observations will probe the population of binary systems of compact objects in connection with kilonovae and short GRBs along with the star formation history and chemical evolution of the Universe.

ET is expected to detect $10^4 - 10^5$ binary neutron star coalescences per year. Thanks to the access at low frequencies, ET will detect binary neutron star mergers before the merger (minutes but also several hours before in the case of close events), and the Earth rotation imprint on the signal will be used to determine the sky localisation. Thus ET, also operating as a single detector will be able to localize a few hundreds detections per year with sky-localization (90%$c.r.$) <100 square degrees. For these events, it will be possible to send early warning alerts. The detection and localization capabilities significantly improve, observing in a network of next generation gravitational-wave observatories (see Fig. 19.3). Thousands of detections per year will have a sky-localization (90%$c.r.$) <10 square degrees, and thousands of detections sky-localization (90%$c.r.$) <1 square degrees for ET observing with Cosmic Explorer [37, 59], and two Cosmic Explorer (one in USA and one in Australia). For recent works on ET and CE detection and localization capabilities see [30, 43, 48, 60]. Since the kilonovae optical emission is intrinsically faint and difficult to detect at redshift larger 0.3, the counterparts at larger redshift will be mainly detected in the high-energy band. A recent comprehensive study [60], starting from simulated binary neutron star population and GRB modelling calibrated and normalized to reproduce properties of observed short GRB samples, has analyzed the joint gravitational wave and gamma and X-ray detections modelling the prompt and afterglow emissions and considering different observational strategies. Almost all detected short GRB will have a gravitational-wave counterparts. Depending on the specific gamma-ray

Sky-localization capabilities: number of detections per years

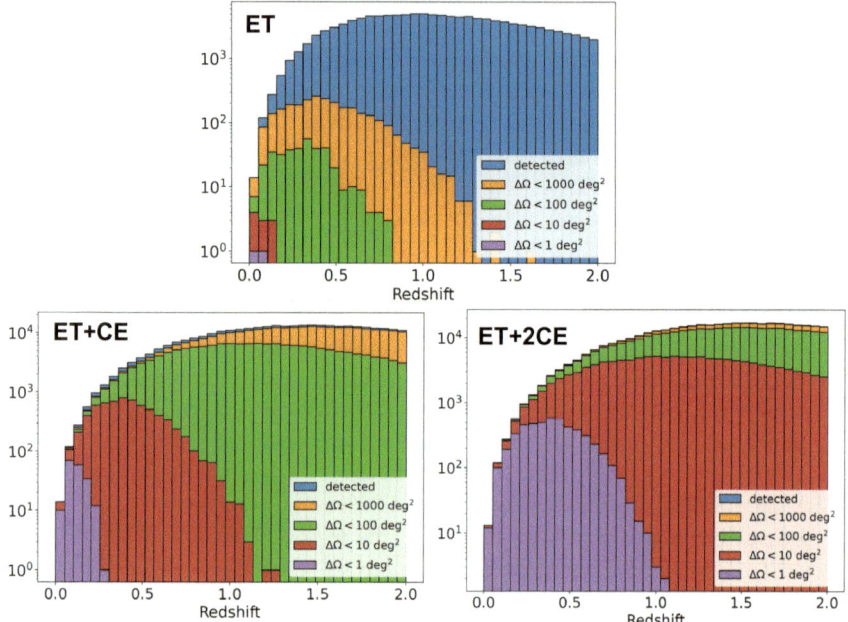

Fig. 19.3 Sky-localization capabilities. The figures (from [61]) show the population of injected binary neutron stars in blue, the number of detections per year localized better than 1000, 100, 10, 1 square degrees in orange, green, red and purple, respectively. Top plot Einstein telescope operating as a single observatory, bottom plots ET in the network of ET and Cosmic Explorer in USA, and ET and two Cosmic Explorer (left plot), one in USA and one in Australia (right plot)

satellites (operating in survey mode), we will have tens to hundreds of gravitational-wave and prompt gamma-ray detections per year. Instead, wide field of view X-ray satellites, such as Einstein Probe, THESEUS, and TAP, are expected to give tens of X-ray afterglow counterparts per year when operating in survey mode. Pointing relatively well localized events (<100 square degrees) by the network given by ET and CE, could increase the detections to hundreds. However, it will be challenging the prioritization of the triggers (based on distance, sky-localization, and viewing angles) to select the ones with a higher chance to be detectable. In summary, ET will make a revolution in our knowledge of the Early Universe, fundamental physics, and transient astrophysics.

References

1. J. Aasi et al., LIGO Scientific Collaboration. Advanced LIGO. Class. Quantum Gravity **32**(7), 074001 (2015)
2. B.P. Abbott, R. Abbott et al., The population of merging compact binaries inferred using gravitational waves through GWTC-3. arXiv:2111.03634
3. B.P. Abbott, R. Abbott et al., Tests of General Relativity with GWTC-3. arXiv:2112.06861
4. B.P. Abbott et al. GWTC-3: Compact Binary Coalescences Observed by LIGO and Virgo During the Second Part of the Third Observing Run. arXiv:2111.03606
5. B.P. Abbott et al., A gravitational-wave standard siren measurement of the Hubble constant. Nature **551**(7678), 85–88 (2017)
6. B.P. Abbott et al., Gravitational Waves and Gamma-Rays from a Binary Neutron Star Merger: GW170817 and GRB 170817A. Astrophys. J. Lett. **848**(2), L13 (2017)
7. B.P. Abbott et al., Multi-messenger observations of a Binary Neutron Star Merger. Astrophys. J. Lett. **848**(2), L12 (2017)
8. B.P. Abbott et al., GW190425: observation of a compact binary coalescence with total mass ~ 3.4 M$_\odot$. Astrophys. J. Lett. **892**(1), L3 (2020)
9. B.P. Abbott et al., Kagra Collaboration, LIGO Scientific Collaboration, and VIRGO Collaboration. Prospects for observing and localizing gravitational-wave transients with Advanced LIGO, Advanced Virgo and KAGRA. Liv. Rev. Relativ. **23**(1), 3 (2020)
10. B.P. Abbott et al., LIGO Scientific Collaboration, and Virgo Collaboration. Astrophysical implications of the binary black-hole Merger GW150914. Astrophys. J. Lett. **818**(2), L22 (2016)
11. B.P. Abbott et al., LIGO Scientific Collaboration, and Virgo Collaboration. Binary black hole mergers in the first advanced LIGO observing run. Phys. Rev. X **6**(4), 041015 (2016)
12. B.P. Abbott et al., LIGO Scientific Collaboration, and Virgo Collaboration. Observation of Gravitational waves from a binary black hole merger. Phys. Rev. Lett. **116**(6), 061102 (2016)
13. B.P. Abbott et al., LIGO Scientific Collaboration, and Virgo Collaboration. GW170817: observation of gravitational waves from a binary neutron star inspiral. Phys. Rev. Lett. **119**(16), 161101 (2017)
14. B.P. Abbott et al., LIGO Scientific Collaboration, and Virgo Collaboration. GW170817: measurements of neutron star radii and equation of state. Phys. Rev. Lett. **121**(16), 161101 (2018)
15. B.P. Abbott et al., LIGO Scientific Collaboration, and Virgo Collaboration. Binary black hole population properties inferred from the first and second observing runs of advanced LIGO and advanced Virgo. Astrophys. J. Lett. **882**(2), L24 (2019)
16. B.P. Abbott et al., LIGO Scientific Collaboration, and Virgo Collaboration. GWTC-1: a gravitational-wave transient catalog of compact binary mergers observed by LIGO and Virgo during the first and second observing runs. Phys. Rev. X **9**(3), 031040 (2019)D
17. B.P. Abbott et al., LIGO Scientific Collaboration, and Virgo Collaboration. Tests of general relativity with the binary black hole signals from the LIGO-Virgo catalog GWTC-1. Phys. Rev. D **100**(10), 104036 (2019)
18. B.P. Abbott et al., LIGO Scientific Collaboration, and Virgo Collaboration. GWTC-2: compact binary coalescences observed by LIGO and Virgo during the first half of the third observing run. Phys. Rev. X **11**(2), 021053 (2021)
19. B.P. Abbott et al., LIGO Scientific Collaboration, and Virgo Collaboration. Population properties of compact objects from the second LIGO-Virgo gravitational-wave transient catalog. Astrophys. J. Lett. **913**(1), L7 (2021)
20. B.P. Abbott et al., LIGO Scientific Collaboration, and Virgo Collaboration. Tests of general relativity with binary black holes from the second LIGO-Virgo gravitational-wave transient catalog. Phys. Rev. D **103**(12), 122002 (2021)
21. R. Abbott et al., LIGO Scientific Collaboration, and Virgo Collaboration. GW190412: observation of a binary-black-hole coalescence with asymmetric masses. Phys. Rev. D **102**(4), 043015 (2020)

22. R. Abbott et al., LIGO Scientific Collaboration, and Virgo Collaboration. GW190521: a binary black hole merger with a total mass of 150 $M_\odot$. Phys. Rev. Lett. **125**(10), 101102 (2020)

23. R. Abbott et al., LIGO Scientific Collaboration, and Virgo Collaboration. GW190814: gravitational waves from the coalescence of a 23 solar mass black hole with a 2.6 solar mass compact object. Astrophys. J. Lett. **896**(2), L44 (2020)

24. R. Abbott et al., LIGO Scientific Collaboration, and Virgo Collaboration. Properties and astrophysical implications of the 150 $M_\odot$ binary black hole merger GW190521. Astrophys. J. Lett. **900**(1), L13 (2020)

25. F. Acernese, et al., Virgo Collaboration. Advanced Virgo: a second-generation interferometric gravitational wave detector. Class. Quantum Grav. **32**(2), 024001 (2015)

26. B.S. Acharya et al., CTA Consortium. Introducing the CTA concept. Astropart. Phys. **43**, 3–18 (2013)

27. K.D. Alexander et al., A decline in the X-Ray through Radio Emission from GW170817 continues to support an Off-axis structured jet. Astrophys. J. Lett. **863**(2), L18 (2018)

28. L. Amati et al., Theseus Consortium. The THESEUS space mission: science goals, requirements and mission concept. Exp. Astron. **52**(3), 183–218 (2021)

29. R.M. Bionta et al., Observation of a neutrino burst in coincidence with supernova 1987A in the Large Magellanic Cloud. Phys. Rev. Lett. **58**(14), 1494–1496 (1987)

30. S. Borhanian, B.S. Sathyaprakash, Listening to the Universe with Next Generation Ground-Based Gravitational-Wave Detectors, Feb. 2022. arXiv:2202.11048

31. J. Camp et al., Transient astrophysics probe, in *Bulletin of the American Astronomical Society*, vol. 51, p. 85, Sept. 2019

32. R. Ciolfi et al., Multi-messenger astrophysics with THESEUS in the 2030s. Exp. Astron. **52**(3), 245–275 (2021)

33. D.A. Coulter et al., Swope Supernova Survey 2017a (SSS17a), the optical counterpart to a gravitational wave source. Science **358**(6370), 1556–1558 (2017)

34. P. D'Avanzo et al., The evolution of the X-ray afterglow emission of GW 170817/ GRB 170817A in XMM-Newton observations. Astron. Astrophys. **613**, L1 (2018)

35. P.E. Dewdney, P.J. Hall, R.T. Schilizzi, T.J.L.W. Lazio, The Square Kilometre Array. IEEE Proc. **97**(8), 1482–1496 (2009)

36. D. Dobie et al., A turnover in the radio light curve of GW170817. Astrophys. J. Lett. **858**(2), L15 (2018)

37. M. Evans et al., A horizon study for cosmic explorer science, observatories, and community. dcc.cosmicexplorer.org/CE-P2100003/public

38. G. Ghirlanda et al., Compact radio emission indicates a structured jet was produced by a binary neutron star merger. Science **363**(6430), 968–971 (2019)

39. R. Gilmozzi, J. Spyromilio, The European extremely large telescope (E-ELT). The Messenger **127**, 11 (2007)

40. A. Goldstein et al., An ordinary short Gamma-Ray burst with extraordinary implications: Fermi-GBM Detection of GRB 170817A. Astrophys. J. Lett. **848**(2), L14 (2017)

41. A. Hajela et al., Two years of nonthermal emission from the binary neutron star merger GW170817: rapid fading of the jet afterglow and first constraints on the Kilonova Fastest Ejecta. Astrophys. J. Lett. **886**(1), L17 (2019)

42. G. Hallinan et al., A radio counterpart to a neutron star merger. Science **358**(6370), 1579–1583 (2017)

43. J. Harms et al., GWFish: a simulation software to evaluate parameter-estimation capabilities of gravitational-wave detector networks, May 2022. arXiv:2205.02499

44. K. Hirata et al., Observation of a neutrino burst from the supernova SN1987A. Phys. Rev. Lett. **58**(14), 1490–1493 (1987)

45. Z. Ivezic et al., LSST Collaboration. Large synoptic survey telescope: from science drivers to reference design. Serbian Astronom. J. **176**, 1–13 (2008)

46. J. Kalirai, Scientific discovery with the James Webb Space Telescope. Contemp. Phys. **59**(3), 251–290 (2018)

47. Y. Li et al., The GECAM and its payload. Scientia Sinica Physica, Mechanica & Astronomica **50**(12), 129508 (2020)
48. Y. Li et al., Exploring the sky localization and early warning capabilities of third generation gravitational wave detectors in three-detector network configurations, Sept. 2021. arXiv:2109.07389
49. J.D. Lyman et al., The optical afterglow of the short gamma-ray burst associated with GW170817. Nat. Astron. **2**, 751–754 (2018)
50. M. Maggiore et al., Science case for the Einstein telescope. J. Cosmol. Astropart. Phys. **2020**(3), 050 (2020)
51. B. Margalit, B.D. Metzger, Constraining the maximum mass of Neutron Stars from Multimessenger observations of GW170817. Astrophys. J. Lett. **850**(2), L19 (2017)
52. R. Margutti et al., The binary neutron star event LIGO/Virgo GW170817 160 days after merger: synchrotron emission across the electromagnetic spectrum. Astrophys. J. Lett. **856**(1), L18 (2018)
53. B.D. Metzger, Kilonovae. Liv. Rev. Relativ. **23**(1), 1 (2019)
54. K.P. Mooley et al., Superluminal motion of a relativistic jet in the neutron-star merger GW170817. Nature **561**(7723), 355–359 (2018)
55. K. Nandra et al., The Hot and Energetic Universe: A White Paper presenting the science theme motivating the Athena+ mission, June 2013. arXiv:1306.2307
56. E. Pian et al., Spectroscopic identification of r-process nucleosynthesis in a double neutron-star merger. Nature **551**(7678), 67–70 (2017)
57. L. Piro et al., Multi-messenger-Athena Synergy White Paper, Oct. 2021. arXiv:2110.15677
58. D. Radice, A. Perego, F. Zappa, S. Bernuzzi, GW170817: joint constraint on the neutron star equation of State from multimessenger observations. Astrophys. J. Lett. **852**(2), L29 (2018)
59. D. Reitze et al., Cosmic explorer: The US contribution to gravitational-wave astronomy beyond LIGO. Bull. Am. Astron. Soc. **51**, 035, 7 (2019)
60. S. Ronchini et al., Perspectives for multi-messenger astronomy with the next generation of gravitational-wave detectors and high-energy satellites, Apr. 2022. arXiv:2204.01746
61. F. Santoliquido et al., The cosmic merger rate density of compact objects: impact of star formation, metallicity, initial mass function, and binary evolution. Mon. Not. R. Astronom. Soc. **502**(4), 4877–4889 (2021)
62. V. Savchenko et al., INTEGRAL detection of the first prompt Gamma-Ray signal coincident with the Gravitational-wave event GW170817. Astrophys. J. Lett. **848**(2), L15 (2017)
63. S.J. Smartt et al., A Kilonova as the electromagnetic counterpart to a gravitational-wave source. Nature **551**(7678), 75–79 (2017)
64. E. Troja et al., The X-ray counterpart to the gravitational-wave event GW170817. Nature **551**(7678), 71–74 (2017)
65. V.A. Villar et al., The combined ultraviolet, optical, and near-infrared light curves of the Kilonova associated with the binary neutron star merger GW170817: unified data set, analytic models, and physical implications. Astrophys. J. Lett. **851**(1), L21 (2017)
66. D. Watson et al., Identification of strontium in the Merger of two Neutron Stars. Nature **574**(7779), 497–500 (2019)
67. J. Wei et al., The Deep and Transient Universe in the SVOM Era: New Challenges and Opportunities—Scientific prospects of the SVOM mission, Oct. 2016. arXiv:1610.06892
68. W. Yuan et al., Einstein Probe—a small mission to monitor and explore the dynamic X-ray Universe, in *Proceedings of Swift: 10 Years of Discovery (SWIFT 10)*, p. 6, Dec. 2014

Chapter 20
High Energy Physics and the European Strategy

Gian Francesco Giudice

Abstract Some remarks about the outcome of the European Strategy for Particle Physics and the future of high-energy physics in Europe.

On 7 March 1960 Bruno Touschek, then a researcher at the INFN laboratory in Frascati, gave a talk proposing the idea of a collider. More than 60 years later, Touschek's idea still underlies the most effective instrument at our disposal to investigate the world of elementary particles and to explore the structure of spacetime at the smallest possible distance scale.

Although I have never met Touschek, he has always been an inspiring figure for me. With his pure scientific genius and ironic sense of humour, he was a worthy successor of Fermi, belonging to a generation of physicists capable of working across the boundaries between theoretical and experimental physics. Indeed, while he is most famous for his accomplishments in experimental physics, he worked on many aspects of quantum field theory and I was told about an anecdote that refers to his studies on CP and time reversal. After a car accident, Touschek was brought to the emergency room of a hospital in Rome. The doctor started his visit with some simple questions to check if the patient suffered from brain damage as a result of the accident. The first question the doctor asked was what he was doing for a living, and Touschek replied: "I am thinking about temporal inversion." The medical examination ended immediately and Touschek was hospitalised with a diagnosis of serious concussion.

Sixty years of colliders have revolutionised our understanding of the microscopic behaviour of the physical world. The field of particle physics went through great discoveries, periods of confusion, unexpected results and brilliant breakthroughs that revealed an order in nature, which is embodied by the elegant conceptual structure now called the Standard Model. The theory is truly a monument of human scientific achievement, since it is able to explain the building blocks of matter and forces in terms of a geometrical principle, which is called gauge symmetry. It appears that gauge symmetry dictates the properties of all fundamental forces, gravity included.

G. F. Giudice (✉)
Theoretical Physics Department, CERN, Geneva, Switzerland
e-mail: Gian.Giudice@cern.ch

L. Bonolis et al. (eds.), *Bruno Touschek 100 Years*,
Springer Proceedings in Physics 287,
https://doi.org/10.1007/978-3-031-23042-4_20

But the story of human discovery is not over. As long as civilisation exists, humans will always ask more and more fundamental questions about the origin of matter, of the universe and of spacetime. These questions still motivate particle physicists to continue their quest. The experimental tools at our disposal have certainly grown enormously since the time Touschek gave his visionary talk about colliders in Frascati, but colliders are still a central instrument for our field. Touschek's legacy is still alive today in the program of future colliders.

Before discussing the future of collider physics, I would like to consider another question: why do we need colliders? Our society, our planet are going through unprecedented challenges: epidemics, poverty, sustainable energy production, clean water supply, global warming and environmental protection. Given these pressing issues, are colliders really a priority for society? The answer to this question lies in the key role that science plays in addressing society urgent challenges. Of course, future scientific effort should focus on the most urgent issues that impact society, but investing only on immediate targets is a short-sighted strategy and, as shown repeatedly in history, it is not an effective way to find radical solutions. Focusing only on applied research simply doesn't work. Fundamental research is a necessary driver of innovation, and investments on fundamental science, which are comparatively small on the macro-economic level, can have a vast impact on the future of humanity. The global emergency caused by the covid-19 pandemic offers a good example. Particle physics cannot produce vaccines, but has produced the world wide web, which was instrumental for society to survive during the covid-19 crisis by allowing for the continuation of economical activities, global coordination, communication and social relations. CERN developed the web having the particle physicists' needs in mind, not the problems of society. But it is scary to think what would have been the impact of the covid-19 pandemic if CERN hadn't developed the web for particle physics.

Many technological advances of great benefit to society have come and will come from fundamental research which is targeted to problems that have nothing to do with society. Just to mention a few examples, research in detector developments led to new technologies for medical imaging and research for accelerators led to innovative therapies for cancer treatment. Future colliders require the development of a new generation of high-field superconducting magnets and research towards high-temperature superconductors. These materials could lead to new ways of storing and transporting energy with virtually no loss from resistance effects, solving one of the greatest limitations of renewable energy, such as wind and solar energy, which is available only for periods of time.

Moreover, collider experiments require the handling of unprecedented sets of big data. Dealing with these problems at the frontier of computing technology will certainly have an impact on our everyday life, when the technology developed for fundamental science is translated into applications for society. Future collider projects require cutting-edge advancements in cryogenics, vacuum, electronics. There is much that technologies developed for particle physics can offer society in the future. Fundamental science channels human talent and creativity towards complex problems, whose solutions lead invariably to unexpected applications. We can rarely predict what they will be, but these applications unfailingly happen.

Another byproduct of large projects at the frontier of science, such as experiments at collider, is scientific training. On average, each year CERN trains thousands of young researchers and PhD students. Not all these people go into academia. The vast majority brings to society and private industry their expertise in dealing with complex problems and in working in close contact with advanced technologies. Scientific training in a place like CERN is an invaluable resource for society. Another, more subtle, aspect that I would like to underline is related to the ethical values that are part of the scientific method. Practicing science helps developing certain principles of tolerance, respect, fairness and justice that contribute to make individuals better fit for society.

Of course, the aspect of fundamental research that is closest to my heart is the advancement of human knowledge. Understanding nature, understanding the universe, understanding physical reality have an immense value for humanity. The scientific exploration of the unknown has an extraordinary inspirational effect on the public and is a powerful driving force for civilisation. Humans simply cannot give up their intellectual curiosity to understand the world they live in.

Indeed, advancing scientific knowledge is CERN's primary mission. CERN is focusing on fundamental research in particle physics, but CERN is not blind to societal issues. Environmental sustainability is the biggest challenge that our society has to face today. As scientists, we cannot remain indifferent to these problems and our way of operating has to reflect the changing attitude. Particle physics should not simply adapt to societal changes, but should lead them. The tradition of particle physics is to be always at the forefront of changes, whether in scientific issues, information technologies, community practices. CERN is well aware of the challenge posed by environmental sustainability and is taking a leading role in the changes that we need to tackle. According to the current budgetary plan, during the period 2016-2026 CERN will invest about 53 MCHF in projects related to environmental protection. Moreover, CERN is carefully scrutinising large collider projects in terms of their environmental impact and studies include R&D on new environmentally friendly gases for the cryogenics of particle detectors. A critical aspect is optimising energy saving and reuse, and each new project is reviewed on the basis of energy consumption. Plans include heat recovery from computing facility, by converting it to heating for buildings during winter, and R&D on efficient power production with potential applications in industry. Another research direction is the development of technologies useful to protect the environment. This involves research with vacuum, high-temperature superconductors for electricity transport, and high-efficiency accelerator techniques.

The ultimate goal of collider research is the exploration of the particle world and the fundamental physical laws. As I mentioned before, today we have consolidated a superb description of nature at the fundamental level, which is given by the Standard Model of particle interactions together with General Relativity. Is this the final chapter of the story? Certainly not. While a perfectly consistent theory, the Standard Model cannot answer more structural questions about the origin of the underlying theory.

Some of the most puzzling aspects of the Standard Model come from the Higgs boson. The Higgs boson was discovered almost 10 years ago, although it was first proposed by theorists at the time that Touschek was pioneering the idea of particle colliders. The experimental data gathered at the LHC match very well the theoretical prediction for the Higgs boson, but the underlying nature of this particle remains a mystery. Its structure does not seem to follow from the same fundamental principles that determine the other particles in the Standard Model. This is why theorists are very much puzzled by the existence of the Higgs boson.

Most particle theorists believe that the Higgs boson must be only the tip of a more complex structure still unknown to us, submerged beyond the present frontier of knowledge. During the last decades, theorists came up with many new creative ideas of what could this submerged world be, and what could lie behind the Higgs boson. Many of these ideas are fascinating, introducing new kinds of forces, new particles, new symmetries, and even a new concept of spacetime. These ideas were put to empirical test with the experiments at the LHC, the CERN particle collider. However, none of these theoretical ideas was shown to be realised in nature.

Some people see in this result a failure of particle physics. I see only a success of the scientific method, which is based on theoretical hypotheses followed by experimental scrutiny. The current results from the LHC give only more reasons to pursue the search because the mysteries in particle physics remain unsolved. Actually, as a theorist, I think that the situation in particle physics is only more interesting today. What the LHC results are telling us is that, once again, nature hides surprises: the next layer of physical reality is very different from what we had imagined so far. The game for theorists is only becoming more intriguing and what we need today is a radical change of paradigm guiding us towards a revolutionary new vision of nature at small distances. This is the great challenge for the young generation of particle theorists. Some new theoretical ideas in this direction are starting to emerge, but it is still too early to tell if any of these ideas is really promising and could help us to resolve some of the mysteries left unexplained by the Standard Model. Of course, theory alone will never be able to tell us whether an hypothesis is correct or not, and we need more experiments to explore unknown territories. Needless to say, CERN has a thriving experimental program aimed at tackling the fundamental open questions in particle physics. In the short-term, the CERN scientific program has five main objectives.

The first objective is a successful Run 3 of the LHC. The LHC operates in alternating phases, with some years of data taking and some years of maintenance and upgrading. Today the Long Shut Down 2 phase has been completed and recently pilot beams have been circulating successfully in the LHC ring, in preparation for the Run 3 phase, in which ATLAS and CMS are expected to collect at least as much luminosity as in the previous run while LHCb and ALICE are expected to increase significantly their data set. After a risk assessment study, it has been decided that the optimal energy for the upcoming LHC run will be 13.6 TeV, slightly higher than in the previous run, but not yet at the ultimate target value of 14 TeV.

The second objective is the completion of preparatory work for the High-Luminosity LHC and the required upgrades of the detectors. The High-Luminosity phase follows Run 3 and will increase the total set of recorded data by a factor of

10. This will allow LHC physics to enter an era of precision measurements that will deliver a lot of important information about the properties of elementary particles. The High-Luminosity LHC is expected to start operating in the late twenties.

The third goal is to reinforce the scientific diversity program. A high-energy collider program is not sufficient to tackle the many open questions in particle physics. More and more, we need a variety of experimental strategies and approaches. A very constructive synergy is building up between particle physics and neighbouring fields, such as observational cosmology, multi-messenger astronomy, underground dark-matter detection, gravitational waves, nuclear physics, and even atomic and condensed-matter physics. New experimental techniques are starting to emerge especially in the search for light dark-matter particles and feebly-interacting particles.

The fourth objective is the support of neutrino experiments in the US and Japan through the Neutrino Platform. In particular, CERN is constructing cryostats for the Dune experiment.

Last, but not least, is theoretical physics, which CERN recognises as an essential objective to open new avenues of exploration and motivate experimental investigation. CERN will continue to support a vast range of theoretical studies, not only related to the laboratory's experimental programme but in a much broader perspective, which will serve as a vehicle of scientific progress and intellectual advancement.

CERN, and the particle-physics community in general, are looking beyond a short-term vision and dream about the future. This dreaming is done in a coherent and comprehensive way through the European Strategy for Particle Physics. This is a community-driven exercise, which first took place in 2005 and has been updated twice, at intervals of about 7 years. The last update took place in 2020 and I had the privilege of participating in the physics preparatory phase as CERN representative. The 2020 European Strategy was a particularly important event because the particle physics community is at a critical moment when decisions about the long-term strategy have to be pondered and debated. The process involved an open call for proposals, a general conference in which the community got together, a preparatory work in which the physics case was outlined in a Briefing Book, and finally a one-week closed session, where representatives from each European country and major labs were present, and where the final document was written.

From the physics point of view, the recommendations made by the European Strategy covered three points. (1) "An electron-positron Higgs factory is the highest-priority next collider." (2) "For the longer term, the European particle physics community has the ambition to operate a proton-proton collider at the highest achievable energy." (3) "A diverse programme that is complementary to the energy frontier is an essential part of the European particle physics Strategy".

The first point refers to the Higgs boson as a priority target. The discovery of the Higgs boson was a milestone in the history of science, but it has left us with many unanswered questions. The true nature of the Higgs boson is a big question mark for particle theorists. Luckily, we have the means to gain further information about its nature and this can be done only by measuring its properties with high precision. At present, we know the couplings of the Higgs to gauge bosons at the level of 10% and the couplings to third generation fermions at the level of 20%. But the goals of future Higgs factories are more ambitious and aim at going below the percent level in precision. This corresponds to testing a possible substructure of the Higgs particle up to distances hundreds of thousand times smaller than the proton radius. This superb probe of the intimate structure of the Higgs boson investigates in depth the question of whether the particle is composite or a truly elementary object.

Besides this fundamental task, future Higgs research aims at probing: *(i)* the Higgs couplings to second-generation fermions, which contain information about the mechanism that feeds mass to matter; *(ii)* invisible decay modes of the Higgs, which contain information about the nature of dark matter; *(iii)* the Higgs self-coupling, which contains information about the nature of the electroweak phase transition; *(iv)* possible rare Higgs decays, which contain information about the symmetry structure of the Standard Model. The experimental program of Higgs precision measurements goes straight into the heart of the many mysteries still enshrouding the origin of electroweak symmetry breaking.

There are several proposals around the world for exploring the properties of the Higgs boson. Other speakers have presented projects in Japan and China. As far as CERN is concerned, we have two projects on the table. The first one is called CLIC and is a linear e^+e^- collider which can be extended in stages, with a first stage in which the tunnel length is 11 km and the machine operates as a Higgs factory. The tunnel can then be extended up to 50 km, with the energy increasing accordingly, to explore the high-energy domain. The justification of this phase could be linked to possible discoveries at the LHC.

The second proposal is called FCC, Future Circular Collider. Essentially, it is a way of repeating the successful story of LEP and the LHC at a larger scale. The plan is to have a new tunnel, about 100 km long, which will host first a circular e^+e^- collider and then a proton-proton collider. In terms of precision, this machine would be a wonder, producing a million Higgs bosons and 10^5 more Z bosons than the full LEP program. The next stage of the FCC would accomplish what the European Strategy defined as "the ambition of the European particle physics community to operate a proton-proton collider at the highest achievable energy." If 16T magnets of Nb3Sn superconducting material are put in the FCC tunnel, one could build a hadron collider operating at about 100 TeV. With high-temperature superconducting material, one could dream of even higher energy, say 150 TeV. But even with regular 6T NbTi magnets, one could already do as well as the SSC.

Another priority identified by the European Strategy is accelerator research with special emphasis on new technologies. Along this line, CERN has doubled the budget on new accelerator projects. The most prominent project is AWAKE, which develops a new plasma wakefield acceleration, and new investments are made also for a far-future muon collider, which could be hosted in the FCC tunnel.

Lacking any precise hint for the scale of new physics from the LHC, we need, on one hand, to explore as deep as possible with the highest possible energy allowed by new technologies and, on the other hand, to broaden the research programme using a variety of different techniques, as recommended by the European Strategy. In this broad landscape of research, it is clear that colliders play an essential role and are irreplaceable tools for exploration. For example, the Higgs boson could have been discovered only through colliders and no other experimental tool or technique could have led to this discovery.

A crucial lesson learned from the LHC is that hadron colliders are not only discovery machines but also excellent precision machines. This result could not have been anticipated at the time the LHC started and was possible only because of the successful interplay between different elements: unprecedented technological advancements, exceptional accelerator performances, excellent detector resolutions, high-performance computing and data handling, higher-order theoretical calculations of background processes with accuracies unthinkable only a few years ago. The merging of different expertise from different scientific communities was the secret behind the success of the LHC precision programme, which brought new knowledge and opened new prospects in research beyond traditional frontiers. Precision has become key for present and future explorations in high-energy physics. There is a lot to learn from precision measurements even without direct access to high-energy process.

A good example of the value of precision measurements are the LHCb results on rare B meson decays, which are showing unexpected discrepancies with the Standard Model predictions. It is too early to tell if these results are real and not only statistical fluctuations or poorly understood systematics, but lots of new data from the LHC and Belle II will come and clarify the situation. If true, these results would be a revolution in particle physics because they cannot be explained by a small deformation of the Standard Model. They would really shatter the basic structure of the Standard Model and imply the existence of a new sector of the theory.

Bruno Touschek was a visionary. His vision is still alive today in present and future CERN scientific projects. Research at future colliders has an impact on society well beyond the boundaries of scientific knowledge, since it can boost technological developments in many areas in ways that are unimaginable without the driving force from fundamental science. But, most of all, it is going to allow us to explore nature at even smaller distance scales and provide humanity with new knowledge about the fundamental principles that govern the physical world. With a rich program in collider physics, CERN is keeping Touschek's dream alive, inspiring today some young girls or boys to say what Touschek wrote in a letter to his father in 1946: "Ich will ein Physiker werden." I want to become a physicist.

Chapter 21
Circular Colliders in China

Yifang Wang

Abstract The project of the Circular Electron-Positron Collider, CEPC, proposed by the Institute for High Energy Physics, Beijing, IHEP, is illustrated.

It is a great honour and pleasure for me to speak at the 100 years Memorial Symposium of Bruno Touschek. The concept of colliders, and the famous Touschek effect are text I learned in books, but now I have the fortune to hear his stories, and give a presentation with his colleagues and students in the same room, although virtual. This is a remarkable memory for me and I appreciate very much the opportunity. Now let me contribute a report about colliders in China.

Accelerator development started in China in early 50s. The first attempt was a 2.5 MeV proton electrostatic accelerator, followed by a 30 MeV electron LINAC in '60s. A number of accelerators for high energy physics, mostly protons on fixed target was proposed in '60–'80s but never approved except a 30 MeV proton LINAC as an exercise.

At the beginning of '80s, the Beijing Electron–Positron Collider (BEPC) was proposed and finally approved. The construction started in 1984 and the first collision was seen in the fall of 1988. BEPC was designed and achieved to have the highest luminosity at the 2–5 GeV energy region for tau and charm physics, a special domain for its abundant resonances, gluon rich environment, being a bridge for pQCD and non-pQCD, and advantages of quantum entanglement of pair production at the threshold. BEPC was a great start for particle physics and synchrotron radiation in China, for its rich physics outcome, and for its training of physicists and engineers on both experimental physics and accelerators. Its success led to a major upgrade in 2004–2008, called BEPCII, which replaced the existing single-ring e^+e^- collider to a factory-type of double-ring machine in the same tunnel to increase the luminosity by a factor of 100. Figure 21.1 shows the luminosity evolution of colliders at 2–5 GeV

Y. Wang (✉)
Institute of High Energy Physics, Beijing 100049, P.R. China
e-mail: yfwang@ihep.ac.cn

© The Author(s) 2023
L. Bonolis et al. (eds.), *Bruno Touschek 100 Years*,
Springer Proceedings in Physics 287,
https://doi.org/10.1007/978-3-031-23042-4_21

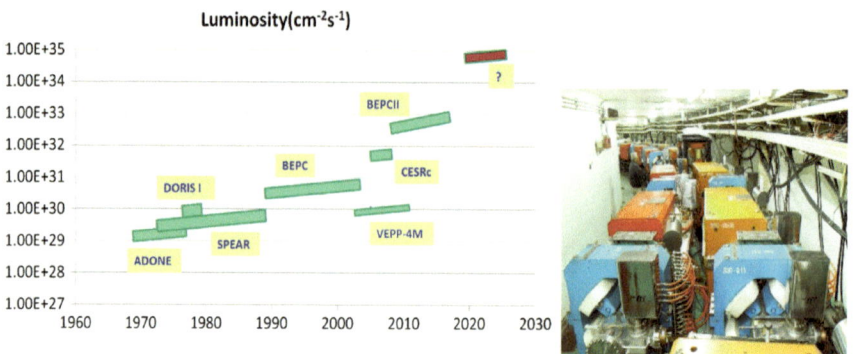

Fig. 21.1 Evolution of the luminosity of e^+e^- colliders at 2–5 GeV energy region, and the BEPCII in the BEPC tunnel (photo IHEP)

energy region and the BEPCII in the BEPC tunnel. It is clear that BEPC and its upgrade, BEPCII, maintained their leadership role in the last 40 years.

The newly built detector, BESIII, has a collaboration with more than 500 members from 17 countries, including 3 institutions from Italy. Its physics program covers light hadron spectroscopy, exotic hadron states, charm and charmonium physics, QCD studies, tau physics and new physics searches [1]. Up to now, more than 380 papers have been published at leading international journals, and a possible 4-quark state, $Z_c^{\pm}(3900)$ was discovered, together with its companion particles, $Z_c^0(3900)$, $Z_c^{\pm}(4020)$, and $Z_c^0(4020)$. Other XYZ particles and their new decay modes, new light hadron resonances and possible glueball candidates were also observed [2].

Even the luminosity of BEPCII reached its design value and after 12 years of data taking, the rich physics program of BESIII still requires 40 fb^{-1} more data, corresponding to another 15 years of data taking [3]. A further upgrade of BEPCII was thus proposed and approved recently by the Chinese Academy of Sciences. The first upgrade item is to increase the luminosity by a factor of 3 at 2.3 GeV for XYZ particle studies (Fig. 21.2), by squeezing the beam size after adding a new RF cavity per beam. The second item is to increase the maximum beam energy from 2.45 to 2.8 GeV for charmed baryons, by replacing the two superconducting quadrupole focusing magnets near the interaction point with higher field strengths. Such an upgrade is also a technology exercise for the future Circular Electron–Positron Collider (CEPC), which will be discussed later. The upgrade is planned to be completed in 2024 and the machine will at least be operational until 2030.

In the mean time we realize that BEPCII cannot be a machine forever and a more ambitious program is desired. Further future of a high energy physics machine after BEPCII has been a topic for discussion since 2005 in the community of China. Various options such as the super-tau-charm factory, super-Flavor factory, even Higgs factory have been talked about in the following years. Joining international projects like ILC was also an option. At a meeting in Sep. 2012, the idea of a Circular Electron–Positron

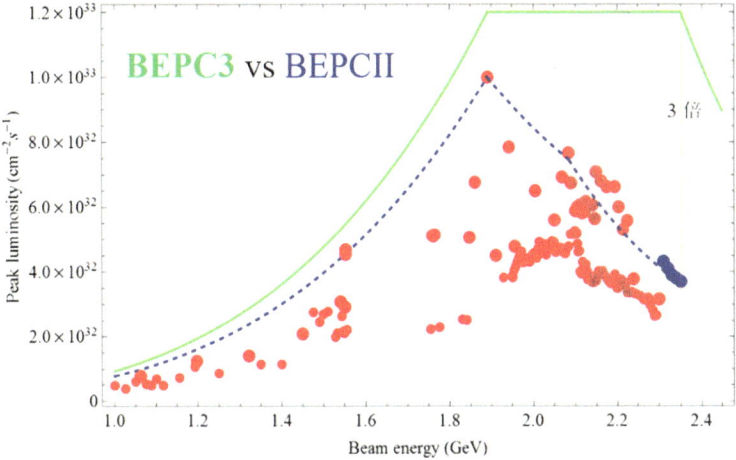

Fig. 21.2 Luminosity upgrade of BEPCII

Collider (CEPC) as a Higgs factory, followed by a very high energy Super Proton-Proton Collider (SPPC) in the same tunnel was proposed. The possibility to re-use the tunnel like LEP-LHC and its advantages over the International Linear Collider (ILC) with a higher luminosity and synchrotron radiation applications quickly gained support in China. The concept, as the first one of its type in the world, was reported in Oct. 2012 at the Fermilab Higgs Factory Workshop [4] and well accepted in the world. Soon after similar ideas such as FCC in Europe appeared and gained momentum.

Higgs factory as the next machine after LHC has a very rich physics potential, as already studied extensively for ILC. If LHC does not find anything new beyond the Standard Model (SM), a Higgs Factory shall be the first choice to discover new physics indirectly beyond the SM. If LHC does find anything new, a Higgs factory is still the first choice to study new physics. Indeed, Higgs is the best window to new physics since it is very special with non-gauge interactions and with a potential similar to that of Landau-Ginzburg which originated from a Cooper pair, an interesting analogy for Higgs being a composite particle. The shape of the Higgs potential also affects the electroweak phase transition at the very beginning of the Universe. Many other inconsistencies and incompleteness of the SM are also Higgs-related, such as the meta-stable vacuum, coupling with dark matter particles, and even the origin of the Higgs mass. An independent study by European physics community also concluded in 2020 that a Higgs Factory is the highest priority for the future of high energy physics [5].

At Higgs factories, couplings of Higgs with fermions and intermediate bosons can be measured to a precision better than 1%, even up to 0.1% with Z. Such a precision can probe new physics up to an energy scale of about 10 TeV, almost a factor of 10 better than that at LHC. If no new physics are found up to this energy scale, the principle of Naturalness is no longer valid which can even be a more important discovery. In addition, comprehensive and high precision tests of the electroweak

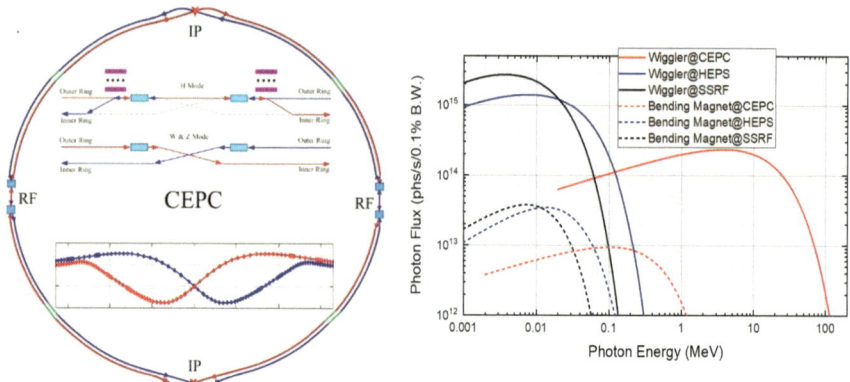

Fig. 21.3 Conceptual design of CEPC, and its photon flux of the synchrotron radiation versus other dedicated synchrotron radiation facilities

theory and QCD studies at Z and W resonances, and flavor physics studies at Z resonances can be performed at circular Higgs factories. Detailed physics potentials are still under study although some have been published [6].

The baseline design of CEPC is 100 km circumference, 30 MW beam power, upgradable to 50 MW beam power and 180 GeV beam energy for ttbar, and compatible with the future pp collider (SPPC) in the same tunnel. Figure 21.3 shows the main ring design and the flux of synchrotron radiation photons. Detailed physics design of the accelerator has been continuously improved since the publication of the "Conceptual Design Report of CEPC" (CDR) in 2018 [7]. Effects such the dynamic aperture with component production errors, beam-beam effects, impedance, electron clouds, etc. have been taken into accounts. Table 21.1 lists the latest key parameters of the CEPC baseline design at Higgs and Z energies, which have been improved dramatically over the CDR with 70% increase of the luminosity at Higgs.

Key components R&D and prototyping have been started since 2014 with funding support from the Ministry of Science and Technology (MoST), Chinese Academy of Sciences (CAS), and National Natural Science Foundation of China (NSFC). Many of the R&D programs were jointly supported with other projects such as the ILC, LHC, High Energy Photon Source (HEPS) in Beijing, China Spallation Neutron Sources (CSNS) as well as startup funds of newly recruited talents and generic R&D. A major progress is the development of Superconducting RF (SRF) cavities. An advanced infrastructure for SRF cavity production, inner surface treatment, QC&QA and testing facilities was established with funding from the Beijing Municipal government and great results have been obtained. Figure 21.4 shows testing results of cavity prototypes for the booster (1.3 GHz) and the main ring (650 MHz) which already satisfied the CEPC design specifications. In fact, the 1.3 GHz cavities which are also applicable to the Shanghai Free Electron Laser Facility and other international projects have already achieved world's best Q-values, thanks to the

Table 21.1 Key parameters of CEPC and its luminosity

	Higgs (high_lum.)	Z (high_lum.)
Number of IPs	2	2
Beam energy (GeV)	120	45.5
Circumference (km)	100	100
Synchrotron radiation loss/turn (GeV)	1.8	0.036
Crossing angle at IP (mrad)	16.5	16.5
Piwinski angle	4.87	18.0
Number of particles/bunch N_e (10^{10})	16.3	16.1
Bunch number (bunch spacing)	214 (0.7us)	10,870 (27 ns)
Beam current (mA)	16.8	841.0
Synchrotron radiation power /beam (MW)	30	**30**
Bending radius (km)	10.2	10.7
Momentum compact (10^{-5})	7.34	2.23
β function at IP β_x^* / β_y^* (m)	0.33/0.001	0.15/0.001
Emittance e_x/e_y (nm)	0.68/0.0014	0.52/0.0016
Beam size at IP σ_x /σ_y (μm)	15.0/0.037	8.8/0.04
Beam-beam parameters ξ_x/ξ_y	0.018/0.115	0.0048/0.129
RF voltage V_{RF} (GV)	2.27	0.13
RF frequency f_{RF} (MHz)	650	650
Natural bunch length σ_z (mm)	2.25	2.93
Bunch length σ_z (mm)	4.42	9.6
Energy spread (%)	0.19	0.12
Energy acceptance requirement (%)	1.7	1.4
Energy acceptance by RF (%)	2.5	1.5
Beamstruhlung lifetime/quantum lifetime (min)	41	–
Lifetime (hour)	21	1.8
Luminosity/IP L (10^{34} cm^{-2} s^{-1})	**5.0**	**101.1**

mid-temperature baking technology. Further R&D to allow the new design of 1-cell 650 MHz applicable to all beam energies of Higgs, Z, and W studies is still on-going.

Other prototypes, including electron guns, all types of magnets, beam diagnostics, vacuum beam pipes with NEG coating, electro-static separators, alignment apparatus, as well as high efficiency klystrons have been in progress. Many were already tested to have satisfied design specifications, as shown in Fig. 21.5.

Design and R&D of detectors have been also progressing well. A new detector concept other than those suggested for ILC, CLIC and FCC have been proposed recently, as shown in Fig. 21.6. A gaseous detector (drift chamber or TPC) in the

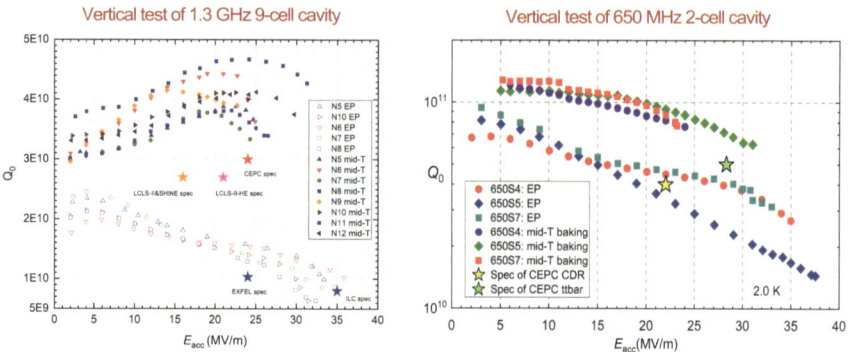

Fig. 21.4 Vertical test results of RF cavities. Left: 1.3 GHz for the booster ring; Right: 650 MHz for the main storage ring

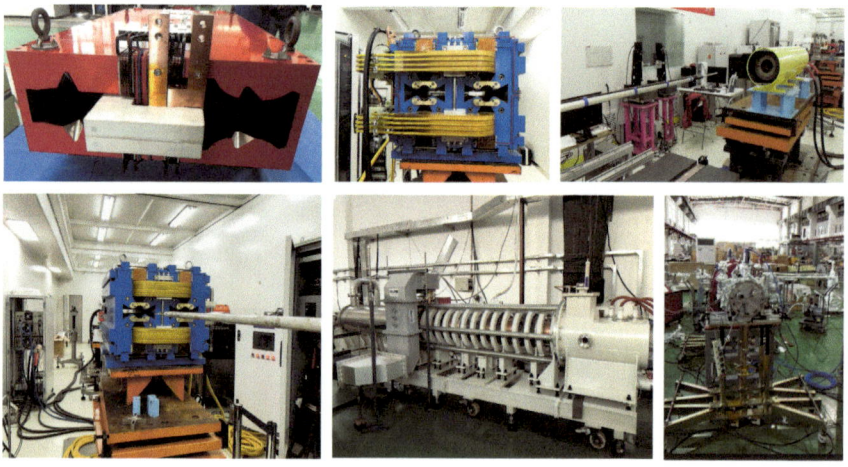

Fig. 21.5 Prototypes of magnets, electro-static separator, Klystrons, vacuum pipes with NEG coating, etc. for CEPC (photos IHEP)

middle of the silicon tracker can improve the track reconstruction, momentum resolution and serve for the particle ID using its dE/dx(or dN/dx) capabilities. A BGO-based crystal electromagnetic (EM) calorimeter with PFA capabilities will have the best energy resolution not only for jets as needed by the Higgs physics, but also for photons needed by the flavor physics. Crystals are arranged in both the x- and y- directions perpendicular to particles from the interaction point. The position of the energy deposition along the crystal bar is obtained from the measured timing difference between two ends of the crystal using SiPMs. Simulation of such a 3D calorimeter shows that ghost hits can be mostly removed and EM showers with a distance more than 4 cm can be well separated. The jet energy and direction can be easily reconstructed and their invariant mass can be obtained with a precision better

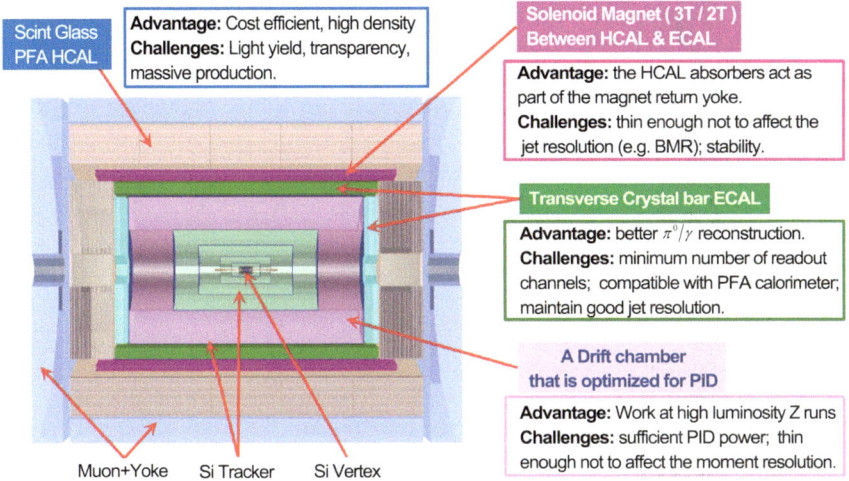

Fig. 21.6 Design of a new detector concept for CEPC

than any other technologies. A very thin solenoid magnet in between the crystal and hadron calorimeter with a field strength of 3 T for Higgs and 2 T for Z physics can be built using high Tc superconducting cables. In order to increase the sampling ratio, the sensitive material of the hadron calorimeter is chosen to use scintillating glass with a density more than 6 g/cm^3, placed in between steel plate. Although it is almost impossible to produce large size glass bars for a total absorption hadron calorimeter, as was suggested years ago, small piece ($\sim$4 $\times$ 4 $\times$ 1 cm^3) of glass with a high light yield (>1000 photons/MeV) is feasible and cost effective. Simulation shows that the stochastic term of the energy resolution for hadrons can be improved from $\sim$50% using traditional technologies to <40% using scintillating glass with a proper sampling ratio.

For the long run, R&D of high Tc superconducting magnets for SPPC is a very important and interesting subject with possible applications to the society. For reasons of cost and applications in the higher magnetic field, Iron-Based Superconducting (IBS) material seems the best choice. A large collaboration with other research institutions, universities and industries has been formed with funding support from CAS, and interesting results have been reported [8]. Another very interesting topic of R&D is the use of the plasma wake-field acceleration (PWA) as the injector of CEPC. By using traditional accelerators before and after the PWA to compensate shortfalls of each other's technology, the injector will be satisfactory and can be very cost-effective. Indeed, an innovative idea to accelerate positrons was proposed [9] recently which may pave the way for e$^+$e$^-$ colliders using PWA technologies.

The CEPC project obtained substantial support from all funding agencies in China, even though the construction is still under discussion. Our plan is to complete the TDR by the end of this year, and the full construction may start at around 2025. The site selection has been on going for almost 10 years, taking into account issues like geology for tunneling, power supply and other infrastructure capabilities, cultural environment and transportation easiness for foreigners, local economy and possible government support, etc. At this moment, 5 cities are running at the front: Qing-Huang-Dao, Chang-Sha, Chang-Chun, Hu-Zhou and Xi'an. Geological investigations and the detailed arrangement of experimental facilities are still under study and the final choice of the site will happen when the project is approved for construction.

We acknowledge that CEPC has a lot of similarities with FCC-ee at CERN, even though the two machines are designed somewhat differently, and running plans are not the same. Their synergies shall be explored more profoundly and the collaboration is much desired. In fact, CEPC will be an international project given its size and the government announced plan to support "China initiated large science projects". We will certainly coordinate with CERN and the international community to move forward with the hope that at least one of the Higgs factories will be realized.

In summary, China has been working on circular e^+e^- colliders for 40 years, and a large science and engineer team has been assembled, together with relevant knowledges and experiences obtained with great efforts. We are eager to make more significant contributions to the high energy physics in the world, and CEPC is a rare opportunity for us. We will work with the international community towards the next phase of the particle physics.

References

1. K.T. Chao, Y.F. Wang (ed.), Physics at BES-III. Inter. J. of Mod. Phys. A **24**(1) (2009) supp
2. Full list of BESIII published papers: http://english.ihep.ac.cn/bes/re/pu/pjp/
3. M. Ablikim et al., Future physics programme of BESIII. Chin. Phys. C **44**, 040001 (2020)
4. A. Blondel et al., Report of the ICFA Beam Dynamics Workshop "Accelerators for a Higgs Factory: Linear versus Circular" (HF2012), arXiv:1302.3318
5. The European Strategy Group, 2020 Update of the european strategy for particle physics. https://home.cern/sites/default/files/2020-06/2020 Update European Strategy.pdf
6. F.P. An et al., Precision higgs physics at the CEPC. Chin. Phys. C **43**, 043002 (2019)
7. The CEPC Accelerator Study Group, Conceptual design report of CEPC-accelerator, arXiv:1809.00285; CEPC Physics-Detector Study Group, Conceptual design report of CEPC-Physics & detector, arXiv:1811.10545
8. D. Wang et al., Supercond. Sci. Technol. **32** (2019) 04LT01
9. S.Y. Zhou et al., Phys. Rev. Lett. **127**, 174801 (2021)

Chapter 22
Linear Colliders

Steinar Stapnes

Abstract This article summarizes the studies for implementation of the Linear Collider projects, CLIC (the Compact LInear Collider) and ILC (the International Linear Collider). The accelerators aim to collide electrons and positrons at 380 and 250 GeV respectively, and both can be extended in length and/or with improved technologies to multi-TeV energies. CLIC is studied for construction at CERN, while ILC is being studied for implementation in Japan. The technical status, expected performances, recent progress and implementation parameters, as schedules, power and costs, are presented. The summary focuses on the accelerator studies for the colliders, but the accelerator studies are accompanied by comprehensive physics and detector studies referred to in the text and references. The future programs including the work for sustainable implementations are briefly summarized at the end. Many of the linear collider studies are common for the two projects and are presented as such. The projects are both implementable at costs similar to LHC and power/energy consumption similar or less than LHC.

22.1 Introduction

The Compact Linear Collider (CLIC) is a multi-TeV high-luminosity linear e^+e^- collider under development by the CLIC accelerator collaboration. The CLIC accelerator has been optimised for three energy stages at centre-of-mass energies 380 GeV, 1.5 and 3 TeV [1]. CLIC uses a novel two-beam acceleration technique, with normal-conducting accelerating structures operating in the range of 70–100 MV/m.

Detailed studies of the physics potential and detector for CLIC, and R&D on detector technologies, have been carried out by the CLIC detector and physics (CLICdp) collaboration. CLIC provides excellent sensitivity to Beyond Standard Model physics, through direct searches and via a broad set of precision measurements of Standard Model processes, particularly in the Higgs and top-quark sectors.

S. Stapnes (✉)
ATS-DO, CERN, Geneva, Switzerland
e-mail: Steinar.Stapnes@cern.ch

© The Author(s) 2023
L. Bonolis et al. (eds.), *Bruno Touschek 100 Years*,
Springer Proceedings in Physics 287,
https://doi.org/10.1007/978-3-031-23042-4_22

The CLIC accelerator, detector studies and physics potential are documented in detail at: http://clic.cern/european-strategy. Information about the accelerator, physics and detector collaborations and the studies in general is available at: http://clic.cern. Since the publication of the reports above for the European Strategy Update in 2018–2019, the baseline luminosity at 380 GeV has been updated according to new studies, new power estimates show a significant reduction, and technical progress and improvements related to X-band technology and klystron design have been achieved. These developments are described in the CLIC input to the 2021 Snowmass process [2].

The International Linear Collider (ILC) is an electron-positron collider with a collision energy of 250 GeV (total length of approximately 20 km). The design study for the ILC for a collision energy of 500 GeV started in 2004, and the Technical Design Report (TDR) [3] was published by the Global Design Effort (GDE) international team in 2013. More than 2,400 researchers contributed to the TDR.

After publication of the TDR, R&D activities regarding linear colliders were organised by the Linear Collider Collaboration (LCC). The 250 GeV ILC for a Higgs factory was proposed and published in the ILC Machine Staging Report 2017 [4].

The International Development Team (IDT) was established [5] by the International Committee for Future Accelerators (ICFA) in August 2020 to prepare to establish the ILC preparatory laboratory (Pre-lab) [6] as the first step towards the construction of the ILC in Japan. The principal accelerator activities of the ILC Pre-lab are foreseen to cover technical preparations, engineering design and documentation for the ILC construction project. The former is summarised in "Technical Preparation and Work Packages (WPs) during ILC Pre-lab" [7]. The ILC Pre-lab activities are expected to continue for approximately four years, and the ILC accelerator construction will require nine years. Currently the Pre-lab activity planning is being revised to start more gradually with a subset of the highest priority technical WPs.

A recent updated and complete summary of the ILC project, including a detailed description of the physics potential, has been submitted to the 2021 Snowmass process. This document can be found at [8], and summarizes also the Pre-lab plans for next phase.

22.2   CLIC Layout

A schematic overview of the accelerator configuration for the first energy stage is shown in Fig. 22.1. To reach multi-TeV collision energies in an acceptable site length and at affordable cost, the main linacs use normal conducting X-band accelerating structures; these achieve a high accelerating gradient of 100 MV/m. For the first energy stage, a lower gradient of 72 MV/m is the optimum to achieve the luminosity goal, which requires a larger beam current than at higher energies.

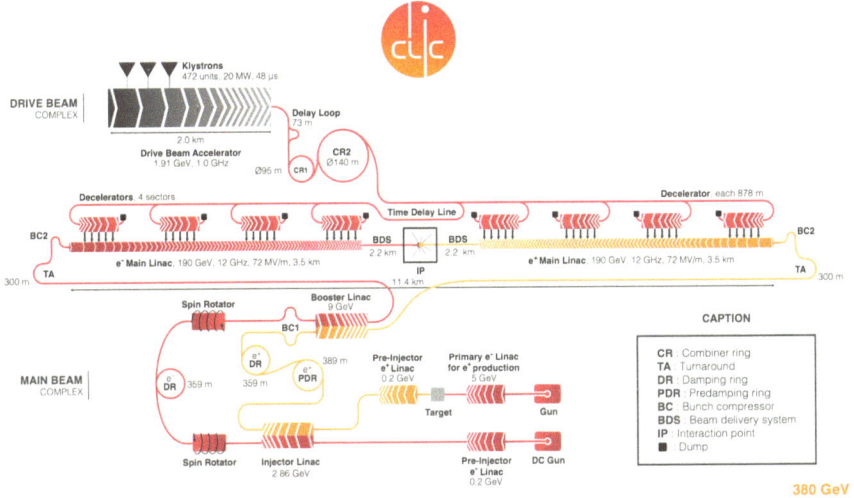

Fig. 22.1 Schematic layout of the CLIC complex at 380 GeV

In order to provide the necessary high peak power, the novel drive-beam scheme uses low-frequency high efficiency klystrons to efficiently generate long RF pulses and to store their energy in a long, high-current drive-beam pulse. This beam pulse is used to generate many short, even higher intensity pulses that are distributed alongside the main linac, where they release the stored energy in power extraction and transfer structures (PETS) in the form of short RF power pulses, transferred via waveguides into the accelerating structures. This concept strongly reduces the cost and power consumption compared with powering the structures directly by klystrons, especially for stages 2 and 3, and is very scalable to higher energies.

The upgrade to higher energies will require lengthening the main linacs. For the RF power the upgrade to 1.5 TeV can be done by increasing the energy and pulse length of the primary drive-beam, while a second drive-beam complex must be added for the upgrade to 3 TeV. An alternative design for the 380 GeV stage has been studied, in which the main linac accelerating structures are directly powered by high efficiency klystrons. The further stages will also in this case be drive-beam based for the reasons mentioned above.

22.3 CLIC Parameter Overview

The parameters for the three energy stages of CLIC are given in Table 22.1. The baseline plan for operating CLIC results in an integrated luminosity per year equivalent to operating at full luminosity for 1.2×10^7 s [9]. Foreseeing 8, 7 and 8 years of running at 380, 1500 and 3000 GeV respectively, and a luminosity ramp up for the first years

Table 22.1 Key parameters of the CLIC energy stages

Parameter	Unit	Stage 1	Stage 2	Stage 3
Centre-of-mass energy	GeV	380	1500	3000
Repetition frequency	Hz	50	50	50
Nb. of bunches per train		352	312	312
Bunch separation	ns	0.5	0.5	0.5
Pulse length	ns	244	244	244
Accelerating gradient	MV/m	72	72/100	72/100
Total luminosity	$1 \times 10^{34} \text{cm}^2\text{s}^1$	2.3	3.7	5.9
Lum. above 99% of $\sqrt{s}$	$1 \times 10^{34} \text{cm}^2\text{s}^1$	1.3	1.4	2
Total int. lum. per year	fb^{-1}	276	444	708
Main linac tunnel length	km	11.4	29.0	50.1
Nb. of particles per bunch	1×10^9	5.2	3.7	3.7
Bunch length	μm	70	44	44
IP beam size	nm	149/2.0	~60/1.5	~40/1
Final RMS energy spread	%	0.35	0.35	0.35
Crossing angle (at IP)	mrad	16.5	20	20

at each stage, integrated luminosities of 1.5, 2.5 and 5.0 ab^{-1} are reached for the three stages. CLIC provides ±80% longitudinal electron polarisation and proposes a sharing between the two polarisation states at each energy stage for optimal physics reach [10].

22.4 Luminosity Margins and Performance

In order to achieve high luminosity, CLIC requires very small beam sizes at the collision point, as listed in Table 22.1. Recent studies have explored the margins and possibilities for increasing the luminosity, operation at the Z-pole and gamma-gamma collisions [11].

The primary beamphysics and luminosity considerations for CLIC are presented in [12]. The impact of static and dynamic imperfections is studied in detail, being the determining factors for the luminosity performance. The dominant imperfections are the static misalignment of beamline elements and ground motion that degrades the initial emittances. Beam-based alignment is used to minimise the impact of static imperfections. For the expected alignment imperfections and with a conservative ground motion model, 90% of the machines achieve a luminosity of 2.3×10^{34} cm^{-2}s^{-1} or greater. This is the value used in Table 22.1. The average luminosity achieved is 2.8×10^{34} cm^{-2}s^{-1}. Future improvements to the technologies used to mitigate imperfections, such as better pre-alignment, active stabilization systems and additional beam-based tuning, will also help further increase this luminosity. A start-to-end simulation of a perfect machine without imperfections shows that a luminosity of 4.3×10^{34} cm^{-2}s^{-1} would be achieved.

At 380 GeV energy also the repetition rate of the facility, and consequently luminosity, could be doubled from 50 to 100 Hz without major changes but with increases in the overall power consumption and cost (at ∼55 and ∼5% levels, respectively).

The CLIC beam energy can be adjusted to meet different physics requirements. In particular, a period of operation around 350 GeV is foreseen to scan the top-quark pair-production threshold. Operation at much lower energies can also be considered. Running at the Z-pole results in an expected luminosity of about 2.3×10^{32} cm^{-2}s^{-1} for an unmodified collider. On the other hand, an initial installation of just the linac needed for Z-pole energy factory, and an appropriately adapted beam delivery system, would result in a luminosity of 0.36×10^{34} cm^{-2}s^{-1} for 50 Hz operation. Furthermore, gamma-gamma collisions at up to ∼315 GeV are possible with a luminosity spectrum interesting for physics.

22.5 Brief Summary of the CLIC Technical Maturity

Accelerating gradients of up to 145 MV/m have been reached with the two-beam concept at the CLIC Test Facility (CTF3). Breakdown rates of the accelerating structures well below the limit of 3×10^{-7} m^{-1} per beam pulse are being stably achieved at X-band test platforms at the foreseen operational gradients of CLIC.

Substantial progress has been made towards realising the nanometre-sized beams required by CLIC for high luminosities: the low emittances needed for the CLIC damping rings are achieved by modern synchrotron light sources; special alignment procedures for the main linac are now available; and sub-nanometre stabilisation of the final focus quadrupoles has been demonstrated. In addition to the results from laboratory tests of components and the experimental studies in ATF2 at KEK, the advanced beam-based alignment of the CLIC main linac has successfully been tested in FACET at SLAC and FERMI in Trieste.

Other technology developments and prototypes include the main linac modules and their auxiliary sub-systems such as vacuum, stable supports, and instrumentation. Beam instrumentation and feedback systems, including sub-micron level resolution beam-position monitors with time accuracy better than 20 ns and bunch-length monitors with resolution better than 20 fs, have been developed and tested with beams in CTF3.

Recent developments, among others of high efficiency klystrons, have resulted in an improved energy efficiency for the 380 GeV stage, as well as a lower estimated cost. For an updated description of the technical developments please see [2].

22.6 CLIC Schedule, Cost Estimate, and Power Consumption

The technology and construction-driven timeline for the CLIC programme is shown in Fig. 22.2 [13]. This schedule has seven years of initial construction and commissioning. The 27 years of CLIC data-taking include two intervals of two years between the stages.

The cost estimate of the initial stage is approximately 5.9 billion CHF. The energy upgrade to 1.5 TeV has an estimated cost of approximately 5.1 billion CHF, including the upgrade of the drive-beam RF power. The cost of the further energy upgrade to 3 TeV has been estimated at approximately 7.3 billion CHF, including the construction of a second drive-beam complex.

The nominal power consumption at the 380 GeV stage is approximately 110 MW. Earlier estimates for the 1.5 and 3 TeV stages yield approximately 370 and 590 MW, respectively [14], however recent power savings applied to the 380 GeV design have not yet been implemented for these higher energy stages. The annual energy consumption for nominal running at the initial energy stage is estimated to be 0.6 TWh. For comparison, CERN's current energy consumption is approximately 1.2 TWh per year, of which the accelerator complex uses approximately 90%.

22.7 The CLIC Programme 2021–2025

The design and implementation studies for the CLIC e^+e^- multi-TeV linear collider are at an advanced stage. The main technical issues, cost and project timelines have been developed, demonstrated and documented.

The CLIC study will submit an updated project description for the next European Strategy Update 2026–2027. Key updates will be related to the luminosity performance at 380 GeV, the power/energy efficiency and consumption at stage 1, but also at multi-TeV energies, and further design, technical and industrial developments of the core-technologies, namely X-band systems, RF power systems, and nano-beams with associated hardware.

The X-band core technology development and dissemination, capitalizing on existing facilities (e.g. X-band test stands and the CLEAR beam facility at CERN), remain a primary focus. More broadly, the use of the CLIC core technologies—primarily X-band RF, associated components and nano-beams—in compact medical, industrial and research linacs has become an increasingly important development and test ground for CLIC, and is destined to grow further [15]. The adoption of CLIC technology for these applications is now providing a significant boost to CLIC related R&D, involving extensive and increasing collaborations with laboratories and universities using the technology, and an enlarging commercial supplier base.

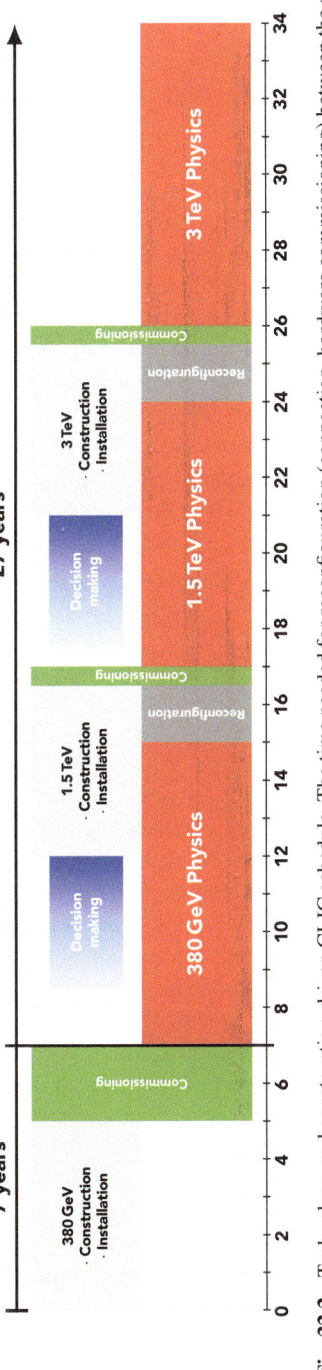

Fig. 22.2 Technology and construction-driven CLIC schedule. The time needed for reconfiguration (connection, hardware commissioning) between the stages is also indicated

On the design side the parameters for running at multi-TeV energies, with X-band or other RF technologies, will be studied further, in particular with energy efficiency guiding the designs.

Other key developments will be related to luminosity performance. On the parameter and hardware side these studies cover among others alignment/stability studies, thermo-mechanical engineering of modules and support systems for critical beam elements, instrumentation, positron production, damping ring and final focus system studies. These technology developments have clear synergies with what is needed for linear colliders using other RF-technologies, and also light sources. Many of the collaboration partners in CLIC involved in these developments are from laboratories with Synchrotron Sources or Free Electron Laser installations, and test components and units in their facilities in view of future use there.

In summary, the CLIC studies foreseen overlap in many areas with challenges for other Higgs-factories or other accelerators, especially with the R&D topics related to high gradient and high efficiency RF systems. CLIC and ILC have for many years had common working groups and workshop sessions on beam-dynamics, sources, damping rings, beam-delivery systems and more. Also the more recent sustainability studies fall into this category. There are also common challenges with the novel accelerator developments concerning linear collider beam-dynamics, drivebeams, nanobeams, polarization and alignment/stability solutions, and also with muon cooling RF systems.

22.8   The ILC Accelerator

The ILC consists of the following domains: (1) electron and positron sources, (2) damping rings (DRs) to reduce the emittance of the e^-/e^+ beams, (3) beam transportation from the damping rings to the main linear accelerators (RTML), (4) the main linear accelerators (MLs), including bunch compressors, that accelerate the e^-/e^+ beams using superconducting RF technology, (5) beam delivery system (BDS) and a final focusing system, to focus and adjust the final beam to increase the luminosity, and the beam interaction region for the machine and detector interface (MDI) where the detectors are installed. After passing through the interaction region, the beams go to the beam dumps. The ILC complex is shown in Fig. 22.3.

Two key technologies are required for ILC, one of which is nano-beam technology applied at DRs, ML and the BDS. The beam is focused vertically to 7.7 nm at the interaction point. The other is SRF technology applied in the MLs. Approximately 8,000 SRF cavities are installed in the MLs and operated at an average gradient of 31.5 MV/m. The accelerator is operated at 5 Hz. In total, 1,312 beam bunches are formed in one RF pulse with a duration of 0.73 ms, and 2×10^{10} electrons and positrons are generated per bunch from the electron source and the positron source, respectively. The high-power output from the klystrons is transferred to the cavities through input couplers generating an electric field of 31.5 MV/m. One klystron's RF power (up to 10 MW) is distributed to 39 cavities. The AC power required to operate

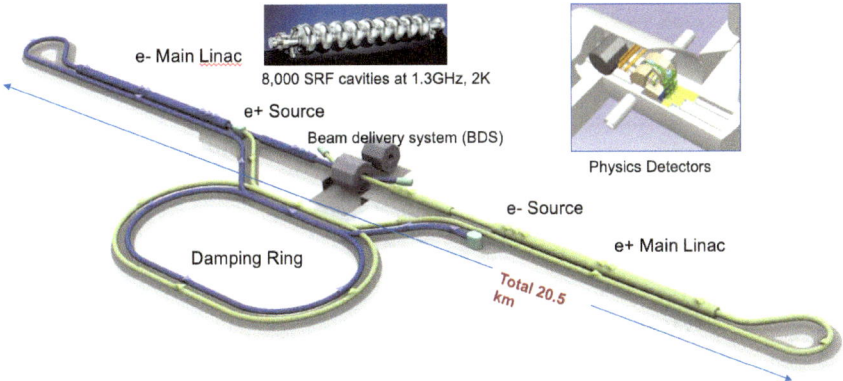

Fig. 22.3 Schematic layout of the ILC complex at 250 GeV

the accelerator will be 111 MW [16]. The spins of the electron and/or positron beams can be maintained during acceleration and collision (polarized sources). This can help significantly improve the precision of measurements. The ILC parameters are summarized in Table 22.2.

The ILC can be upgraded in energy by extending the tunnel or increasing the acceleration gradient. An important feature of linear colliders is that the energy can be increased without being affected (limited) by synchrotron radiation, allowing to adjust the facilities to emerging new physics. The beam delivery system (BDS) and beam dump of the ILC can handle collision energies up to 1 TeV. Another upgrade scenario is a luminosity upgrade. By increasing the high-power RF system, the luminosity can be doubled as compared to the current scenario discussed in the TDR. It might also be possible to re-use the tunnel, infrastructure and other facility resources for a future multi-TeV linear collider based on further improved or novel accelerator RF-technologies. Some of these options are described in [8].

22.9 Status of the ILC Accelerator Developments

22.9.1 Positron Source

There are two options for ILC positron sources: undulator and electron driven. The undulator scheme provides polarization (30%), but is a novel method for a collider. The electron-driven scheme is conventional and technically more proven. Considering the physical potential of the polarized positron, the undulator and electron-driven schemes are being developed in parallel. A superconducting helical undulator has been put into operation at APS (ANL, USA) and long undulators are also operated at European XFEL. Concerning the undulator scheme, the necessary techniques for

Table 22.2 Parameters for ILC250 GeV and future 500 GeV and 1 TeV upgrades. See [8] for detailed explanations

Parameter	Symbol	Unit	Option					
			Higgs			500 GeV		TeV
			Baseline	Lum. Up	L Up, 10 Hz	Baseline	Lum. Up	Case B
Center-of-Mass Energy	E_{CM}	GeV	250	250	250	500	500	1000
Beam Energy	E_{beam}	GeV	125	125	125	250	250	500
Collision rate	f_{col}	Hz	5	5	10	5	5	4
Pulse interval in electron main linac		ms	200	200	100	200	200	200
Number of bunches	n_b		1312	2625	2625	1312	2625	2450
Bunch population	N	10^{10}	2	2	2	2	2	1.737
Bunch separation	Δt_b	ns	554	366	366	554	366	366
Beam current		mA	5.79	8.75	8.75	5.79	8.75	7.6
Average power of 2 beams at IP	P_B	MW	5.26	10.5	21	10.5	21	27.3
RMS bunch length at ML & IP	σ_z	mm	0.3	0.3	0.3	0.3	0.3	0.225
Emittance at IP (x)	γe_x^*	mm	5	5	5	10	10	10
Emittance at IP (y)	γe_y^*	nm	35	35	35	35	35	30
Beam size at IP (x)	σ_x^*	mm	0.515	0.515	0.515	0.474	0.474	0.335
Beam size at IP (y)	σ_y^*	nm	7.66	7.66	7.66	5.86	5.86	2.66
Luminosity	L	10^{34} cm^{-2}s^{-1}	1.35	2.7	5.4	1.79	3.6	5.11
AC power	P_{site}	MW	111	138	198	173	215	300
Site length	L_{site}	km	20.5	20.5	20.5	31	31	40

undulator positron sources such as installation precision and orbit correction have been established. The durability test of the titanium alloy target was carried out and good results were obtained. For the electron drive system, the rotating target with magnetic fluid vacuum sealing was tested for degradation of the sealing part by irradiation and for long-term running of the simulated target, and the stable rotation and sufficient vacuum sealing performance were confirmed. For the magnetic convergence circuit, the electromagnetic design of the flux concentrator was completed based on the results at BINP, and the thermal design is now in progress.

22.9.2 BDS and Interaction Point

Nanobeam technology has been demonstrated at the ATF-2, hosted at KEK as an international collaboration, and is close to satisfying the requirements of the ILC. The ATF-2 has two goals. One is the generation of a small 37 nm beam, which is equivalent to 7.7 nm at the ILC-250 final focus at the IP. Until now 41 nm has been achieved. The other is to demonstrate precise position feedback. A feedback latency of 133 ns has satisfied the ILC requirement of less than 366 ns. Evaluation of the effect of the wakefield on the beam size at the ATF has led to studies aiming at suppressing wakefield effects at the ILC. The ATF programme, results, status and future opportunities, has recently been reviewed by an international committee, and the importance of continuing the research for detailed design and performance studies, of the ILC (and CLIC) final focus systems was highlighted.

22.9.3 SRF Technology

The SRF technology readiness has been proved by the successful operation of the European XFEL, where approximately 800 superconducting cavities (one-tenth the scale of the ILC SRF cavities) have been installed. A distributed and collaborative construction model was also successfully demonstrated. Following the European XFEL, the LCLS-II at SLAC and SHINE in Shanghai are under construction. Two major R&D programs are underway to improve the performance and reduce the cost of superconducting cavities. One is a new surface treatment for high Q and gradients, and the other is a new approach for niobium (Nb) material processes. New cavity surface treatments, such as two-step baking developed at FNAL, improve both the acceleration gradient and Q. Such surface treatments lead to a higher beam energy and/or cost reduction by shortening the length of the SRF linac and reducing the cryogenic heat load. Nb material R&D aims to reduce material costs during the production of Nb discs and sheets, including direct slicing and tube formation. Automation in a clean environment is important for the mass production of high-performance SRF cavities. The equipment for the automation of activities such as dust removal, is under development. Cryomodule assembly of a collection of 38 MV/m cavities significantly exceeding ILC specifications is in progress at FNAL in the USA with international cooperation.

22.10 Technical Preparation of ILC with a Pre-lab

Although significant work has already been done and described in the TDR and its addendum, it is necessary to revisit all the items to examine whether improvements or further developments are needed. The technical preparations during the Pre-lab phase, i.e. accelerator work necessary for producing the final engineering design and documentation, are anticipated to be a starting point to discuss the international cooperation and technical efforts to be shared as in-kind contributions among the participating laboratories worldwide. A total of 18 work packages (WPs) have been proposed covering five accelerator domains.

A dominant part of the Pre-lab plans are related to SRF development, and three of the WPs are related to this topics. The technical preparations for the SRF include cavity industrial production readiness (WP-1), demonstration of cryomodule (CM) production readiness and global transfer while maintaining specified performance (WP-2), and crab cavity (WP-3). In WP-1, a total of 120 cavities will be produced (40 cavities per region, Europe, the Americas, and Asia), and successful production yields ($\geq$ 90%) are to be demonstrated in each region. Recent high-performance cavity preparation will be included. In WP-2, six CMs (two CMs per region) will be fabricated, and their performance will be qualified within each region. Thus, 48 of the 120 produced cavities will be used in the six CM assemblies. The compatibility of the CMs from different regions will be confirmed.

If the cavity is to be operated at a 10% higher gradient of 35 MV/m, it is necessary to confirm that the input coupler is compatible with the high gradient, and the introduction of a high-efficiency klystron is expected to reduce the electric power consumption. These are in line with the development of high-performance SRF cavities, input couplers, and high-efficiency klystrons.

WP-2 will also demonstrate readiness for the cost-effective production of other cryomodule components, such as couplers, tuners, and superconducting magnets. Overall CM testing after assembling these components into the CM is the last step for confirming the performance of the CM as a primary accelerator component unit.

The Americas and Europe have already developed significant expertise in cavity and CM production for their large SRF accelerators, including the formulation of countermeasures against performance degradation after cryomodule assembly, as well as degradation during ground transport of modules. As part of WP-2, the resilience of CMs to intercontinental transport will be established. In WP-3 (crab cavity), the first down-selection of the crab cavity will be carried out before Pre-lab to narrow down the choices from four to two, and then one of the two will be selected after the performance test during the Pre-lab.

The other WPs concerns the electron and positron sources, the damping rings, the beam-delivery and final focus system, and the dump. Overall their address the key elements needed for providing high luminosity with the ILC nano-beams. All the Pre-lab work-packages are described in detail in [6].

22.11 Sustainability of Linear Colliders

Power and energy efficiency studies will continue, covering accelerator structures and cavities, but also very importantly high efficiency RF power system with optimal system designs using high efficiency klystrons and modulators. It is expected that the

CLIC and ILC power consumptive can be further consolidated and possibly reduced. In particular for stages 2 and 3 of CLIC many technical developments affecting the power have not been included in the current power estimates. For ILC the wall-plug power is minimized making use of the small surface resistance of the SRF accelerating structures (cavities). Future SRF cavity studies can further improve the power efficiency, being particularly important for potential upgrades towards and into the TeV region. For ILC further improvements in energy efficiency are anticipated as part of the Green ILC concept, which aims to establish a sustainable laboratory [17] in a wide perspective, as part of the local region and economy.

Sustainability studies in general, e.g. power/energy efficiency, using power predominantly in low cost periods as is possible for a linear collider, use of renewable energy sources, and energy/heat recovery where possible, will therefore be a priority for further studies for both LC projects. Such studies were already made with initial parameters for the CLIC Implementation Plan (see Chap. 7 in [13]), but for example a complete carbon footprint analysis has not been made. Similar studies will be made for ILC. Both machines can benefit from use of permanent magnets and several studies and prototype have been successfully made. Future work in the area of sustainability will be synergetic with any future large accelerator study. In particular there are clear plans for future work common work between CLIC and ILC regarding sustainability and power/energy optimisation.

Acknowledgements This summary is made on behalf of the CLIC and ILC collaborations and communities. Many researchers work across and studies are carried out in common between the two collider concepts. The written resources used for this summary are primarily the CLIC and ILC sections being part of the European LDG roadmap report [18] and the CLIC and ILC input documents to the Snowmass 2021 process [2, 8]. The authors and editors of these reports are specially acknowledged.

References

1. P.N. Burrows, P. Lebrun, L. Linssen, D. Schulte, E. Sicking, S. Stapnes, M.A. Thomson (eds.), *Updated Baseline for a Staged Compact Linear Collider*. CERN Yellow Reports: Monographs. CERN, Geneva (2016). https://doi.org/10.5170/CERN-2016-004
2. O. Brunner, et al., *The CLIC Project: Report to Snowmass 2021* (2022). arXiv:2203.09186 [physics.acc-ph]
3. T. Behnke, et al. (eds.), *The International Linear Collider Technical Design Report—Volume 1: Executive Summary* (2013). arxiv:1306.6327
4. L. Evans, S. Michizono, *The International Linear Collider Machine Staging Report* (2017). arXiv:1711.00568 [physics.acc-ph]. https://arxiv.org/pdf/1711.00568.pdf
5. *ICFA announces a new phase towards preparation for the International Linear Collider* (2020). https://www.interactions.org/press-release/icfa-announces-new-phase-towards-preparation-international
6. ILC International Development Team, *Proposal for the ILC Preparatory Laboratory (Pre-lab)* (2021). https://arxiv.org/pdf/2106.00602.pdf

7. ILC International Development Team Working Group 2, *Technical Preparation and Work Packages (WPs) during ILC Pre-lab* (2021). http://doi.org/10.5281/zenodo.4742018
8. A. Aryshev et al., *The International Linear Collider: Report to Snowmass 2021* (2022). arXiv:2203.07622 [physics.acc-ph]
9. F. Bordry et al., *Machine Parameters and Projected Luminosity Performance of Proposed Future Colliders at CERN* (2018). arXiv:1810.13022 [physics.acc-ph]
10. P.G. Roloff, A. Robson, Updated CLIC luminosity staging baseline and Higgs coupling prospects. Technical report, CERN, Geneva (Oct. 2018). 9 pages, 6 figures. https://cds.cern.ch/record/2645352
11. C. Gohil, A. Latina, D. Schulte, S. Stapnes, High-Luminosity CLIC Studies. Technical report, CERN, Geneva, Aug. 2020. https://cds.cern.ch/record/2687090
12. C. Gohil, P.N. Burrows, N. Blaskovic Kraljevic, A. Latina, J. Ögren, D. Schulte, Luminosity performance of the Compact Linear Collider at 380 GeV with static and dynamic imperfections. Phys. Rev. Accel. Beams **23**, 101001 (2020). https://doi.org/10.1103/PhysRevAccelBeams.23.101001
13. P.N. Burrows, et al. (eds.), *CLIC Project Implementation Plan*. CERN Yellow Reports: 4/2018 (2018). https://doi.org/10.23731/CYRM-2018-004. https://edms.cern.ch/document/2053292/
14. M. Aicheler, P. Burrows, M. Draper, T. Garvey, P. Lebrun, K. Peach, N. Phinney, H. Schmickler, D. Schulte, N. Toge (eds.), *A Multi-TeV Linear Collider Based on CLIC Technology: CLIC Conceptual Design Report* vol. CERN-2012-007 (2012). https://doi.org/10.5170/CERN-2012-007
15. G. D'Auria et al., *Status of the CompactLight Design Study*, 078 (2019). https://doi.org/10.18429/JACoW-FEL2019-THP078
16. ILC International Development Team, *Updated power estimate for ILC-250* (2021). https://agenda.linearcollider.org/event/8389/contributions/45111/attachments/35278/54677/ILC-CR-0018.pdf
17. Green-ILC Project Team: *Green ILC project* (2017). http://green-ilc.in2p3.fr/home/
18. C. Adolphsen et al., *European Strategy for Particle Physics—Accelerator R&D Roadmap*. CERN Yellow Rep. Monogr. 1, 1–270 (2022). arXiv:2201.07895 [physics.acc-ph]. https://doi.org/10.23731/CYRM-2022-001

Chapter 23
Remembering Bruno Touschek

Giovanni Battimelli, Franco Buccella, Luisa Cifarelli, Carlo Di Castro, Giovanni Gallavotti, and Luciano Pietronero

Abstract Participants' recollections about their relations with Bruno Touschek.

23.1 Remembering the Founders: Bruno Touschek's Papers, *Giovanni Battimelli*

I cannot say that I ever came to really know Bruno Touschek, when he was still active at the Physics Institute in Rome. Back then, around 1970, I was a student there, and almost all I knew of him was that he was in charge of the course of

G. Battimelli (✉)
INFN Sezione di Roma, c/o Dipartimento di Fisica, Università di Roma "La Sapienza", P.le Aldo Moro, 5, 00185 Rome, Italy
e-mail: giovanni.battimelli@uniroma1.it

F. Buccella (✉)
INFN, Sezione di Napoli, Via Cintia, Napoli, Italy
e-mail: buccella@roma1.infn.it

L. Cifarelli (✉)
University of Bologna, Bologna, Italy
e-mail: luisa.cifarelli@bo.infn.it

C. Di Castro (✉)
Dipartimento di Fisica, Università di Roma "La Sapienza", P.le Aldo Moro 5, 00185 Roma, Italy
e-mail: carlo.dicastro@roma1.infn.it

G. Gallavotti (✉)
Dipartimento di Fisica, INFN Università di Roma "La Sapienza", P.le A. Moro 2, 00185 Roma, Italia
e-mail: giovanni.gallavotti@uniroma1.it

L. Pietronero (✉)
Enrico Fermi Research Center, Via Panisperna 89a, 00184 Roma, Italy
e-mail: luciano.pietronero@roma1.infn.it

L. Bonolis et al. (eds.), *Bruno Touschek 100 Years*,
Springer Proceedings in Physics 287,
https://doi.org/10.1007/978-3-031-23042-4_23

"Metodi matematici della fisica", and, in his official role of professor, one of the "counterparts" of the frequent and confused agitations shaking in that turbulent period our students' life. We had some hints about his being a brilliant theorist that had been involved with the creation in Frascati of a new kind of clever experimental tool, but what made him famous to us were rather the humorous and seemingly incongruous puns that he scattered liberally across his course notes, which left us at the same time dumbfounded and pondering how close he was to the stereotype of the crazy scientist.

It was about ten years later when I had the first real interaction with Touschek's legacy. It must have been someday early in 1982. I was chatting with my friend and colleague Michelangelo De Maria in the office we shared at the first floor of the Institute, when the door opened and the head of Amilcare Bietti poked in. Amilcare had been Bruno's assistant for quite some time, in my student years. "Hey guys, they are cleaning Bruno's office upstairs. You'd rather have a look on what's going on".

(Which shows two things: one, that back then there must have been no pressing demand for space in the Institute, given that Touschek's papers still were in his old office, almost four years after his death and over seven years after his actually no longer coming to the university; and, two, that back then it still was a current practice, in order to make room in an office previously occupied by a retired, or deceased professor, to get rid of old books and papers just throwing everything away, thus paying tribute, probably unknowingly, to Alfred Whitehead's famous sentence "a science that hesitates to forget its founders is lost").

Well, we were, or pretended to be, historians of physics, and did not hesitate to act in the opposite direction to the one suggested by a strict interpretation of Whitehead's prescription. Urged by Amilcare's intervention, upstairs we went and we found out that the current practice referred above was being duly performed, and already a good portion of the papers left in the office had been discharged in the large garbage can on the back of the building. We ran down and started searching through the box like hungry homeless desperate for leftover food, extracting from the overall mess quite a bit of correspondence, including letters to and from Werner Heisenberg, Max von Laue and the like, lab logbooks, drawings, original sketches and notes related to the early days of the AdA project, and so on. Luckily, we could stop the "current practice" just in time to prevent that valuable documentation from getting lost.

And so was born the first block of what was going to become, in the course of the following years, the richest collection of physicists' personal papers in Italy, now duly preserved in the basement of the department's library. A few days after our first intervention, we went to see Francis Touschek at the family house in via Pola, and he lent us more papers and documents that his father had kept at home, thus allowing the building up of a substantial archive that has proved to be, in the course of time, a unique and most valuable source for those who have researched, documented, and written about, Touschek's scientific life and his impact on the course of twentieth century physics.

Sometimes I ask myself the seemingly silly question "what if". What if Bietti had not knocked on our door that day? What if he went to see us and we had not been there? What if we had dismissed his warnings? Silly questions maybe, that leave us pondering on the fortuitous contingencies impending upon so many of our endeavours. Be it as it may, it gives us pleasure to know that, among the several possibilities open at that moment for the course of events to be, the one that actually materialized gave us the chance to keep the door open for our science to not forget (one of) its founders.

23.2 Touschek: A Great Master of Quantum Electrodynamics and Statistical Mechanics, *Franco Buccella*

In the summer of 1963 Guido Altarelli and I were trying to compute the differential cross-section for the emission of a photon in electron–positron scattering, the issue proposed for our thesis by our tutor Raffaele Gatto. The numerical evaluation gave conflicting results, negative (!) or very large values. To account for this last case, we told Prof. Touschek that the amplitude with all the final particles in the same direction had a very small denominator. Immediately he replied with his nice Austrian accent: "Denominatore piccolo, numeratore zero." In fact, the transverse polarization vector of the photon is orthogonal to all the longitudinal momenta of the particles. This led us to perform the ultrarelativistic approximation for the final fermions, which allowed us to complete the analytical evaluation. The comparison of the formula with the experimental measurements at AdA proved that the machine worked. Our paper (1) was quoted on the book of quantum field theory by Landau and his collaborators.

Few months before I followed the course taught by Prof. Touschek on Statistical Mechanics and I was impressed by the mathematical elegance of the derivation of the Maxwell–Boltzmann, Fermi–Dirac and Bose–Einstein formulas. This allowed me to propose Fermi–Dirac and Planck formulas, respectively, for the valence partons and gluons distributions, as boundary conditions to the DGLAP equations.

These formulae are in agreement with the shapes of the distributions and with the isospin and spin asymmetries of the proton sea (2). More recently, the gluon distribution measured by ATLAS has been well described with a value equal to the adimensional variable, which plays the role of the temperature and fixes the behaviour of valence partons (3).

(1) G. Altarelli and F. Buccella, Nuovo Cimento 34 (1964) 1337
(2) F. Buccella, F. Tramontano and Sozha Sohaily, J. Stst. Mech. (2019) (7) 073,302
(3) L. Bellantuono, R. Bellotti and F. Buccella, arxiv::2201.07640v2 [hep-ph].

23.3   AdA as Historic Site of the European Physical Society to Pay a Tribute to Bruno Touschek, *Luisa Cifarelli*

The European Physical Society (EPS) was founded in Geneva, Switzerland in 1968 through the visionary leadership of Gilberto Bernardini (then CERN Research Director) "as a further demonstration of the determination of scientists to collaborate as close as possible in order to make their positive contribution to the strength of European cultural unity".

In line with this "cultural unity", the Historic Sites initiative of the European Physical Society was launched at the end of 2011, when I had the honour of being president of the EPS. The initiative was inspired by an analogous initiative on the other side of the Atlantic by the American Physical Society. A dedicated EPS Historic Sites Committee was created, which has been actively operating since then.

The EPS Historic Site awards commemorate places in Europe, sometimes outside geographical Europe, with national or international significance for the development and the history of physics. Examples of sites to be considered are laboratories, buildings, institutions, universities, towns, etc., each associated with an event, a discovery, a research or body of work, by one or more individuals, that made long lasting contributions to physics.

Until now, more than 100 proposals of Historic Sites were received, either spontaneous or channelled through national member societies of the EPS. The Historic Sites Committee examines the proposals typically three times per year. Almost 70 EPS Historic Sites have been inaugurated up to 2022 in 25 different countries (even outside geographical Europe): Austria, Belgium, Bulgaria, Czech Republic, Denmark, France, Germany, Hungary, India, Israel, Ireland, Italy, Lithuania, The Netherlands, Norway, Poland, Portugal, Romania, Russia, Serbia, Spain, Sweden, Switzerland, United Kingdom, United States.

On December 5, 2013, the Frascati National Laboratories (LNF) of the Italian National Institute of Nuclear Physics (INFN) hosted the naming ceremony of AdA (*Anello di Accumulazione*/Storage Ring) as a Historic Site of the European Physical Society (EPS).

AdA was built in 1961 by a small group of Italian physicists under the brilliant leadership of Austrian physicist Bruno Touschek. It was the world's first prototype electron–positron storage ring. AdA was later moved to Orsay, to the *Laboratoire de l'Accélérateur Linéaire* (LAL), in order to operate with higher intensity beams. AdA was by far the forerunner of several generations of e^+e^- colliders of gradually increasing energy and luminosity, in Italy and around the world. In Frascati, in particular, its successors were ADONE and DAPHNE.

As for each Historic Site inauguration "fest", a plaque was unveiled in the presence of the local representatives and authorities. The ceremony was chaired by Umberto Dosselli, then LNF Director, and the speakers included, in particular: Stefano Di Tommaso, then Mayor of Frascati; Giorgio Salvini, Director of LNF in 1961; Fernando Ferroni, then President of INFN. The many distinguished participants included Samuel C. C. Ting, who was also invited for a special seminar the same day on the AMS (Anti Matter in Space) experiment as part of the traditional "Bruno Touschek Memorial Lectures". The establishment of the AdA Historic Site was not only meant as a recognition of a glorious past but also as a wish for a brilliant future of the LNF.

So far, the Historic Site initiative of the EPS has been a series of success stories implying the improvement of mutual relations between the EPS, its national member societies, and local institutions and official bodies. Therefore, while stamping important and meaningful places for the history and the progress of physics, the EPS Historic Sites provide visibility to physics and to the physics community and, at the same time, enhance a sense of belonging to the EPS.

This initiative has also the (maybe) ambitious objective to create the awareness that not only artistic cultural heritage and natural heritage should be preserved for humankind, but also scientific cultural heritage. AdA is indeed part of it.

23.4 Touschek's Approach to Students, *Carlo Di Castro*

I entered the university as a freshman in 1956. All people at the Physics Institute in that period are in debt in a way or another to Bruno Touschek, in Rome since 1952. I was not an exception. In my personal studies I became interested in thermodynamics, statistical mechanics and in the theoretical physics of condensed matter—largely ignored in Rome, at a time when everyone was engaged in the study of elementary particle physics. The course in statistical mechanics was given by Bruno Touschek. Even though statistical mechanics was not his field, his course was brilliant and stimulating. For him physics was a unifying vision, the basic notions were given following E. Schrödinger's *Statistical Thermodynamics*, but he would also extemporize on specific topics, not teaching, strictly speaking professional statistical mechanics, but rather how a theoretical physicist should approach problems with technique, imagination and enthusiasm. When time came to select the argument for my Laurea thesis, I had to use this imagination to find my way. The Institute in Rome had little to offer in terms of my interests in condensed matter physics and statistical mechanics. Obviously, there was also a problem of cultural legacy. Under the Fascist dictatorship (the Racial Laws of '38, the war, etc.), Italian physics was destroyed, and after the war, Edoardo Amaldi had the difficult task of rebuilding the field. Obviously, capable young physicists, at least in the theory group, wanted to pursue the physics of the moment, i.e. elementary particles. Giorgio Careri however, at that time in Padua, had obtained brilliant results with his experiments on Superfluid Helium four and was supposed to come back to Roma. So, I started to study superfluidity. According to the newly proposed BCS theory (1957) of superconductivity, below a certain temperature, electrons may couple in Cooper pairs and then condense, like bosons. I therefore decided, for my thesis, to introduce the pairing approximation in superfluidity. Marcello Cini was my advisor, and my outside examiner was of course my Statistical Mechanics teacher, who returned the thesis with a comment written in his typical misspelled Germano-Italian style: "Con complimenti del avvocato del diavolo" ("With best wishes from the devil's advocate").

In the same period, we all were in a classroom waiting for a seminar on the Landau quasiparticle spectrum of superfluid helium and the speaker didn't arrive. Touschek, the theoretician present, was asked to extemporize a short talk. He drew the quasiparticle spectrum energy versus momentum which starts linearly (the so-called phonon part), goes through a maximum and then has a minimum (the so-called roton part) at a wave vector inversely proportional to the average distance between the Helium particles. Touschek then paradoxically presented superfluid helium as a missed solid. According to him the rotonic minimum was the sign of the missed periodicity of the solid Helium when the average distance between the helium particles is substituted with the lattice constant. Actually, I realized that after all Touschek was not far from the famous Feynman explanation of those few excited states compatible with superfluidity of a system of bosonic interacting particles. In short, the ground state function is a real positive totally symmetric function of the positions of the well separated and evenly spaced atoms. Phonons are the only low-lying excited states compatible with the Bose statistics because variation in the density cannot be accomplished by just permuting atoms starting from a homogeneous configuration. All other states either are equivalent to the ground state by permutations or involve movements of atoms on distances less than the average atomic distance, i.e., are rotons separated from the ground state by parabolic excitation energies with an effective mass and wavevector proportional to the inverse average distance between Helium particles. Bruno, with his approach to physics, was a continuous unintentional teacher for all of us.

23.5 Memories of Bruno Touschek, *Giovanni Gallavotti*

In 1963 I asked Professor Touschek to accept to follow my work towards my "Laurea". He assigned me a problem on quantum electrodynamics and soon he realized that I was not ready to work on such a subject. I still feel gratitude that he simply did not insist to deal with the problem and changed it into a more technical study on the lifetime of electrons in the storage rings at the time under construction (ADONE) or already operative (AdA).

The electrons of one packet collide with the light emitted from the positrons of another packet and as a consequence are expelled from the orbit. The question is to estimate how long a packet can stay on the ring in which it circulates, i.e. which is the half-life of a packet. The problem is relevant for the storage rings and was checked independently, while being useful to a student to learn not to hesitate over very long calculations.

This was for me a very difficult task although it did not require particular originality, but it trained me to consider computations as a minor problem. So, I kept asking regularly suggestions by daring to go to his office: the explanations were given on a blackboard (still there in the office that has become a room for visiting scholars) which was densely covered by ever changing formulae: during the several months of my work I vividly recall that there, essentially in the center, was written with white chalk, immutable, "*amice diem perdidi*".

The work for the graduation was over around November: I cannot think that he was happy with it and I thank him for letting me, nevertheless, go through the final exam. I regard that a sign of his confidence that my future work might be of better quality.

I then left Roma for about 10 years without further interaction with him: until in 1971 he chaired an Italian Physical Society meeting where I was a speaker. He listened to my work and, returning to Roma, he mentioned it enthusiastically to some of the senior professors (so I was told). I warmly thank him for this as, from that moment, I was accepted as a "physicist". Although my work was not on a subject of his typical interests, still he was open minded to publicly appreciate it: his open mindedness remains for me as a permanent example of the attitude that senior researchers have to take in dealing with the new generations. I remember it with deep gratitude.

Eventually, in the eighties, I obtained a position at "La Sapienza" in the Physics Institute (now department): but it was too late to interact regularly with Touschek and I can only regret that his departure had happened too early.

23.6 Memories of an Extraordinary Person: Bruno Touschek, *Luciano Pietronero*

It was 1970 when I attended the course of Mathematical Methods for Physics by Bruno Touschek and Amilcare Bietti as assistant. The cour*se* was very original, all based on his personal and mostly handwritten notes and, even the response of a harmonic oscillator, already studied in other courses, became a fascinating scientific adventure: "at high frequency the oscillator trembles but it does not oscillate". So, after this course, I went to him asking for some possible subject for a Laurea thesis. I was 20 and pretty ignorant and he was the great scientist. He started discussing with me almost on any subject in physics and beyond and manifested a certain scepticism about the situation of high energy physics which he considered a bit stuck at the time. In that period, he was intrigued by a chapter of Pauli's book on General Relativity which he found unclear. In his words: "It is always difficult to nail down Pauli". The question was about Mach's principle, the equivalence principle and the meaning of the inertial mass. The gravitational mass is a local property of particles like the electric charge, while the inertial mass is the resistance to acceleration, quite a different property. Mach's principle states that inertial forces should be due to the interaction of a body with all the other masses in the universe. It was never stated in a mathematically rigorous form, but it was very influential to Einstein. In General Relativity this principle is not fully present and is replaced by the equivalence principle, according to which inertial and gravitational masses are linked by the gravitational constant, which is a fixed number. However, according to a strict interpretation of Mach's principle, the inertial mass should be non local and depend on all the other masses of the universe and their positions. In this perspective the gravitational constant cannot be just a fixed number. All this may appear almost philosophical but it took a very concrete perspective in 1918 with two papers. In one Lens and Thirring computed the so-called frame dragging of the earth that induces an inertial dragging and precession to the orbit of satellites. This has been accurately measured in the past years with the LARES satellites. This effect does not touch directly the problem of the inertial mass but it begins to show that the motion of masses induces inertia like effects. The second paper by Thirring was more intriguing. It studied the metric inside a rotating cylindrical mass shell and it was clearly inspired by the conceptual problem of Mach's principle. This study revealed the appearance of a force with the structure of Coriolis force. No centrifugal like force was present but there was also a curious vertical force. This was the paper discussed by Pauli that intrigued Touschek. He immediately realized that the vertical force was spurious and due to the spherical shell that should be substituted by a cylindrical one to have the correct rotational symmetry. Also Thirring had resorted to the linearization of the GR equations and clearly this cannot lead to quadratic effects that are necessary for the centrifugal force. So the problem was clear, use a cylindrical geometry and solve the GR equations to second order in the gravitational constant. This implied the construction of a novel mathematical scheme to go to second order and then consider the cylinder as the source of the field. Well, after quite some calculations, we found that a rotating cylinder leads to a metric which gives exactly the Coriolis term and the centrifugal one with the correct relations. Touschek was extremely excited because this result paved the way for a more concrete implementation of Mach's principle and possibly a generalization of GR. However, he did not want to sign the paper because, in his opinion, I had done all the calculations, but he had given me all the ideas. The paper was published in *Annals of Physics* in 1973. After this enthusiastic period things became problematic with his health and I realized that this beautiful and exciting experience was not going to have a continuation. So I moved to the field of condensed matter and went first to US and then to Switzerland. I never missed to visit him when I could, up to the last days in 1978. I was so influenced by his sparkling originality that, on many occasions, in front of a difficult problem, I asked myself what would Touschek do with this problem? Grazie Bruno, una luce brillante nella mia vita.